T0396459

Springer Proceedings in Physics

Volume 261

Indexed by Scopus

The series Springer Proceedings in Physics, founded in 1984, is devoted to timely reports of state-of-the-art developments in physics and related sciences. Typically based on material presented at conferences, workshops and similar scientific meetings, volumes published in this series will constitute a comprehensive up-to-date source of reference on a field or subfield of relevance in contemporary physics. Proposals must include the following:

- name, place and date of the scientific meeting
- a link to the committees (local organization, international advisors etc.)
- scientific description of the meeting
- list of invited/plenary speakers
- an estimate of the planned proceedings book parameters (number of pages/ articles, requested number of bulk copies, submission deadline).

Please contact:

For Americas and Europe: Dr. Zachary Evenson; zachary.evenson@springer.com
    For Asia, Australia and New Zealand: Dr. Loyola DSilva; loyola.dsilva@springer. com

More information about this series at http://www.springer.com/series/361

Prafulla Kumar Behera · Vipin Bhatnagar ·
Prashant Shukla · Rahul Sinha

Editors

XXIII DAE High Energy Physics Symposium

Select Proceedings

Volume 2

 Springer

Editors
Prafulla Kumar Behera
Department of Physics
Indian Institute of Technology Madras
Chennai, India

Prashant Shukla
Bhabha Atomic Research Center
Mumbai, India

Vipin Bhatnagar
Panjab University
Chandigarh, India

Rahul Sinha
The Institute of Mathematical Sciences
Chennai, India

ISSN 0930-8989   ISSN 1867-4941 (electronic)
Springer Proceedings in Physics
ISBN 978-981-33-4407-5   ISBN 978-981-33-4408-2 (eBook)
https://doi.org/10.1007/978-981-33-4408-2

This Springer imprint is published by the registered company Springer Nature Singapore Pte Ltd.
The registered company address is: 152 Beach Road, #21-01/04 Gateway East, Singapore 189721, Singapore

Preface

Particle Physics has been at the forefront of all Physical sciences since the advent of the electron. Over the years, this field of Particle Physics has given more verities or types of particles than the different types of animals, generally found, in a city zoo! Such things are possible in this field due to the advancement in the theoretical understanding and the technological improvements happening all the time and up to some extent over a similar timescale. Many predictions done by the theoreticians were tested time and again in the experimental labs. Some of these were discovered, inferred or negated. This process is still on, but a bit slowed down due to challenges faced by the technology. For example, the present-day technology allows us to probe a distance of the order of 10-18 m (i.e., a decimal followed by 18 zeros) and at such tiny dimensions the entities that are seen, and are not further resolvable, are Electron and Quarks. These two types of entities, falling under a Generalized category called Fermions, are responsible for all the Matter in the visible Universe!

Moving on to the hunt for the "most sought-after" Higgs Boson ended on 4th July, 2012 by CERN-based Mega-Particle smasher aka LHC, the field appeared to be settling down for a moment for the trust in Standard Model of Particle Physics working extremely well. But, as witnessed with earlier discoveries, as usual it opened up questions on even deeper symmetries to be explored which bring out the various properties of the elementary particles once the "mass" source is accounted for. This in one sense translates to having more and more Center of Mass energy or Even packing more particles in the colliding bunches. These are technological challenges (along with other associated technologies) which attract a large number of manpower in these fields.

From the Particle Physics point of view, which is somewhat synonymous with LHC Physics these days, Super-Symmetry is the next level of improvement in the Standard Model most probed ever. Primordial soup has been extensively probed by many Heavy-ion accelerating instruments with very good outcomes on a new phase of matter—quark–gluon plasma, matter density probes, and hadron freeze-outs. The techniques and tools developed for such physics analyses are the sure shot straight outcomes applicable in many diverse fields in the society.

The "elusive neutrino" appears to be daunting the field heavily with the abrupt rise in the facilities appearing across the globe and opening up of the Astroparticle regime. The beyond SM signature shown by neutrinos has opened up the whole field now with connections being made from Supernovae, Galactic, Solar, Atmospheric, and Geo neutrinos. Interesting, but not many, properties from this sector have far-reaching deep impacts on the cosmological evolution of the Universe, like: Mass hierarchy, Study of CPV in neutrinos, Searches beyond the 3-neutrino framework, and Neutrino cross-sections. Even applications of these weakly interacting neutral leptons are emerging on the horizon having geopolitical and strategic linkage. Experimental setups like LIGO—searching for Gravitational Waves have initiated a complete new revolution in unraveling the mysteries of the Universe with the discovery of the Gravity waves. Thus, the linking of such experiments, with Particle Physics experiment has become stronger. The applications based on the technologies developed in the field were also presented, ranging from Medical Imaging,

The DAE-BRNS symposium, as usual, attracted 450 plus participants and assembled at the sprawling campus of IITM making "Particles" a buzzword for few days! There were invited talks by the experts, followed by sectional talks and parallel talks besides the famous poster sessions for the young minds. The DAE-BRNS-HEP symposium also witnessed participation from the industry and technocrats who have local to global footprints in the field. This led to the development of many Institute–Industry collaborative partnerships for the future of the field and embedded technologies. The proceedings book contains the selected papers covering almost all the aspects of particle physics highlighting the achievements from the past to the present and further illuminating the future path for the field.

Chennai, India   Prafulla Kumar Behera
Chandigarh, India   Vipin Bhatnagar
Mumbai, India   Prashant Shukla
Chennai, India   Rahul Sinha

Contents

Editors and Contributors

About the Editors

Dr. Prafulla Kumar Behera is currently an associate professor at the Department of Physics, Indian Institute of Technology Madras, Chennai. He obtained his Ph.D. from Utkal University, Bhubaneswar. He was a postdoctoral researcher at the University of Pennsylvania, USA and a research scientist at the University of Iowa, USA. His major research interests include experimental particle physics, detector building and data analysis. Dr. Behera was a member of BABAR experiment, Stanford, USA and ATLAS experiment, Geneva, Switzerland. Currently, he is a member of CMS and INO experiments, and he is also a member of the international advisory committee of VERTEX conference. He has published more than 1000 articles in respected international journals. He serves as a referee for several American Physical Society journals.

Dr. Vipin Bhatnagar is currently a professor at the Department of Physics, Panjab University, Chandigarh. He obtained his Ph.D. from Panjab University, Chandigarh, following which he was a postdoctoral fellow at LAL Orsay, France and CERN associate, CERN, Geneva. His major areas of research include experimental particle physics, detector building, computational physics and data analysis. He is a member of NOvA experiment, Fermilab, USA and CMS experiment, Geneva, Switzerland. He has published more than 700 journal articles in international peer-reviewed journals.

Dr. Prashant Shukla is a professor at the Nuclear Physics Division, Bhabha Atomic Research Center (BARC), Mumbai. He obtained his Ph.D. from Mumbai University, India. His research interests include high energy nuclear collisions, quark gluon plasma, and cosmic ray physics. Currently, he is a member of CMS and INO collaboration. He has published more than 150 articles in international journals of repute.

Dr. Rahul Sinha is currently a professor at the Institute of Mathematical Sciences, Chennai. He obtained his M.A and Ph.D from Rochester, USA, which was followed by a postdoctoral fellowship at the University of Alberta, Canada. His research primarily focuses on theoretical particle physics. Dr. Sinha is a member of BELLE and BELLE II experiment at KEK, Japan. He has supervised 11 doctoral students, and has also published more than 100 articles in international peer-reviewed journals.

Contributors

Sandeep Aashish Department of Physics, Indian Institute of Science Education and Research, Bhopal, India

Aman Abhishek Physical Research Laboratory, Navrangpura, Ahmedabad, India

J. Ablinger RISC, Johannes Kepler University, Linz, Austria

Debabrata Adak Department of Physics, Government General Degree College, West Bengal, India

Souvik Priyam Adhya Variable Energy Cyclotron Centre, Kolkata, India

Madan M. Aggarwal Panjab University, Chandigarh, India

Ritu Aggarwal Savitribai Phule Pune University, Pune, India;
Department of Science and Technology, New Delhi, India

Zubayer Ahammed Variable Energy and Cyclotron Centre, Kolkata, India

Asar Ahmed University of Delhi, Delhi, India

Rizwan Ahmed University of Delhi, Delhi, India

C. L. Ahmed Rizwan Department of Physics, National Institute of Technology Karnataka (NITK), Mangaluru, India

K. M. Ajith Department of Physics, National Institute of Technology Karnataka, Mangalore, India

Sampurn Anand Physical Research Laboratory, Ahmedabad, India

Richa Arya Physical Research Laboratory, Ahmedabad, India;
Indian Institute of Technology Gandhinagar, Gandhinagar, India

Abhishek Atreya Center For Astroparticle Physics and Space Sciences, Bose Institute, Kolkata, India

Anjali Attri Panjab University, Chandigarh, India

N. Ayyagiri Electronics Division, Bhabha Atomic Research Centre, Trombay, Mumbai, India

Partha Bagchi Variable Energy Cyclotron Centre, Kolkata, India

Bindu A. Bambah School of Physics, University of Hyderabad, Hyderabad, India

Aritra Bandyopadhyay Departamento de Física, Universidade Federal de Santa Maria, Santa Maria, RS, Brazil

Triparno Bandyopadhyay Department of Theoretical Physics, Tata Institute of Fundamental Research, Mumbai, India

Avik Banerjee Saha Institute of Nuclear Physics, HBNI, Kolkata, India

Pinaki Banerjee ICTS-TIFR, Bengaluru, India

Monika Bansal DAV College, Chandigarh, India

Sunil Bansal UIET, Panjab University, Chandigarh, India

R. C. Baral NISER, HBNI, Jatni, Bhubaneswar, Odisha, India

W. Bari University of Kashmir, Srinagar, India

Mitesh Kumar Behera University of Hyderabad, Hyderabad, India

Nirbhay Kumar Behera Department of Physics, Inha University, Incheon, Republic of Korea

Prafulla Kumar Behera Indian Institute of Technology Madras, Chennai, India

A. Behere Electronics Division, Bhabha Atomic Research Centre, Trombay, Mumbai, India

Rajkumar Bharathi Department of High Energy Physics, Tata Institute of Fundamental Research, Colaba, Mumbai, India

Akanksha Bhardwaj Physical Research Laboratory (PRL), Ahmedabad, Gujarat, India;
IIT Gandhinagar, Gandhinagar, India

Ashutosh Bhardwaj Centre for Detector and Related Software Technology, Department of Physics and Astrophysics, University of Delhi, Delhi, India

Shankita Bhardwaj Department of Physics and Astronomical Science, Central University of Himachal Pradesh, Dharamshala, India

Vishal Bhardwaj Department of Physical Sciences, IISER, Mohali, India

Vipin Bhatnagar Panjab University, Chandigarh, India

A. D. Bhatt Tata Institute of Fundamental Research, Mumbai, India;
Institute of Nuclear Physics Polish Academy of Sciences, Krakow, Poland

Jitesh R. Bhatt Theoretical Physics Division, Physical Research Laboratory, Ahmedabad, India

Sukannya Bhattacharya Theory Divison, Saha Institute of Nuclear Physics, Bidhannagar, Kolkata, India;
Theoretical Physics Division, Physics Research Laboratory, Ahmedabad, India

Gautam Bhattacharyya Saha Institute of Nuclear Physics, HBNI, Kolkata, India

Rik Bhattacharyya School of Physical Sciences, National Institute of Science Education and Research, HBNI, Jatni, India

Debabrata Bhowmik Saha Institute of Nuclear Physics, HBNI, Kolkata, India

P. S. Bhupal Dev Department of Physics and McDonnell Center for the Space Sciences, Washington University, St. Louis, MO, USA

M. Biswal Institute of Physics, Sachivalaya Marg, Bhubaneswar, India

Ambalika Biswas Department of Physics, Vivekananda College, Thakurpukur, India

S. Biswas Department of Physics and CAPSS, Bose Institute, Kolkata, India;
Department of Physics, National Institute of Technology, Durgapur, West Bengal, India

J. Blümlein DESY, Zeuthen, Germany

Debasish Borah Department of Physics, Indian Institute of Technology Guwahati, Guwahati, Assam, India

Ankita Budhraja Indian Institute of Science Education and Research, Bhopal, MP, India

Alexandra Carvalho Estonian Academy of Sciences, Tallinn, Estonia

Aleena Chacko Indian Institute of Technology Madras, Chennai, India

Kaustav Chakraborty Theoretical Physics Division, Physical Research Laboratory, Ahmedabad, India;
Discipline of Physics, Indian Institute of Technology, Gandhinagar, India

S. Chakraborty Department of Physics and CAPSS, Bose Institute, Kolkata, India

H. C. Chandola Department of Physics (UGC-Centre of Advanced Study), Kumaun University, Nainital, India

Sinjini Chandra Variable Energy Cyclotron Centre, Kolkata, India;
Homi Bhabha National Institute, Mumbai, India

Vinod Chandra Indian Institute of Technology Gandhinagar, Gandhinagar, Gujarat, India

V. B. Chandratre Electronics Division, Bhabha Atomic Research Centre, Trombay, Mumbai, India

Akshay Chatla School of Physics, University of Hyderabad, Hyderabad, India

Arindam Chatterjee Indian Statistical Institute, Kolkata, India

Bhaswar Chatterjee Department of Physics, Indian Institute of Technology Roorkee, Roorkee, India

S. Chatterjee Department of Physics and CAPSS, Bose Institute, Kolkata, India

Subhasis Chattopadhyay Variable Energy Cyclotron Centre, HBNI, Kolkata, India

Prakrut Chaubal Physical Research Laboratory, Ahmedabad, India

Ankur Chaubey Department of Physics, Institute of Science, Banaras Hindu University, Varanasi, India

Geetanjali Chaudhary Panjab University, Chandigarh, India

B. C. Chauhan Department of Physics and Astronomical Science, School of Physical and Material Sciences, Central University of Himachal Pradesh (CUHP), Dharamshala, Kangra, HP, India

Bhavesh Chauhan Physical Research Laboratory, Ahmedabad, India;
Indian Institute of Technology, Gandhinagar, India

Garv Chauhan Department of Physics and McDonnell Center for the Space Sciences, Washington University, St. Louis, MO, USA

Sushil Singh Chauhan Department of Physics, Panjab University, Chandigarh, India

Sandhya Choubey Harish-Chandra Research Institute, Jhunsi, Allahabad, India;
Department of Physics, School of Engineering Sciences, KTH Royal Institute of Technology, AlbaNova University Center, Stockholm, Sweden;
Homi Bhabha National Institute, Mumbai, India

Debajyoti Choudhury Department of Physics and Astrophysics, University of Delhi, Delhi, India

S. Choudhury Indian Institute of Technology Hyderabad, Sangareddy, Telangana, India

Maria Agness Ciocci Department of Physics, University of Pisa, Pisa, Italy

Ranjeet Dalal Centre for Detector and Related Software Technology, Department of Physics and Astrophysics, University of Delhi, Delhi, India

Sanskruti Smaranika Dani Institute of Physics, PO: Sainik School, HBNI, Bhubaneswar, India

Arpan Das Theory Division, Physical Research Laboratory, Navrangpura, Ahmedabad, India

Dipankar Das Department of Astronomy and Theoretical Physics, Lund University, Lund, Sweden

Mrinal Kumar Das Department of Physics, Tezpur University, Tezpur, India

Pritam Das Department of Physics, Tezpur University, Tezpur, India

S. Das Department of Physics and CAPSS, Bose Institute, Kolkata, India

Ashutosh Dash School of Physical Sciences, National Institute of Science Education and Research, Jatni, India

V. M. Datar Department of High Energy Physics, Tata Institute of Fundamental Research, Colaba, Mumbai, India

S. De Department of Physics, Indian Institute of Technology Indore, Simrol, Indore, India

K. N. Deepthi Mahindra Ecole Centrale, Hyderabad, India

P. S. Bhupal Dev Department of Physics, McDonnell Center for the Space Sciences, Washington University, St. Louis, MO, USA

Mayuri Devee Department of Physics, University of Science and Technology Meghalaya, Ri-Bhoi, Baridua, India

Ram Krishna Dewanjee Laboratory of High Energy and Computational Physics, KBFI, Tallinn, Estonia

Atri Dey Regional Centre for Accelerator-based Particle Physics, Harish-Chandra Research Institute, HBNI, Jhunsi, Allahabad, India

Ujjal Dey Asia Pacific Center for Theoretical Physics, Pohang, Korea

Lobsang Dhargyal Harish-Chandra Research Institute, HBNI, Jhusi, Allahabad, India

S. Digal The Institute of Mathematical Sciences, Chennai, India

A. K. Dubey Variable Energy Cyclotron Centre, Kolkata, India

Sandeep Dudi Department of Physics, Panjab University, Chandigarh, India

Juhi Dutta Regional Centre for Accelerator-based Particle Physics, Harish-Chandra Research Institute, HBNI, Jhusi, Allahabad, India

Nirupam Dutta National Institute of Science Education and Research Bhubaneswar, Odisha, India

Rupak Dutta National Institute of Technology, Silchar, India

Keval Gandhi Department of Applied Physics, Sardar Vallabhbhai National Institute of Technology, Surat, Gujarat, India

S. Ganesh Department of Physics, Birla Institute of Technology and Science, Pilani, India

Mayukh Raj Gangopadhyay Theory Divison, Saha Institute of Nuclear Physics, Kolkata, India;
Centre for Theoretical Physics, Jamia Millia Islamia, New Delhi, India

Avijit K. Ganguly Department of Physics (MMV), Banaras Hindu University, Varanasi, India

Ila Garg Department of Physics, Indian Institute of Technology Bombay, Powai, Mumbai, India

Renu Garg Department of Physics, Panjab University, Chandigarh, India

V. Gaur Virginia Polytechnic Institute and State University, Blacksburg, Virginia, India

Rajiv V. Gavai Tata Institute of Fundamental Research, Colaba, Mumbai, India

Elizabeth George Department of Physics, Indian Institute of Technology Bombay, Mumbai, India

Pulkit S. Ghoderao Indian Institute of Technology Bombay, Powai, Mumbai, India

C. Ghosh Variable Energy Cyclotron Centre, Kolkata, India;
Homi Bhabha National Institute, Mumbai, India

S. K. Ghosh Department of Physics and CAPSS, Bose Institute, Kolkata, India

Snigdha Ghosh IIT Gandhinagar, Gandhinagar, Gujarat, India

Sovan Ghosh Post Graduate, Department of Physics, Vijaya College, Bangalore, India

Anjan K. Giri Indian Institute of Technology, Hyderabad, Kandi, India

U. Gokhale Department of High Energy Physics, Tata Institute of Fundamental Research, Colaba, Mumbai, India

Mohit Gola University of Delhi, Delhi, India

Srubabati Goswami Theoretical Physics Division, Physical Research Laboratory, Ahmedabad, India

Rajat Gupta Panjab University, Chandigarh, India

Najmul Haque School of Physical Sciences, National Institute of Science Education and Research, HBNI, Jatni, Khurda, India

Nikhil Hatwar Department of Physics, Birla Institute of Technology and Science, Pilani, India

Honey Tata Institute of Fundamental Research, Mumbai, India

D. Indumathi The Institute of Mathematical Sciences, Tharamani, Chennai, Tamil Nadu, India;
Homi Bhabha National Institute, Anushaktinagar, Mumbai, Maharashtra, India

Vijay Iyer School of Physical Sciences, National Institute of Science Education and Research, HBNI, Jatni, India

Ambar Jain Indian Institute of Science Education and Research, Bhopal, MP, India

Chakresh Jain Centre for Detector and Related Software Technology, Department of Physics and Astrophysics, University of Delhi, Delhi, India

Geetika Jain Centre for Detector and Related Software Technology, Department of Physics and Astrophysics, University of Delhi, Delhi, India

Manoj K. Jaiswal Department of Physics, University of Allahabad, Prayagraj, India

Bharti Jarwal Department of Physics and Astrophysics, University of Delhi, Delhi, India

Abhik Jash School of Physical Sciences, National Institute of Science Education and Research, HBNI, Jatni, Odisha, India

Pawan Joshi Department of Physics, Indian Institute of Science Education and Research Bhopal, Bhopal, India

S. R. Joshi Department of High Energy Physics, Tata Institute of Fundamental Research, Colaba, Mumbai, India

Anjan S. Joshipura Theoretical Physics Division, Physical Research Laboratory, Ahmedabad, India

Jan Kalinowski Faculty of Physics, University of Warsaw, Warsaw, Poland

A. B. Kaliyar Indian Institute of Technology Madras, Chennai, India

D. Kalra Department of Physics, Panjab University, Chandigarh, India

Bithika Karmakar Theory Division, Saha Institute of Nuclear Physics, HBNI, Kolkata, India

Saikat Karmakar Tata Institute of Fundamental Research, Mumbai, India

Monal Kashav Department of Physics and Astronomical Science, Central University of Himachal Pradesh, Dharamshala, India

Varchaswi K. S. Kashyap School of Physical Sciences, National Institute of Science Education and Research, HBNI, Jatni, Odisha, India

Daljeet Kaur S.G.T.B. Khalsa College, University of Delhi, New Delhi, India

Manjit Kaur Panjab University, Chandigarh, India

P. K. Kaur Department of High Energy Physics, Tata Institute of Fundamental Research, Colaba, Mumbai, India

Najimuddin Khan Centre for High Energy Physics, Indian Institute of Science, Bangalore, India

Anisa Khatun Department of Physics, Aligarh Muslim University, Aligarh, India

Virendrasinh Kher Applied Physics Department, Polytechnic, The M S University of Baroda, Vadodara, Gujarat, India

Bharti Kindra Physical Research Laboratory, Ahmedabad, India; Indian Institute of Technology, Gandhinagar, India

H. Kolla Electronics Division, Bhabha Atomic Research Centre, Trombay, Mumbai, India

Jyothsna Rani Komaragiri Indian Institute of Science, Bengaluru, India

Partha Konar Theoretical Physics Group, Physical Research Laboratory, Ahmedabad, India

Paweł Kozów Faculty of Physics, University of Warsaw, Warsaw, Poland

A. Kumar Variable Energy Cyclotron Centre, Kolkata, India; Homi Bhabha National Institute, Mumbai, India

Abhass Kumar Physical Research Laboratory, Ahmedabad, India; Harish-Chandra Research Institute, Jhunsi, Allahabad, India

Ajit Kumar Homi Bhabha National Institute, Mumbai, India

Anil Kumar Saha Institute of Nuclear Physics, Kolkata, India; Homi Bhabha National Institute, Trombay, Mumbai, India

Arvind Kumar Dr. B R Ambedkar National Institute of Technology, Jalandhar, Punjab, India

Ashok Kumar Department of Physics and Astrophysics, University of Delhi, Delhi, India

Hemant Kumar Department of Physics & Astrophysics, University of Delhi, Delhi, India

J. Kumar Variable Energy Cyclotron Centre, Kolkata, India

M. Kumar Malaviya National Institute of Technology, Jaipur, India

Nilanjana Kumar Department of Physics and Astrophysics, University of Delhi, Delhi, India

Priyanka Kumar Department of Physics, Cotton University, Guwahati, Assam, India

Rajesh Kumar Dr. B R Ambedkar National Institute of Technology, Jalandhar, Punjab, India

Ramandeep Kumar Akal University Talwandi Sabo, Punjab, India

Sanjeev Kumar Department of Physics and Astrophysics, University of Delhi, New Delhi, India

Sunil Kumar Department of Physics, Panjab University, Chandigarh, India

Utkarsh Kumar Department of Physics, Ariel University, Ariel, Israel

V. Kumar Saha Institute of Nuclear Physics, Kolkata, India;
Homi Bhabha National Institute, Trombay, Mumbai, India

Ajay Kumar Rai Department of Applied Physics, Sardar Vallabhbhai National Institute of Technology, Surat, Gujarat, India

Santosh Kumar Rai Regional Centre for Accelerator-based Particle Physics, Harish-Chandra Research Institute, HBNI, Jhusi, Allahabad, India

Priyanka Kumari Panjab University, Chandigarh, India

Suman Kumbhakar Indian Institute of Technology Bombay, Mumbai, India

Sourav Kundu School of Physical Sciences, National Institute of Science Education and Research, HBNI, Jatni, India

Manu Kurian Indian Institute of Technology Gandhinagar, Gandhinagar, Gujarat, India

Apurba Laha Department of Electrical Engineering, Indian Institute of Technology Bombay, Mumbai, India

Amitabha Lahiri SNBNCBS, Kolkata, India

Anirban Lahiri Fakultät für Physik, Universität Bielefeld, Bielefeld, Germany

Jayita Lahiri Regional Centre for Accelerator-based Particle Physics, Harish-Chandra Research Institute, HBNI, Jhunsi, Allahabad, India

S. M. Lakshmi Indian Institute of Technology Madras, Chennai, India

K. Lalwani Malaviya National Institute of Technology, Jaipur, India

Gaetano Lambiase Dipartimento di Fisica "E.R. Caianiello", Universitá di Salerno, Fisciano (SA), Italy

James F. Libby Indian Institute of Technology Madras, Chennai, India

Manisha Lohan IRFU, CEA, Université Paris-Saclay, Gif-sur-Yvette, France

A. Lokapure Department of High Energy Physics, Tata Institute of Fundamental Research, Colaba, Mumbai, India

Kallingalthodi Madhu Department of Physics, BITS-Pilani, Zuarinagar, India

Namit Mahajan Physical Research Laboratory, Ahmedabad, India

Rajesh K. Maiti IISER Mohali, Punjab, India

Snehanshu Maiti Indian Institute of Technology Madras, Chennai, India

Rudra Majhi University of Hyderabad, Hyderabad, India

P. Maji Department of Physics, National Institute of Technology Durgapur, Durgapur, West Bengal, India

Debasish Majumdar Astroparticle Physics and Cosmology Division, Saha Institute of Nuclear Physics, HBNI, Kolkata, India

Nayana Majumdar Saha Institute of Nuclear Physics, HBNI, Kolkata, India

Devdatta Majumder University of Kansas, Lawrence, Kansas, USA

Gobinda Majumder DHEP, Tata Institute of Fundamental Research, Mumbai, India

B. Mallick Institute of Physics, HBNI, Bhubaneswar, India

Debasish Mallick National Institute of Science Education and Research, HBNI, Jatni, India

Rusa Mandal IFIC, Universitat de València-CSIC, València, Spain

A. Manna Electronics Division, Bhabha Atomic Research Centre, Trombay, Mumbai, India

P. Marquard DESY, Zeuthen, Germany

A. Maulik Department of Physics and CAPSS, Bose Institute, Kolkata, India

Arindam Mazumdar Physical Research Laboratory, Ahmedabad, India

Alberto Messineo Department of Physics, University of Pisa, Pisa, Italy

Aditya Nath Mishra Instituto de Ciencias Nucleares, UNAM, CDMX, Mexico

Arvind Kumar Mishra Theoretical Physics Division, Physical Research Laboratory, Ahmedabad, India;
Indian Institute of Technology Gandhinagar, Gandhinagar, India

Dheeraj Kumar Mishra The Institute of Mathematical Sciences, Chennai, India;
Homi Bhabha National Institute, Mumbai, India

Hiranmaya Mishra Theory Division, Physical Research Laboratory, Navrangpura, Ahmedabad, India

M. Mishra Department of Physics, Birla Institute of Technology and Science, Pilani, India

Subhasmita Mishra IIT Hyderabad, Hyderabad, India

Akhila Mohan Department of Physics, BITS-Pilani, Zuarinagar, India

Lakshmi S. Mohan Indian Institute of Technology Madras, Chennai, India

Rukmani Mohanta School of Physics, University of Hyderabad, Hyderabad, Telangana, India

Bedangadas Mohanty School of Physical Sciences, National Institute of Science Education and Research, HBNI, Jatni, Bhubaneswar, Odisha, India; Department of Experimental Physics, CERN, Geneva, Switzerland

G. B. Mohanty Tata Institute of Fundamental Research, Mumbai, India

Subhendra Mohanty Physical Research Laboratory, Ahmedabad, India

Manas K. Mohapatra Indian Institute of Technology, Hyderabad, Kandi, India

R. N. Mohapatra Department of Physics, Maryland Center for Fundamental Physics, University of Maryland, College Park, MD, USA

Ranjita K. Mohapatra Department of Physics, Indian Institute of Technology Bombay, Mumbai, India; Institute of Physics, Bhubaneswar, India; Homi Bhabha National Institute, Mumbai, India

S. Moitra Electronics Division, Bhabha Atomic Research Centre, Trombay, Mumbai, India

Mitali Mondal Variable Energy and Cyclotron Centre, Kolkata, India

N. K. Mondal HENPPD, Saha Institute of Nuclear Physics, Kolkata, India

S. Mondal Tata Institute of Fundamental Research, Mumbai, India

Suryanarayan Mondal NPD, Homi Bhaba National Institute, Mumbai, India; DHEP, Tata Institute of Fundamental Research, Mumbai, India

Sanjib Muhuri Variable Energy Cyclotron Centre, Kolkata, India

Ananya Mukherjee Department of Physics, Tezpur University, Tezpur, India

Arghya Mukherjee Saha Institute of Nuclear Physics, Kolkata, India

Tamal K. Mukherjee Department of Physics, School of Sciences, Adamas university, Kolkata, India

Sourav Mukhopadhyay Bhabha Atomic Research Centre, Mumbai, India

Supratik Mukhopadhyay Saha Institute of Nuclear Physics, HBNI, Kolkata, India

Upala Mukhopadhyay Astroparticle Physics and Cosmology Division, Saha Institute of Nuclear Physics, HBNI, Kolkata, India

Biswarup Mukhopadhyaya Regional Centre for Accelerator-based Particle Physics, Harish-Chandra Research Institute, HBNI, Jhunsi, Allahabad, India

Lakshmi P Murgod Department of Physics, Central University of Karnataka, Kalaburagi, India

M. V. N. Murthy The Institute of Mathematical Sciences, Chennai, India

Munshi G. Mustafa Theory Division, Saha Institute of Nuclear Physics, HBNI, Kolkata, India

P. Nagaraj Department of High Energy Physics, Tata Institute of Fundamental Research, Colaba, Mumbai, India

Srishti Nagu Department of Physics, Lucknow University, Lucknow, India

Bharati Naik Indian Institute of Technology Bombay, Mumbai, India

Md. Naimuddin Department of Physics and Astrophysics, University of Delhi, Delhi, India

Dibyendu Nanda Department of Physics, Indian Institute of Technology Guwahati, Assam, India

Ekata Nandy Variable Energy Cyclotron Centre, HBNI, Kolkata, India;
Homi Bhabha National Institute, Mumbai, India

Ashish Narang Physical Research Laboratory, Ahmedabad, India;
Indian Institute of Technology Gandhinagar, Gandhinagar, India

Nimmala Narendra Indian Institute of Technology Hyderabad, Kandi, Sangareddy, Telangana, India

Newton Nath Institute of High Energy Physics, Chinese Academy of Sciences, Beijing, China;
School of Physical Sciences, University of Chinese Academy of Sciences, Beijing, China

A. Naveena Kumara Department of Physics, National Institute of Technology Karnataka (NITK), Mangaluru, India

Surya Narayan Nayak Jyoti Vihar, Burla, Sambalpur, Odisha, India

T. K. Nayak Variable Energy Cyclotron Centre, Kolkata, India;
CERN, Geneva, Switzerland

P. Nayek Department of Physics, National Institute of Technology, Durgapur, West Bengal, India

V. Negi Variable Energy Cyclotron Centre, Kolkata, India

M. Nizam Homi Bhabha National Institute, Mumbai, India;
Tata Institute of Fundamental Research, Mumbai, India

Manjunath Omana Kuttan Department of Physics, Central University of Karnataka, Kalaburagi, India

Abhilash Padhy Department of Physics, Indian Institute of Science Education and Research, Bhopal, India

A. Padmini Electronics Division, Bhabha Atomic Research Centre, Trombay, Mumbai, India

Rita Paikaray Department of Physics, Ravenshaw University, Cuttack, India

N. Panchal Homi Bhabha National Institute, Mumbai, India;
Tata Institute of Fundamental Research, Mumbai, India

Sukanta Panda Department of Physics, Indian Institute of Science Education and Research Bhopal, Bhopal, India;
Department of Physics, Ariel University, Ariel, Israel

Susil Kumar Panda Department of Physics, Ravenshaw University, Cuttack, India

H. C. Pandey Department of Physics, Birla Institute of Applied Sciences, Bhimtal, India

Madhurima Pandey Astroparticle Physics and Cosmology Division, Saha Institute of Nuclear Physics, HBNI, Kolkata, India

Sujata Pandey Discipline of Physics, Indian Institute of Technology Indore, Indore, India

Sudhir Pandurang Rode Indian Institute of Technology Indore, Indore, Madhya Pradesh, India

J. N. Pandya Faculty of Technology and Engineering, Applied Physics Department, The Maharaja Sayajirao University of Baroda, Vadodara, Gujarat, India

Lata Panwar Indian Institute of Science, Bengaluru, India

Priyank Parashari Physical Research Laboratory, Ahmedabad, India;
Indian Institute of Technology Gandhinagar, Gandhinagar, India

Bibhuti Parida Tomsk State University, Tomsk, Russia

M. K. Parida CETMS, SOA University, Bhubaneswar, India

Sonia Parmar Panjab University, Chandigarh, India

Avani Patel Indian Institute of Science Education and Research Bhopal, Bhopal, India

Shesha D. Patel Faculty of Science, Physics Department, The Maharaja Sayajirao University of Baroda, Vadodara, Gujarat, India

Vikas Patel Department of Applied Physics, Sardar Vallabhbhai National Institute of Technology, Surat, Gujarat, India

Mahadev Patgiri Department of Physics, Cotton University, Guwahati, Assam, India

Pathaleswar Department of High Energy Physics, Tata Institute of Fundamental Research, Colaba, Mumbai, India

Sourav Patra IISER Mohali, Punjab, India

Avik Paul Astroparticle Physics and Cosmology Division, Saha Institute of Nuclear Physics, HBNI, Kolkata, India

S. Pethuraj NPD, Homi Bhaba National Institute, Mumbai, India; DHEP, Tata Institute of Fundamental Research, Mumbai, India

Aman Phogat Department of Physics and Astrophysics, University of Delhi, Delhi, India

Shailesh Pincha Department of Applied Physics, Sardar Vallabhbhai National Institute of Technology, Surat, Gujarat, India

Soumita Pramanick Department of Physics, University of Calcutta, Kolkata, India; Harish-Chandra Research Institute, Jhunsi, Allahabad, India

S. K. Prasad Department of Physics and CAPSS, Bose Institute, Kolkata, India

Bhabani Prasad Mandal Department of Physics, Institute of Science, Banaras Hindu University, Varanasi, India

Massimiliano Procura Fakultät für Physik, Universität Wien, Wien, Austria; Theoretical Physics Department, CERN, Geneva, Switzerland

M. Punna Electronics Division, Bhabha Atomic Research Centre, Trombay, Mumbai, India

Moh. Rafik Department of Physics and Astrophysics, University of Delhi, Delhi, India

S. Raha Department of Physics and CAPSS, Bose Institute, Kolkata, India

K. V. Rajani Department of Physics, National Institute of Technology Karnataka (NITK), Mangalore, India

N. Rajeev National Institute of Technology, Silchar, India

S. Rajkumarbharathi Tata Institute of Fundamental Research, Mumbai, India

Harishankar Ramachandran Indian Institute of Technology Madras, Chennai, India

P. Ramadevi Indian Institute of Technology Bombay, Powai, Mumbai, India

Arun Rana Department of Physics, Indian Institute of Science Education and Research, Bhopal, India

N. Rana DESY, Zeuthen, Germany; INFN, Milano, Italy

Raghavan Rangarajan School of Arts and Sciences, Ahmedabad University, Ahmedabad, India

Kirti Ranjan Centre for Detector and Related Software Technology, Department of Physics and Astrophysics, University of Delhi, Delhi, India

R. Rath Department of Physics, Indian Institute of Technology Indore, Simrol, Indore, India

Haresh Raval Department of Physics, Institute of Science, Banaras Hindu University, Varanasi, India;
Department of Physics, Indian Institute of Technology Delhi, New Delhi, India

K. C. Ravindran Department of High Energy Physics, Tata Institute of Fundamental Research, Colaba, Mumbai, India

Deependra Singh Rawat Department of Physics (UGC-Centre of Advanced Study), Kumaun University, Nainital, India

Atasi Ray School of Physics, University of Hyderabad, Hyderabad, India

Rajarshi Ray Department of Physics & Center for Astroparticle Physics & Space Science, Bose Institute, Kolkata, India

Amitava Raychaudhuri Department of Physics, University of Calcutta, Kolkata, India

Karaparambil Rajan Rebin Indian Institute of Technology Madras, Chennai, India

P. K. Resmi Indian Institute of Technology Madras, Chennai, India

Niharika Rout Indian Institute of Technology, Madras, India

Promita Roy Saha Institute of Nuclear Physics, HBNI, Kolkata, India

S. Roy Department of Physics and CAPSS, Bose Institute, Kolkata, India

Sahithi Rudrabhatla Department of Physics, University of Illinois at Chicago, Chicago, IL, USA

Samrangy Sadhu Variable Energy Cyclotron Centre, Kolkata, India;
Homi Bhabha National Institute, Mumbai, India

Soumya Sadhukhan Department of Physics and Astrophysics, University of Delhi, Delhi, India

Rajib Saha Indian Institute of Science Education and Research, Bhopal, MP, India

Debashis Sahoo TIFR, Mumbai, India

Pragati Sahoo Discipline of Physics, School of Basic Sciences, Indian Institute of Technology Indore, Indore, India

Raghunath Sahoo Discipline of Physics, School of Basic Sciences, Indian Institute of Technology Indore, Indore, India

S. Sahoo Department of Physics, National Institute of Technology, Durgapur, West Bengal, India

Sarita Sahoo CETMS, SOA University, Bhubaneswar, India; Institute of Physics, HBNI, Bhubaneswar, India

Suchismita Sahoo Theoretical Physics Division, Physical Research Laboratory, Ahmedabad, India

Narendra Sahu Indian Institute of Technology Hyderabad, Kandi, Sangareddy, Telangana, India

P. K. Sahu Institute of Physics, HBNI, Bhubaneswar, India

S. Sahu Institute of Physics, HBNI, Bhubaneswar, India

S. K. Sahu Institute of Physics, HBNI, Bhubaneswar, India

Jogender Saini Variable Energy Cyclotron Centre, Kolkata, India

Jyoti Saini Indian Institute of Technology Jodhpur, Jodhpur, India

Amrutha Samalan Department of Physics, Central University of Karnataka, Kalaburagi, India

Rome Samanta Physics and Astronomy, University of Southampton, Southampton, UK

Subhasis Samanta School of Physical Sciences, National Institute of Science Education and Research, HBNI, Jatni, Bhubaneswar, India

Deepak Samuel Department of Physics, Central University of Karnataka, Kalaburagi, India

Kaur Sandeep Panjab University, Chandigarh, India

S. Sandilya University of Cincinnati, Cincinnati, Ohio, India

M. N. Saraf Department of High Energy Physics, Tata Institute of Fundamental Research, Colaba, Mumbai, India

Pradeep Sarin Department of Physics, Indian Institute of Technology Bombay, Mumbai, India

Sandip Sarkar Saha Institute of Nuclear Physics, HBNI, Kolkata, India

B. Satyanarayana Department of High Energy Physics, Tata Institute of Fundamental Research, Colaba, Mumbai, India

P. S. Saumia Bogoliubov Laboratory of Theoretical Physics, JINR, Dubna, Russia

H. Saveetha Institute of Mathematical Sciences, Chennai, India

S. Sawant Tata Institute of Fundamental Research, Mumbai, India

C. Schneider RISC, Johannes Kepler University, Linz, Austria

J. Selvaganapathy Theoretical Physics Group, Physical Research Laboratory, Ahmedabad, India

Pritam Sen The Institute of Mathematical Sciences, Tharamani, Chennai, Tamil Nadu, India;
Homi Bhabha National Institute, Anushaktinagar, Mumbai, Maharashtra, India

Wadut Shaikh Saha Institute of Nuclear Physics, HBNI, Kolkata, India

Anjali Sharma Panjab University, Chandigarh, India

Ashish Sharma IIT Madras, Chennai, Tamil Nadu, India

Gazal Sharma Department of Physics and Astronomical Science, School of Physical and Material Sciences, Central University of Himachal Pradesh (CUHP), Dharamshala, Kangra, HP, India

Meenakshi Sharma Department of Physics, University of Jammu, Jammu and Kashmir, Jammu, India

Umesh Shas Department of High Energy Physics, Tata Institute of Fundamental Research, Colaba, Mumbai, India

R. R. Shinde Department of High Energy Physics, Tata Institute of Fundamental Research, Colaba, Mumbai, India

Anup Kumar Sikdar Indian Institute of Technology Madras, Chennai, India

G. Sikder University of Calcutta, Kolkata, India

S. Sikder Electronics Division, Bhabha Atomic Research Centre, Trombay, Mumbai, India

D. Sil Department of High Energy Physics, Tata Institute of Fundamental Research, Colaba, Mumbai, India

Dipankar Sil Tata Institute of Fundamental Research, Mumbai, India

R. N. Singaraju Variable Energy Cyclotron Centre, Kolkata, India

Captain R. Singh Department of Physics, Birla Institute of Technology and Science, Pilani, India

J. B. Singh Department of Physics, Panjab University, Chandigarh, India

Jagbir Singh Panjab University, Chandigarh, India

Janardan P. Singh Faculty of Science, Physics Department, The Maharaja Sayajirao University of Baroda, Vadodara, Gujarat, India

Jaydip Singh Department of Physics, Lucknow University, Lucknow, India

Jyotsna Singh Department of Physics, Lucknow University, Lucknow, India

Lakhwinder Singh Institute of Physics, Academia Sinica, Taipei, Taiwan

Ravindra Singh Department of Electrical Engineering, Indian Institute of Technology Bombay, Mumbai, India

S. Somorendro Singh Department of Physics and Astrophysics, University of Delhi, Delhi, India

V. Singhal Variable Energy Cyclotron Centre, Kolkata, India

Shivaramakrishna Singirala Indian Institute of Technology Indore, Simrol, Madhya Pradesh, India;
University of Hyderbad, Hyderabad, India

Roopam Sinha Saha Institute of Nuclear Physics, Kolkata, India

N. R. Soni Faculty of Technology and Engineering, Applied Physics Department, The Maharaja Sayajirao University of Baroda, Vadodara, Gujarat, India

C. Soumya Institute of Physics, Bhubaneswar, India

Vipin Sudevan Indian Institute of Science Education and Research, Bhopal, MP, India

M. Sukhwani Electronics Division, Bhabha Atomic Research Centre, Trombay, Mumbai, India

V. Sunilkumar Department of Physics, UC College, Aluva, Kerala, India

S. Swain Institute of Physics, HBNI, Bhubaneswar, India;
School of Physics, Sambalpur University, Sambalpur, India

Sagarika Swain Institute of Physics, PO: Sainik School, HBNI, Bhubaneswar, India

Michał Szleper National Center for Nuclear Research, High Energy Physics Department, Warsaw, Poland

S. H. Thoker University of Kashmir, Srinagar, India

M. Thomas Electronics Division, Bhabha Atomic Research Centre, Trombay, Mumbai, India

Swatantra Kumar Tiwari Discipline of Physics, School of Basic Sciences, Indian Institute of Technology Indore, Indore, India

Sławomir Tkaczyk Fermi National Accelerator Laboratory, Batavia, IL, USA

K. Trabelsi Laboratory of the Linear Accelerator (LAL), Orsay, France

Jyoti Tripathi Department of Physics, Panjab University, Chandigarh, India

A. Tripathy Institute of Physics, HBNI, Bhubaneswar, India;
Utkal University, Bhubaneswar, India

Sushanta Tripathy Department of Physics, Indian Institute of Technology Indore, Simrol, India

L. Umesh Tata Institute of Fundamental Research, Mumbai, India

S. S. Upadhya Department of High Energy Physics, Tata Institute of Fundamental Research, Colaba, Mumbai, India

Deepak Vaid Department of Physics, National Institute of Technology Karnataka (NITK), Mangalore, India

Ton van den Brink Utrecht University, Utrecht, The Netherlands

P. Verma Tata Institute of Fundamental Research, Mumbai, India

Surender Verma Department of Physics and Astronomical Science, School of Physical and Material Sciences, Central University of Himachal Pradesh (CUHP), Dharamshala, Kangra, HP, India

K. N. Vishnudath Theoretical Physics Division, Physical Research Laboratory, Ahmedabad, India;
Discipline of Physics, Indian Institute of Technology, Gandhinagar, India

A. S. Vytheeswaran Department of Theoretical Physics, University of Madras, Chennai, India

Dinesh Yadav Department of Physics (UGC-Centre of Advanced Study), Kumaun University, Nainital, India

M. Younus Department of Physics, Nelson Mandela University, Port Elizabeth, South Africa

E. Yuvaraj Department of High Energy Physics, Tata Institute of Fundamental Research, Colaba, Mumbai, India

Yongchao Zhang Department of Physics and McDonnell Center for the Space Sciences, Washington University, St. Louis, MO, USA

Chapter 101
Characterization of Metal by GEM Detector Using Ion Beam Facility at IOP

A. Tripathy, P. K. Sahu, S. Swain, S. Sahu, and B. Mallick

Abstract A Triple GEM prototype of area 10×10 cm^2 is fabricated and characterized using Fe^{55} source at the Institute of Physics, Bhubaneswar, India. The same GEM detector is used to characterize different metals by using the ion beam facility at the Institute of Physics. Proton beam generated from a 3 MV Tandem Pelletron has been used to emit X-rays from different metal targets, which irradiate on GEM. X-ray yield of the metals is directly proportional to the proton beam current. The spectra obtained from the detector have been used to detect the sample based on their energy output. The spectrum is taken with a brass sample shows two peaks corresponding to two X-ray energies.

101.1 Introduction

In recent years, there have been significant advances in the field of Micro Pattern Gas Detector (MPGD) [1] due to their features of high rate capability, excellent spatial and time resolution, stability, radiation hardness, etc. These evolving micropattern technologies are already introduced in different mega projects of nuclear and high energy physics experiments [2]. The Gas Electron Multiplier (GEM) is one of these micropattern detectors, which is the first choice of researcher whenever operation in a high luminosity environment, stability over performance and high radiation resistance are required. GEM detectors are already employed in several particle trackers in high energy experiments [3], imaging instruments with optical and electrical readouts [4], cryogenic detectors for dark matter and neutrino search and fast single-photon detectors, and many more applications are foreseen. Apart from the above features, GEM has one more advantage that it can be used in multiple layers, i.e., multi-GEM structure. The main objective to use the multi-GEM configuration is to get

A. Tripathy (✉) · P. K. Sahu · S. Swain · S. Sahu · B. Mallick
Institute of Physics, HBNI, Sachivalaya Marg, Bhubaneswar 751005, India
e-mail: alekhika.t@gmail.com

A. Tripathy
Utkal University, Vani Vihar, Bhubaneswar 751004, India

© Springer Nature Singapore Pte Ltd. 2021
P. K. Behera et al. (eds.), *XXIII DAE High Energy Physics Symposium*,
Springer Proceedings in Physics 261,
https://doi.org/10.1007/978-981-33-4408-2_101

high detector gain (order of 10^3–10^4), suppression of ion feedback, low discharge probability, reduced aging issue, optimum spatial and time resolution, etc.

GEMs were first introduced by Fabio Sauli in the Gas Detector Development Group at CERN in 1997 [5]. GEM consists of 50 μm Kapton foil with 5 μm copper cladding on both sides. Kapton is generally used because of its stability over a wide range of temperature (−269 °C–400 °C). With the acid etching process and specific photolithography technique, an array of holes are created throughout the foil. The typical hole diameter is 70 μm at the surface and 50–55 μm at the center with hole-to-hole distance, i.e., pitch is 140 μm. A magnified view of a standard double mask GEM foil with hole diameter and pitch is given in Fig. 101.1. Potential difference is applied across the copper metals and high-intensity electric field is generated through the holes. The reduced diameter concentrates the field lines within the holes through out the foil. The typical field configuration of a GEM foil with field lines is shown in Fig. 101.2.

The simplest GEM detector, often called as single GEM is formed with a single GEM foil, a drift plane (Kapton with one-sided copper) placed above the GEM and a readout plane below the GEM for collection of electrons. In this configuration when external radiation enters, electrons are ionized through primary and secondary ionization in the drift gap and then carried toward the holes by the provided drift field. Inside the holes, electron multiplication occurs due to the avalanche process and then all the electrons are collected over the readout plane. For R&D purposes, we have

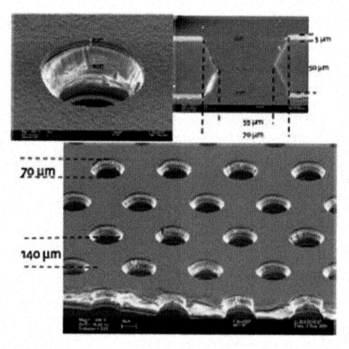

Fig. 101.1 Magnified view of GEM foil from electron microscope

Fig. 101.2 Cross section diagram of electric field lines and equipotential lines for two holes in a GEM foil

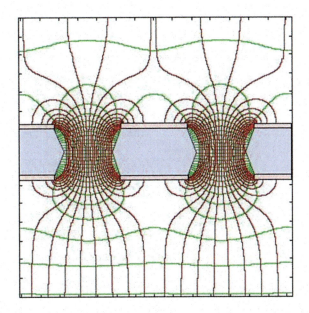

assembled a triple-layered GEM detector and obtained its characteristic properties by using ion beam facility. The full detector description with beam facility at IOP is discussed below.

101.2 Triple GEM Detector Geometry

The triple GEM detector consists of a gas-filled chamber with a drift plane, three cascaded GEM foils and a read-out plane. All the components are procured from CERN and assembled here in IOP, High Energy Detector laboratory. The negatively biased drift plane is placed at 3 mm distance above the top of the first GEM, acting as a cathode. Three $10 \times 10 \, cm^2$ standard stretched double mask GEM foils are kept one above another, separated by 2 mm distance forming two transfer gaps. Finally, a read-out plane is placed below the third GEM, forming an induction gap of 2 mm. When any external radiation enters into the detector and deposits its energy, ionization occurs, thus electron-ion pairs are generated in the drift volume. The field applied in this drift region is just high enough to push the primarily generated electrons into the holes of GEM 1. The electrons, which reached near the hole, experience a huge field due to a high potential difference around 60 kV/cm across the hole. Here electrons undergo proportional amplification and avalanche multiplication occurs. Appropriate fields are applied between the GEMs to transfer the avalanche generated electrons from one layer to another. Once the electrons come out from the holes of GEM1, they enter GEM 2 and again multiplications occur. Again, an avalanche occurs and electrons are transferred into holes of GEM 3 for additional multiplication. Finally,

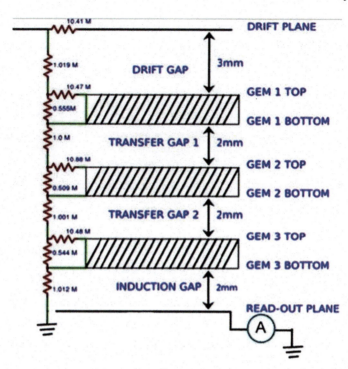

Fig. 101.3 Voltage divider circuit for triple GEM detector designed for voltage application to each layer

all the amplified electrons are collected over the readout and the induced signal is processed to further electronics. The readout used here has two-dimensional strips with variable width. The voltage distribution to the detector is done with a high-voltage divider circuit and the corresponding design is given in Fig. 101.3. Continuous flow of gas is done throughout the detector during the experiment. Here a mixture of Ar and CO_2 gas is used in a 70:30 ratio. As the detector performance is significantly affected by temperature and pressure ratio, the gas flow rate with ambient parameters such as temperature, pressure, and relative humidity is monitored by a data logger built in-house [6].

101.3 Ion Beam Facility at IOP

The Ion Beam laboratory of IOP consists of an NEC model 9SDH-2 tandem Pelletron accelerator that can provide a proton beam of energy range 1–6 meV [7]. The maximum terminal voltage is 3 MV and it can be set as per the user's requirement from 0.5 MV to 3.0 MV. Two negative ions sources Alphatross and MC-SNICS are employed within the accelerator. Alphatross is exclusively used for producing H^- and

He⁻ ions, whereas MC-SNICS produces all other types of negative ions. Commonly used ion beams are that of H, He, Li, C, Si, Cu, Ag, Au, etc. When energetic negative ions (order of 55 keV) are injected into the accelerator, they lose their electrons due to several collisions with stripper (Ar) gas atoms. At high terminal voltage, positive ions are emitted with a huge amount of energy. These positive beams are incident on the target with current varying from a few nano Ampere (nA) to a few tens of nano Ampere (nA). Here, 3 meV proton beams generated from 3 MV tandem Pelletron are used with different targets. The emitted X-rays from those targets are then detected by the detector for further characterization purposes.

101.4 Experimental Setup

The total setup consists of a triple GEM detector with high-voltage power supply, 3 MeV Proton beam as external radiation to be incident on $2 \times 1 \times 0.2$ cm^3 Fe, brass, and Cu sample and a Keithley Pico-ammeter for anode current measurement. The high-voltage power supply provides the negative potential to each layer through a divider circuit for the detector operation. The readout i.e. the anode plane is connected to the Picoammeter. A premixed gas mixture of Ar/CO$_2$ is continuously passed through the detector in a 70/30 ratio at an optimum flow rate value. The distance between the target to GEM detector is around 20 cm. The samples are placed at 45° on the beamline such that diffracted X-rays are incident on the detector perpendicularly. The schematic diagram of the setup and a picture of the setup in IBL are shown in Figs. 101.4 and 101.5.

The terminal voltage is set to 3 MV and variable beam current from 50 to 200 nA is applied during the experiment. Anode current from the detector was observed with different GEM voltage starting from 3500 to 4200 V.

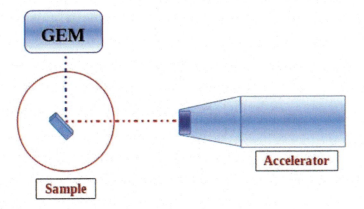

Fig. 101.4 Schematic view of detector setup with Ion beam

Fig. 101.5 Experimental setup in IBL

101.5   Results

The anode current was observed with increasing GEM voltage keeping beam current fixed for three different samples, Fe, Cu, and Brass. This is done for beam currents 150 nA and 200 nA and the same pattern is obtained as shown in Fig. 101.6.

The energy spectrum obtained for the brass sample is shown in Fig. 101.7. We can clearly distinguish the two energy peaks for brass using the GEM detector. The GEM detector can characterize the unidentified metal samples, which is very cost-effective and cheap.

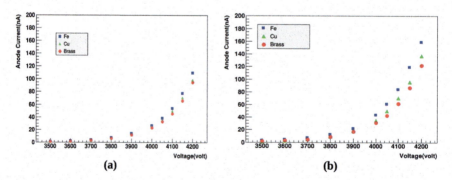

Fig. 101.6 Variation of anode current with GEM voltage for target currents (**a**) 150 nA and (**b**) 200 nA

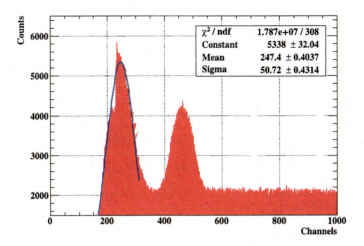

Fig. 101.7 Energy Spectrum for brass sample showing two energy peaks

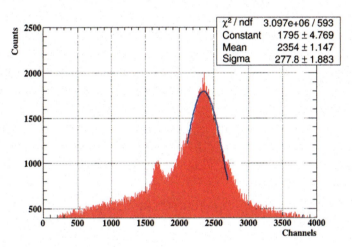

Fig. 101.8 Fe^{55} spectrum at 4200 V for the GEM detector

The detector is tested with a Fe^{55} X-ray source which provides 5.9 keV X-rays. The pulse height spectrum at 4200 V for this source is shown in Fig. 101.8.

Acknowledgements We would like to thank the members of the IOP workshop and IBL for their invaluable cooperation.

References

1. S. Pinto, Micropattern gas detector technologies and applications The work of the RD51 collaboration, in *IEEE Nuclear Science Symposium 2010 Conference Record*. https://arxiv.org/abs/1011.5529
2. TESLA Technical Design Report, Part IV, A Detector for TESLA (2001)
3. Technical Design Report for the Upgrade of the ALICE Time Projection Chamber, ALICE -TDR-016 (2016)
4. S. Bachmann et al., High rate X-ray imaging using multi- GEM detectors with a novel readout design. Nucl. Instr. Meth. A **478**, 104 (2002)
5. F. Sauli, GEM: A new concept for electron amplification in gas detectors. Nucl. Instr. Meth. A **386**, 531 (1997)
6. S. Sahu et al., Design and fabrication of data logger to measure the ambient parameters in gas detector R&D. JINST **12**, C05006 (2017)
7. www.pelletron.com

Chapter 102
Muon Momentum Spectra with mini-ICAL

A. D. Bhatt, Gobinda Majumder, V. M. Datar, and B. Satyanarayana

Abstract The upcoming 50 kt magnetized iron calorimeter (ICAL) detector at the India-based Neutrino Observatory (INO) is designed to study the atmospheric neutrinos and antineutrinos separately over a wide range of energies and path lengths. A prototype magnet (mini-ICAL) detector made of 11 layers of $4\,m \times 4\,m \times 5.6\,cm$ iron layer with interlayer gap of 4.5 cm is set up at IICHEP campus, Madurai (9.9 ° N, 78.1 ° E), where maximum magnetic field is 1.3 T. The RPCs are placed in the interlayer gap and act as sensitive detectors. The data are collected using coincidence signal in four layers, which are mainly due to cosmic ray muon. The momentum distribution as well as charge ratio of cosmic ray muons at the sea level will be presented.

102.1 Introduction

INO will be an underground experiment with at least 1 km rock overburden in all directions [1]. To carry out the neutrino physics experiment, an Iron Calorimeter (ICAL) has been proposed for installation at the INO cavern to observe the neutrino oscillation pattern at least over one full period. The main goals of this experiment are precise measurement of neutrino oscillation parameters including the sign of the

A. D. Bhatt, G. Majumder, V. M. Datar, B. Satyanarayana on behalf of INO mini-ICAL group

A. D. Bhatt (✉)
Institute of Nuclear Physics Polish Academy of Sciences, PL-31342 Krakow, Poland
e-mail: apoorva.bhatt@tifr.res.in

A. D. Bhatt · G. Majumder · V. M. Datar · B. Satyanarayana
Tata Institute of Fundamental Research, Mumbai, India
e-mail: gobinda@tifr.res.in

V. M. Datar
e-mail: vivek.datar@tifr.res.in

B. Satyanarayana
e-mail: bsn@tifr.res.in

© Springer Nature Singapore Pte Ltd. 2021   739
P. K. Behera et al. (eds.), *XXIII DAE High Energy Physics Symposium*,
Springer Proceedings in Physics 261,
https://doi.org/10.1007/978-981-33-4408-2_102

2–3 mass-squared difference, $\Delta m^2_{32} \left(= m^2_3 - m^2_2\right)$ through matter effects, the value of the leptonic CP phase and, last but not the least, the search for any non-standard effect beyond neutrino oscillations. The Resistive Plate Chamber (RPC) [3] is chosen as the active detector element for ICAL detector due to its low cost and excellent timing resolution.

As part of the ICAL detector R & D programme, two detector stacks have been successfully built to study the performance and long-term stability of the RPCs using cosmic ray muons. But, these tests were done with different electronics including the final design for the INO-ICAL experiment [1]. Also, these were tested in absence of magnetic field. To validate the magnet design and all the electronics in the fringe field of magnet, the miniICAL detector is constructed.

Other than the test of electronics in presence of magnetic field, these following items are also being tested in this setup (i) magnetic field measurement using pickup coil and Hall probes and compare with 2D simulation by MAGNET [2] software, (ii) performance of RPC including DC–DC power supply, (iii) feasibility study of using Muon Spin Rotation to measure B-field, complementary to sense wire loop and Hall probe data, (iv) measurement of the charge-dependent muon flux upto $\sim$ 1.5 GeV and compare with simulation, (v) proof of principle test of the cosmic muon veto detector for the feasibility of a shallow ICAL detector, etc.

102.2 Detector

The detector is $\sim 1/600$th the size of final ICAL detector in terms of volume and around 0.1 %interms of RPC and related electronics. A view of the detector is shown in Fig. 102.1. The detector mainly consists of 11 layers of soft iron plates having high magnetic permeability and low carbon percentage. The iron plate has material density of 7850 kg/m^3 and Young's modulus of 200 GPa. These iron plates have a knee point at 1.5 T magnetic field. Spacers, made of non-magnetic material (SS304), are used to create gap of 45 mm between two iron layers for placing resistive plate chamber detectors. Pure copper (purity 99.99%) OF grade with high electrical and thermal conductivity is used for mini-ICAL coil manufacturing. There are 2 coils with 18 turns each and induction of 24,000 ampere-turns. For cooling the coil, DM water with low conductivity is used. The Magnet Power Supply with rating 30 V, 1500 A, made by "Danfysik" is used for energizing the mini ICAL Magnet to achieve required magnetic field. A closed-loop low-conductivity ($\leq 10 \mu$ mho/cm) chilled water system is used to cool the magnet power supply system and the conducting copper coils.

The RPCs made of 1.74×1.87 m $\times 8$ mm glass gap are used as the sensitive detectors in miniICAL. The signal is readout from two orthogonally placed copper pickup panels on either side of the gas gaps, labelled as X- and Y-palnes. The strips are of width 2.8 cm with an inter-strip gap of 0.2 cm. Thus, there are $\sim$ 60 strips on X-side and $\sim$ 63 strips on Y-side. They can record a 2D position information of the charged particle traversing through the gas gap. The signal from the RPCs is first

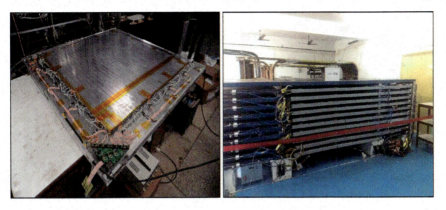

Fig. 102.1 (left) A RPC with complete readout electronics and (right) A view of miniICAL detector with 10 layers of RPC detector

passed through front-end electronics which includes 128 channels of fast amplifier and discriminator. Out of 10 layers of RPCs, the second layer uses ANUSPARSH [4] front-end amplifiers, designed by BARC. The other nine layers uses NINO [5] preamplifier-discriminator chip which is an ultrafast, low noise, 8-channel front-end developed at CERN to be used mainly in the ALICE time-of-flight detector. Also, there is DC–DC high voltage (HV) supply for providing bias voltage for RPC. An FPGA based DFE module with most of the data acquisition functionality built into it is used for on board DAQ [6]. The front-end electronics, RPCs, power supplies and gas lines are packed in a steel-reinforced epoxy tray. The back end electronics consists of multilevel trigger system, calibration unit and pulse shaper. The RPCs use a gas mixture of R134a, iso-butane and SF_6 with a proportion of 95.5:4.3:0.2. A closed-loop gas system, which maintains a flow rate of 6 SCCM with 2–3 mbar over pressure with respect to atmospheric pressure for individual chamber, is used for the circulation of this mixture [7]. The backend is formed with data concentrator, event builder, data storage system, etc. along with slow controls and monitoring of gas, magnet, power supplies, etc.

102.3 Simulation Code for MiniICAL

A GEANT4 [8] based simulation code for mini-ICAL detector is developed. The detector geometry in the simulation is designed as the actual detector, along with room and building in which mini-ICAL detector is placed to account for all the materials in the muons path till it reaches the detector. The muon events obtained from the simulation of primary cosmic rays using CORSIKA [9] generator are used in the simulation. These events were simulated in the CORSIKA with the geomagnetic location of the detector in Madurai. An INO specific digitization algorithm

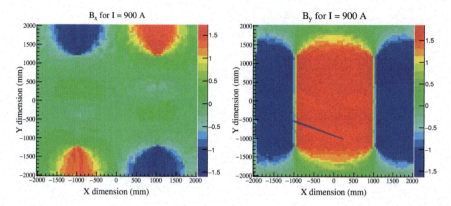

Fig. 102.2 Magnetic field map at 900 A current, (left) B_x and (right) B_y in Tesla. Length in X- & Y-axis are in mm

is developed to digitize the data. This helps in simulating various possible detector properties like detector inefficiencies, strip multiplicity, time resolution, detector noise, hardware trigger, etc. The magnetic field simulated using MAGNET [2] software is also an useful input in the simulation. The magnetic field in a layer of iron obtained through simulation is presented in Fig. 102.2.

102.4   Reconstruction of Track

The cosmic ray muons incident on the detector registers clean tracks with just about one cluster (combination of nearby hits) per RPC layer in the detector. In the absence of magnetic field, the hits from muon (being minimum ionizing particle) will present a straight line pattern with small kinks appearing due to multiplie scattering and strip multiplicity effect in the RPCs. Hence these tracks are fitted with straight line and various detector properties like efficiencies, strip multiciplties, electronic offsets for time measurements, etc. which are estimated. The mechanical alignment of the RPCs is also determined and perfected from these observations. These properties are further used as correction parameters and as inputs to the full detector simulation.

In presence of magnetic field, the hits from muon (experiencing Lorentz force) will show a curvature representing the bending of muon trajectory. The standard Kalman fit algorithm developed for the ICAL detector is used in these trajectories. The momentum, zenith angle and the azimuthal angle of the incident muon can be estimated through this fit. The track reconstruction is performed in two stages: track finding and track fitting. The track finding algorithm analyses the topology of the strips in order to identify seed tracks for fitting algorithm. A Kalman-filter-based algorithm is used to fit the tracks based on the bending of the tracks in the magnetic field. Every track is identified by a starting vector $X_0 = (x, y, dx/dz, dy/dz, q/p)$ which contains the position of the earliest hit (x, y, z) as recorded by the finder, with the charge-weighted inverse momentum q/p taken to be zero. Since the tracks are virtually straight in the starting section, the initial track direction is calculated from the first two layers. This initial state vector is then extrapolated to the next layer using a standard Kalman-filter based algorithm. The track fitting algorithm is explained in detail in [10].

102.5   Cosmic Muon Spectrum at Madurai

Figure 102.3 shows some cosmic muon events observed in mini-ICAL detector at 900 A current which corresponds to magnetic field of ∼1.5 T in the central region. The magnetic field in the central region of the mini-ICAL detector is along the Y-direction and the particle is moving along Z-direction. Hence, the bending of the track is supposed to be observed only along X-coordinate. It is expected that the track will not show any deviation in Y-coordinate except for multiple scattering.

The cosmic-ray muons events recorded using the mini-ICAL detector were passed through the same reconstruction algorithm, where the top four layers are used in the trigger criteria. To compare the observed data with MC, discrete muons from the CORSIKA [9] event generator were passed through the GEANT4-based detector simulation. In the simulation, the muons are generated above the ceiling of the building. The various detector parameters like uncorrelated and correlated inefficiencies, trigger efficiencies, strip multiplicity, layer residuals and hardware trigger criterias

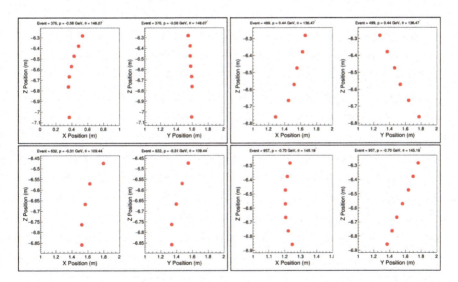

Fig. 102.3 Typical cosmic muon events in miniICAL

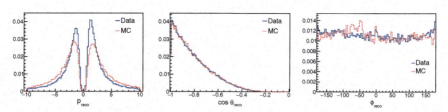

Fig. 102.4 Comparison of muon spectrum in data and CORSIKA simulation

were incorporated during the digitisation process of simulation. The detector parameters were calculated with the cosmic-ray muon events recorded in absence of magnetic field. The normalized momentum, polar angle and azimuthal angle distribution of data and MC sample are shown in Fig. 102.4.

Due to limited number of layer and poorer position resolution ($\sigma_x \sim 7–8$ mm) with respect to conventional tracking device ($\sigma_x \leq 100 \mu m$), the momentum resolution is poor. But, the discrepancy between the data and MC is mainly due to small uncertainty of magnetic field measurement and improper prediction of momentum spectrum of muon at Madurai. The measured momentum spectra of cosmic ray muons is distorted due to the finite resolution, limited trigger acceptance and other systematic effects of the detector. This spectrum needs to be unfolded to get the true muon spectrum at Madurai.

102.6   Summary

The mini-ICAL detector is commissioned with 10 layers of RPCs and is operational since May with 1.3 T field. The magnetic field measurement is closely matched with the simulation using MAGNET software. There is no unexpected noise in the RPC electronics due to fringe field. The observed muon spectrum is also closely matched with CORSIKA predictions. The measurement of momentum spectrum can be used to improve the neutrino flux at Theni.

References

1. A. Kumar et al., Physics potential of the ICAL detector at the India-based Neutrino Observatory (INO). Pramana J. Phys. **89**, 79 (2017)
2. Infolytica Corp, Electromagnetic field simulation software. http://www.infolytica.com/en/products/magnet/
3. M. Bhuyan et al., Development of 2m Œ 2m size Glass RPCs for INO. Nucl. Instrum. Meth. A **661**, S64–S67 (2012)
4. V.B. Chandratre et al., ANUSPARSH-II frontend ASIC for avalanche mode of RPC detector using regulated cascode trans-impedance amplifier, in *Proceedings of the DAE-BRNS Symposium on Nucl. Phys.*, vol. 60 (2015), pp. 928–929

5. P.K. Kaur et al., Development of Fast, Low Power 8-Channel Amplifier-Discriminator Board for the RPCs, in *XXII DAE High Energy Physics Symposium*, Springer Proceedings in Physics, vol. 203 (2018), pp. 571–574
6. Soft-Core processor based data acquisition module for ICAL RPCs with network interface, in *XXI DAE High Energy Physics Symposium*, Springer Proceedings in Physics vol. 174 (2015), pp. 565–570
7. M. Bhuyan et al., Performance of the prototype gas recirculation system with built-in RGA for INO RPC system. Nucl. Instrum. Meth. A **661**, S234–S240 (2012)
8. GEANT4 Collaboratio, GEANT4: a simulation toolkit. Nucl. Instrum. Meth. A **506**, 250–303 (2003)
9. D. Heck et al., CORSIKA: a monte carlo code to simulate extensive air showers (1998)
10. K. Bhattacharya et al., Error propagation of the track model and track fitting strategy for the Iron CALorimeter detector in India-based neutrino observatory. Comput. Phys. Commun. **185**, 3259–3268 (2014)

Chapter 103
Modeling Neutron Damage in Silicon Detectors for High Energy Physics Experiments

Chakresh Jain, Geetika Jain, Ranjeet Dalal, Ashutosh Bhardwaj, Kirti Ranjan, Alberto Messineo, and Maria Agness Ciocci

Abstract Silicon detectors are expected to experience an unprecedented neutron flux in the future upgrades of the detectors at the Large Hadron Collider (LHC). The challenging radiation environment of these experiments will severely affect the performance of silicon detectors. An irradiation campaign is generally carried out, followed by measurements, to develop radiation-hard silicon detectors. Device modeling complements the measurement results for the detailed understanding of the silicon detectors. Our Group at the University of Delhi successfully developed the radiation damage model for proton irradiation. However, a similar model for neutron irradiation has been missing. In the present work, a Technology Computer-Aided Design (TCAD) simulation software—Silvaco, has been used to study the effects of neutron irradiation on silicon detectors. The effects of radiation damage are incorporated using an effective two-trap model. A trap level is characterized by a number of trap parameters, e.g. trap type, trap energy level, introduction rate of acceptors and donors, and carrier (electrons and holes) capture cross section for that particular trap level. A systematic study of the sensitivity of various macroscopic parameters of silicon detectors to these trap parameters has been performed. The simulation results on leakage current (I_{LEAK}), full depletion voltage (V_{FD}) and charge collection efficiency (CCE), using the neutron damage model, are found to be in good agreement with the measurement results.

C. Jain (✉) · G. Jain · R. Dalal · A. Bhardwaj · K. Ranjan
Centre for Detector and Related Software Technology, Department of Physics and Astrophysics, University of Delhi, Delhi 110007, India
e-mail: interuniversechakresh15@gmail.com

A. Messineo · M. A. Ciocci
Department of Physics, University of Pisa, Pisa, Italy

© Springer Nature Singapore Pte Ltd. 2021
P. K. Behera et al. (eds.), *XXIII DAE High Energy Physics Symposium*,
Springer Proceedings in Physics 261,
https://doi.org/10.1007/978-981-33-4408-2_103

747

103.1 Introduction

The interaction of radiation with the detector causes the formation of trap levels in the forbidden energy gap of silicon. These trap levels modify the macroscopic properties of the detector namely, V_{FD}, I_{LEAK} and CCE. In simulations, we parameterize the effects of all such real trap levels by using an effective radiation damage model consisting of a few bulk and surface traps. Earlier, the University of Delhi developed a proton radiation damage model [1], and the effects of proton irradiation on the macroscopic properties of the detector were studied. In this work, a model for neutron irradiation which is extremely important for predicting the behavior of Si detectors in an irradiated environment is developed. A systematic study for the sensitivity of macroscopic parameters towards various trap parameters has been done in order to develop an effective neutron radiation damage model.

103.2 Sensitivity of Macroscopic Parameters of the Detector

The two-trap bulk damage model for proton irradiation has been used as a starting point and the trap parameters, viz. Introduction rate (G_{int}) and capture cross sections for electrons and holes (σ_e, σ_h), are varied one at a time, to study the variation in macroscopic properties of the detector viz. V_{FD}, I_{LEAK} and CCE.

103.2.1 Simulation Structure and Parameters

A plane parallel p-on-n structure with a physical thickness of 300 μm and a resistivity of 1.8 KΩ cm has been used to study V_{FD} and I_{LEAK}. While for CCE simulations, a similar structure has been used with p-type bulk (with resistivity of 4.6 KΩ cm) and a physical thickness of 320 μm.

103.2.2 Sensitivity of V_{FD}

The variation in V_{FD} is linked to the change in the density and properties of trap levels which are responsible for a change in the effective bulk doping density (N_{eff}). Similarly, variation in σ_e, σ_h may vary the ionization ability of the traps and may also affect the trapping and de-trapping time of free charge carriers. Figure 103.1 shows the effect of variation of some of these trap parameters on V_{FD} of p-on-n detectors. It can be seen that V_{FD} depends on G_{int} (Donor) but not on σ_e (Donor).

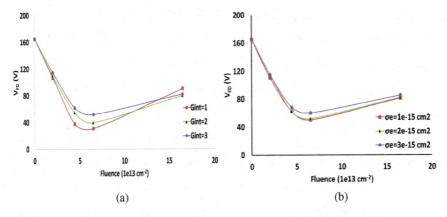

Fig. 103.1 Effect of trap parameters, **a** G_{int} (Donor), **b** σ_e (Donor) on V_{FD} values of n-type detectors

103.2.3   Sensitivity of I_{LEAK}

The leakage current of a Si detector depends upon temperature, volume of the detector and recombination of free charge carriers through recombination centers or trap levels. This recombination rate further depends upon lifetime of the free charge carriers which in turn depends on density and capture cross sections of trap levels.

Further, according to Shockley–Read–Hall (SRH) theory, recombination of charge carriers depends upon the position of recombination centers within the forbidden energy gap of Si and is maximum for mid gap levels. Therefore, the change in the detector leakage current on varying the trap parameters is attributed to the change in density and other parameters of these trap levels. Figure 103.2 shows the effect of variation of some of the trap parameters on I_{LEAK} of the Si detectors. It is observed

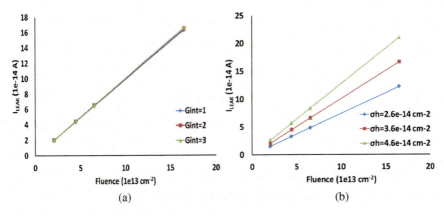

Fig. 103.2 Effect of trap parameters, **a** G_{int} (Donor), **b** σ_h (Acceptor) on I_{LEAK} values of n-type detectors

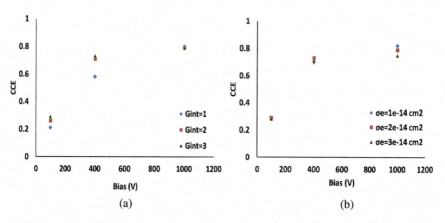

Fig. 103.3 Effect of trap parameters, **a** G_{int} (Donor), **b** σ_e (Acceptor) on CCE values of p-type detector. Simulations are performed at 3 bias levels: 100 V, 400 V and 1000 V

that I_{LEAK} has a strong dependence on σ_h (Acceptor) and is independent of G_{int} (Donor).

103.2.4 Sensitivity of CCE

The deep levels, generated due to irradiation, trap the free charge carriers and may have de-trapping time larger than the detector electronics peaking time. So, the collected charge is less resulting in decrease of CCE. Figure 103.3 shows the effect of variation of some of the trap parameters on CCE of the p-type silicon detectors. The CCE is found to be a function of σ_e (Acceptor) in fully depleted detectors.

103.2.5 Results of Sensitivity Study of Macroscopic Parameters

In the previous subsections, the variations of macroscopic parameters have been shown only for two-trap parameters. However, similar studies are performed by varying other trap parameters and the conclusions related to sensitivity of V_{FD}, I_{LEAK} and CCE with respect to these parameters are given below.

1. V_{FD} has a strong dependency on G_{int} (donor trap) and σ_e, σ_h (Acceptor traps).
2. I_{LEAK} depends only on σ_h (Acceptor) and is independent of other parameters.
3.   CCE depends mainly on G_{int} (donor) before full depletion, σ_e (Acceptor) after full depletion and slightly on other parameters.

Table 103.1 Parameters of the two-trap neutron bulk damage model

Trap type	Energy level (eV)	Introduction rate (cm^{-1})	σ_e (cm^2)	σ_h (cm^2)
Acceptor	$E_C-0.51$	4	7.2e−15	2.8e−14
Donor	$E_V + 0.48$	1	2e−15	2e−15

103.3   Development of Two-Trap Neutron Bulk Damage Model

To construct a radiation damage model for neutron irradiation, the developed proton bulk damage model is considered as the base model and the results of various experimental studies related to neutron irradiation are incorporated to tune the model.

It has been observed that neutron irradiation produces only one-sixth of the donor traps in comparison to proton irradiation [2]. Also, the V_{FD} values, after type inversion, for neutron-irradiated n-type bulk are found to be higher in comparison to proton irradiation [3]. I_{LEAK} is observed to be independent of proton and neutron irradiation [3]. Table 103.1 shows the proposed neutron radiation damage model with values of various trap parameters based on the understanding gained from experimental studies and the sensitivity studies performed in the previous section.

103.4   Results and Comparisons: Simulation Versus Measurements

103.4.1 V_{FD} Variation with Fluence

Figure 103.4a shows the comparison of simulation and measurement results of V_{FD} for an n-type detector for two different bulk doping: 2.37e12 cm^{-3} and 1e12 cm^{-3}. From these plots, it can be seen that simulation and measurement results [4] are in good agreement. For n-type detectors, V_{FD} first decreases with an increase in fluence reaches the minima and thereafter it starts increasing again. This is because the neutron irradiation mainly creates acceptor kind of traps. Therefore, for an originally donor-rich bulk the N_{eff} first decreases with increase in fluence, eventually leading to the type inversion of the bulk from n-type to p-type and after that N_{eff} and hence the V_{FD} keep on increasing with increasing neutron fluence.

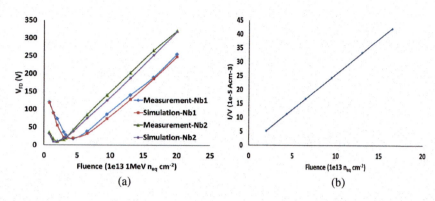

(a)   (b)

Fig. 103.4 **a** Variation of simulated and measured V_{FD} values with fluence for two different bulk doping: Nb1 = 2.37e12 cm^{-3} and Nb2 = 1e12 cm^{-3}. **b** Variation of I_{LEAK} with fluence

103.4.2 I_{LEAK} Variation with Fluence

Figure 103.4b shows the variation of the saturation value of I_{LEAK} with fluence for a n-type detector. As expected, the saturation value of I_{LEAK} (Simulated) increases with increase in fluence. The slope of this plot gives the value of the current related damage parameter α. The simulated value of α is coming out to be 3.97e$-$17 A/cm which is very close to the measured value of 4e$-$17 A/cm, at a temperature of 293 K.

103.4.3 CCE Variation with Fluence

Figure 103.5 shows the CCE variation with bias voltage both for the non-irradiated and irradiated p-type detector at a fluence of 9e14 n_{eq} cm^{-2}. From this plot we can see that for a particular fluence CCE increases with an increase in bias voltage and for a fixed bias voltage, CCE decreases significantly for a fluence of 9e14 n_{eq} cm^{-2}. Simulation results are in good agreement with the measurement results [5].

Fig. 103.5 CCE variation
with fluence for a
non-irradiated and irradiated
p-type detector with a
fluence of 9e14 n_{eq} cm^{-2}

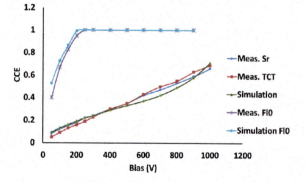

103.5 Summary

Neutron radiation damage model is developed by using the results of experimental measurements and the sensitivity studies of V_{FD}, I_{LEAK} and CCE. This model is applied to the n- and p-type Si detectors and the simulation results are found to be in good agreement with the measurements.

Acknowledgements The authors acknowledge the financial support from the DST, UGC, and the University of Delhi.

References

1. R. Dalal, Simulation of irradiated Si detectors. PoS, Vertex-2014 030
2. I. Pintilie et al., NIMA **611**, 52–68 (2009)
3. F. Hartmann, Evolution of silicon sensor technology in particle physics. Springer Tracts Mod. Phys. (2009)
4. G. Lindstrom, University of Hamburg + CERN-RD48, PIXEL 2000 Genoa 05–09 June 2000
5. E. Curras, Nuclear instruments and methods in physics research A, Proceeding 2016

Chapter 104
Cut-Based Photon ID Tuning of CMS Using Genetic Algorithm

Debabrata Bhowmik

Abstract Whenever we expect a photon as a final state particle of a collider search, it is essential to identify whether the photon is coming from a hard scattering at the primary vertex of the collision(prompt photons) or from decays of π^0 and fragmentation processes within a jet depositing their energies in the electromagnetic calorimeter. The goal of photon identification(ID) is to accept prompt photons at a high efficiency while rejecting the non-prompt photons maximally, for a given efficiency. Identifying prompt photons and measuring the energy deposited by the photons accurately, play a crucial role for analyses which have photon(s) in the final state. In this talk, we will discuss the optimization of the photon identification criteria obtained by using genetic algorithm. Using Monte Carlo samples of CMS corresponding to run conditions of 2017, different working points of the ID corresponding to signal efficiencies 90%, 80%, and 70% are derived for both barrel and endcap.

104.1 Introduction

In a $p - p$ collision at LHC, multiple number of photons can be observed in a single event. All of them may not come from the primary vertex as there are tens of soft collision vertices produced in every branch crossing. Specially for photons we do not get any track associated with them and so it is not trivial to know the vertex a photon is coming from. Also particles from these pile up vertices can deposit their partial energies inside the isolaton cone of the photon candidates we are interested in. In this analysis, we first removed such pile-up effects using a median energy density technique. We used γ-Jet sample where a prompt photon and a jet are present in every event. Jet may have several photons inside it coming from π_0 decay and fragmentation

Debabrata Bhowmik on behalf of CMS Collaboration.

D. Bhowmik (✉)
Saha Institute of Nuclear Physics, HBNI, 1/AF Bidhannagar, Kolkata 700064, India
e-mail: debabrata.bhowmik@cern.ch

© Springer Nature Singapore Pte Ltd. 2021
P. K. Behera et al. (eds.), *XXIII DAE High Energy Physics Symposium*,
Springer Proceedings in Physics 261,
https://doi.org/10.1007/978-981-33-4408-2_104

processes leading the jet to fake as a photon. These jets are treated as background candidates and our aim is to distinguish these jets from the prompt photons which are treated as signals.

104.2 Analysis

For this study, we have chosen five variables as discriminators of signal from background which are $\sigma_{i\eta i\eta}$,[1] H/E and three isolations (Photon, charged, and neutral-hadron), where $\sigma_{i\eta i\eta}$ is shower shape variable. H/E is the ratio of energy deposited of the particle in the Hadron calorimeter and electromagnetic calorimeter. We want to optimize the cut values of these variables to separate signal from background.

104.2.1 Necessity of Pile-Up Correction

In collision, we will detect the particles not only coming from primary vertices but from the soft collisions too. If we are looking for isolated photon then the momentum of other particles coming from pile up may be counted. For charged particles, we have tracks in the tracker and can identify the vertices they are originating from and remove their contributions. For neutral hadrons and photons, it is necessary to remove the effect of pileup. We have used a median energy density(ρ)-based technique to correct pile-up effect.

104.2.2 Pile-Up Correction

If whole Ecal $\eta - \phi$ plane can be thought of N patches and area of the i-th patch is A_i. Then ρ is defined as

---

$$\sigma_{i\eta i\eta} = \left(\frac{\sum(\eta_i - \bar{\eta})^2\omega_i}{\sum\omega_i}\right)^{1/2} ; \quad \bar{\eta} = \frac{\sum\eta_i\omega_i}{\sum\omega_i} ; \quad \omega_i = \max\left(0, 4.7 + \log\frac{E_i}{E_{5\times5}}\right)$$

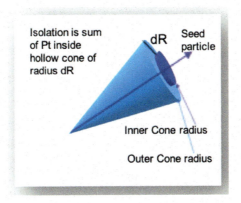

$$\rho = \underset{i \in patches}{\mathrm{median}} \left\{ \frac{p_{ti}}{A_i} \right\}$$

ρ is thus an effective contribution in P_t sum from pile-up per unit area. If isolation sum is defined as the sum of pt(excluding the seed particle) inside a hollow cone, it should vary linearly with ρ.

$$Isolation = EA \times \rho + Isolation_{Corr}$$
$$(Y = m \times X + C)$$

We now plot ρ versus isolation and get the slope which is called effective area.

Figure 104.1 shows such plots for three (charged hadron, neutral hadron, and photon) isolations in one eta region ($1.0 < \eta < 1.479$) and the table in Fig. 104.2 contains effective area (EA) values for different eta regions. Having the EA values pile-up corrected isolations can be calculated easily as

$$Isolation_{Corr} = Isolation - EA \times \rho$$

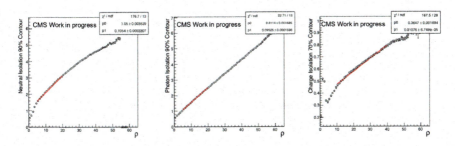

Fig. 104.1 Effective area for $1.0 < \eta < 1.479$

bin	EA charged hadrons(70% cont)	EA neutral hadrons(90% cont)	EA photons(90% cont)
abs(η)<1.0	0.0112	0.0668	0.1113
1.0<abs(η)<1.479	0.0108	0.1054	0.0953
1.479<abs(η)<2.0	0.0106	0.0786	0.0619
2.0 <abs(η) < 2.2	0.01002	0.0233	0.0837
2.2 <abs(η) < 2.3	0.0098	0.0078	0.1070
2.3 <abs(η) < 2.4	0.0089	0.0028	0.1212
abs(η) > 2.4	0.0087	0.0137	0.1466

Fig. 104.2 Effective area for isolations in different η regions

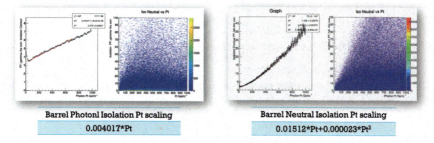

Barrel PhotonI Isolation Pt scaling	Barrel Neutral Isolation Pt scaling
0.004017*Pt	0.01512*Pt+0.000023*Pt²

Fig. 104.3 P_t scaling for photon and neutral hadron isolations in barrel

104.2.3 Pt Scaling

Pile-up corrected isolations has a dependency on the candidate photon Pt as shown
in Fig. 104.3. Leakage of energy from the particle cone to the isolation cone with
the increase in Pt mainly causes this. The dependency with photon Pt for photon
isolation is linear and for neutral hadron isolation is quadratic in nature.

104.2.4 Genetic Algorithm

Aim of the analysis is to find the optimized cut values in a five-dimensional (for 5
variables) space. Every variable can take at least 1000 different values leaving the
choice of at least 10^{15} solutions. Pure canonical way is to apply every solution on
a large number of signal and background MC samples and find out which one has
maximum signal efficiency and maximum background rejection. For that, a large
number of samples has to be processed 10^{15} times and providing the amount of time
and computing it will demand is next to impossible. Most importantly, can we not
have a smart way to reach close enough to the right solution. Genetic algorithm is
one of the effective approaches towards it.

The algorithm has few steps as described below, and a chart flow is also shown in the right side.

1. To **initialize**, forget right solution, make a population of solutions. For us that is set of values of the variables of the events in the training samples. These solutions are called chromosomes and the set of chromosomes is considered as initial population. Every solution (Chromosome) consists of five values corresponding to five variables and each value is named as gene.

2 & 3. Based on the probability (which is one form of **fitness assignment**), randomly **select** two chromosomes as parents. Advantage is that the solutions which have less background rejection ability will contribute less in the next generation. Thus, a faster way toward right solution.

In the previous step, parent chromosomes have been selected that will produce off-springs.

4. One point **crossover**. A random crossover point is selected and the tails of both the chromosomes are swapped to produce a new off-springs.

5. There is some change in the genes of children which makes them different from its parents. A random tweak in the chromosome, which also promotes the idea of diversity in the population. A simple method of **mutation** is randomly replacing one particular gene of a chromosome by that of another.

6. There are different **termination conditions**, which are listed below:

(i)   There is no improvement in the population for over x iterations.
(ii) We have already predefined an absolute number of generation for our algorithm.
(iii) When our fitness function has reached a predefined value.

We train a genetic algorithm based network with MC samples using Root TMVA package.

104.2.5 Results

After the training, the network gets ready with the optimized values of all the five variables for particular signal efficiencies with maximum possible background rejection. This means we actually obtain the cut values of the variables from which given an unknown mixed sample of signal and background, eventually which will be the case of data, we will be able to know for a working point (loose/medium/tight) corresponding to a particular signal efficiency how much the background rejection will be. This set of optimized cut values is called an photon ID. Optimized ID (for barrel) to be used for 2017 analyses is presented in Fig. 104.4. Results in Fig. 104.5 verifies that the optimized ID is consistent with different Pt values, with number of vertices and in different eta region (inside barrel) of the detector.

BARREL	Loose (90.08%)	Medium (80.29%)	Tight (70.24%)
Background Rejection	Loose (86.25%)	Medium (89.36%)	Tight (90.97%)
HoverE	0.04596	0.02197	0.02148
$sigma_{ietaieta}$	0.0106	0.01015	0.00996
Rho corrected PF charged hadron isolation	1.694	1.141	0.65
Rho corrected PF neutral hadron isolation	24.032 + 0.01512*pho_pt + 2.259e-05*pho_pt^2	1.189 + 0.01512*pho_pt + 2.259e-05*pho_pt^2	0.317 + 0.01512*pho_pt + 2.259e-05*pho_pt^2
Rho corrected PF photon isolation	2.876 + 0.004017*pho_pt	2.08 + 0.004017*pho_pt	2.044 + 0.004017*pho_pt

Fig. 104.4 Optimized ID for barrel

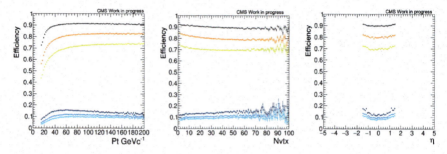

Fig. 104.5 Dependency of signal and background efficiencies on P_t (left panel), number of vertices (middle panel), and η (right panel)

References

1. CMS collaboration "*Performance of photon reconstruction and identification with the CMS detector in proton-proton collisions at sqrt(s) = 8 TeV. JINST 10*", (2015) P08010, https://doi.org/10.1088/1748-0221/10/08/P08010, https://arxiv.org/abs/1502.02702, 1502.02702
2. CMS collaboration, "*Particle-flow reconstruction and global event description with the CMS detector*". JINST **12** (2017) P10003, https://doi.org/10.1088/1748-0221/12/10/P10003, https://arxiv.org/abs/1706.049651706.04965
3. CMS Collaboration, "*The CMS electromagnetic calorimeter project: Technical Design Report*" (Technical Design Report CMS. CERN, Geneva, 1997)

Chapter 105
Design and Development of Various Cooling Arrangements for Muon Chamber Detector Electronics

C. Ghosh, A. K. Dubey, J. Kumar, A. Kumar, and Subhasis Chattopadhyay

Abstract The Compressed Baryonic Matter (CBM) experiment at FAIR is being designed to explore the QCD phase diagram of high baryon density matter using high-energy fixed target nucleus–nucleus collisions [1]. Muon Chamber (MUCH) is the detector to be used in CBM to detect low-momentum muons. The final design of the muon detector system consists of six hadron absorber layers and 18 Gaseous tracking chambers located in triplets behind each absorber slab [2]. Collaboration has decided to perform a mini CBM (mCBM) experiment before commencing the main CBM experiment. The mCBM setup consists of mMUCH system which contains 3 GEM prototype stations consisting of real-size modules with 2304 pads each. Each module has 18 MUCH-XYTER asic; almost 7K readout channels in total for 3 modules. In total 135 W heat is deposited by the 3 GEM modules. The main CBM will consist of 48 detector modules in first station resulting in total heat generation of 2.2 KW; for extraction of this heat continuously, we have to design some proper cooling arrangement. Here we have discussed about the design and development of various cooling arrangement.

105.1 Design of Aluminum Plate Water Cooling System

A 10 mm thick aluminum plate with water channels grooved inside (Fig. 105.1a) is designed and fabricated at VECC for cooling purpose. A 2 mm aluminum strip was welded on the top of the groove (Fig. 105.1b) and the entire plate was surface finished.

C. Ghosh (✉) · A. K. Dubey · J. Kumar · A. Kumar · S. Chattopadhyay
Variable Energy Cyclotron Centre, 1/AF Bidhannagar, Kolkata 700064, India
e-mail: c.ghosh@vecc.gov.in

A. K. Dubey
e-mail: anand@vecc.gov.in

C. Ghosh · A. K. Dubey · A. Kumar · S. Chattopadhyay
Homi Bhabha National Institute, Mumbai 400094, India

© Springer Nature Singapore Pte Ltd. 2021
P. K. Behera et al. (eds.), *XXIII DAE High Energy Physics Symposium*,
Springer Proceedings in Physics 261,
https://doi.org/10.1007/978-981-33-4408-2_105

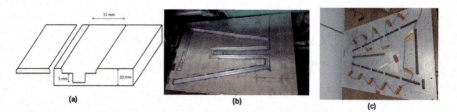

Fig. 105.1 **a** Schematic of water channel groove **b** 5 mm groove is made before welding of 2 mm strip **c** Surface finished aluminum plate with water channel grooved inside

105.1.1 Cooling Performance Test

For testing the performance of the cooling plate, we used 18 heating resistors each of 2.5 W heating capability. The resistors were placed on to the surface of the aluminum plate to replicate the heat sink of Front End Boards (FEBs) electronics (Fig. 105.1c).

A DC voltage is applied across the series connection of all the resistors such that each element deposits a heat of 2.5 W. Simultaneously, water is pumped using a submersible pump into the aluminum plate from a water chiller and the outlet hot water from aluminum plate is looped back into the chiller. We used an aurdino-based micro controller board to control and monitor the temperature of the plate, Fig. 105.2a.

If the plate is heated without any water circulation, then the plate temperature reaches 50 °C. We need to keep our FEBs below 25 °C for accurate data taking, so we need to cool down the plate continuously as long as the FEBs are taking data. To measure the uniformity of temperature across the entire surface we placed 18 temperature sensors (DS18B20) at different points on the plate. Figure 105.2b shows the temperature variation with time. The blue curve shows the chiller water temperature, black curve is ambient temperature, and the red curve is cooling plate temperature. One can see the stability of the plate temperature (red curve) which remains stable for 100 h continuous operation [3].

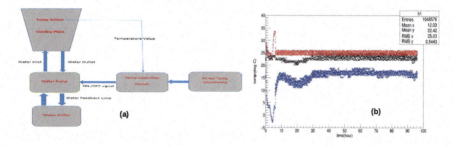

Fig. 105.2 **a** Schematic of the cooling arrangement **b** Temperature profile of three different sensors

105.1.2 Cooling Plate Arrangement in CERN SPS H4 Test Beam Line

The cooling plate fabricated at VECC workshop was used at CERN SPS H4 test beam for cooling of the detector electronics [3]. The detector was mounted on one side of the plate and on the other side the FEBs were placed, and these FEBs were connected by flexible cable to the detector. The plates were successful in maintaining the temperature of all FEBs within 25 °C.

105.2 Air Cooling Setup for MUCH Electronics

After demonstrating the aluminum plate water cooling system we tried to develop an alternative cooling system which can work without any liquid like water. In long run duration of a few years, water cooling system is prone to leakage, so to eliminate this issue we tried some other cooling arrangements.

Here we have developed a dummy of the actual trapezoidal GEM detector with 18 FEBs attached to it. The entire dummy is made out of G10 sheet. To simulate the heat sink of each FEB we have used 18 resistors of each $4\,\Omega$. A voltage of $57\,V$ is applied across the series connection of the resistors such that each resistor deposits $2.5\,W$ of heat. One high flow rate DC fan is used to extract the heat of each element as shown in Fig. 105.3a. Ambient temp= 20 °C, R.H= 47%. At first the resistors were heated without any cooling, when all the resistor temperature gets stable at higher value then the fan was switched on and the resistor temperatures went down and got stable at lower value.

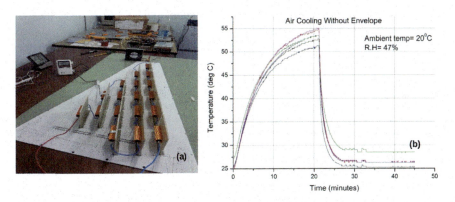

Fig. 105.3 **a** Schematic of the cooling arrangement **b** Temperature profile of three different sensors at three different location

From Fig. 105.3b, it can be seen that if no cooling is applied then the heating element temperature rises above 50 °C. From the plots we can infer that after cooling the final temp value lies between 25 °C and 28 °C. It is a nice observation that almost all the heating element temp goes down below 28 °C.

105.3   Cooling of MUCH Electronics Using Peltier Module

We have used a 2 mm thick Aluminum plate on which 14 heating resistors of each 10 Ω is placed in series and a voltage of 70 V is applied across the series connection of the resistors such that each resistor deposits 2.5 W of heat. Three Peltier cooling unit (TEC1-12706) is kept in contact of the aluminum plate as shown in Fig. 105.4a.

We observed almost 8 °C temperature variation across different position of the aluminum plate. Peltier modules are good for localized cooling. The area within few cm of the peltier module gets cooled below dew point temperature and water droplets are formed on the aluminum plate, which is detrimental for the readout electronics. Another issue with peltier module is the extraction of heat from the hot side of the module, and we tried to extract the heat with a special arrangement of heat sink and small cooling fan, but this procedure is difficult to implement in main CBM environment.

105.4   Water Cooling with Minimum Material Budget

mini CBM experiment will be a platform for detector integrity test along with all electronics and data acquisition chain. It is decided to have three real-size Muon Chamber (MUCH) modules with 10 mm thick aluminum cooling plate in mCBM. The effect of the aluminum plate on momentum distribution of primary and secondary particles was simulated and it was found that the momentum distribution changes

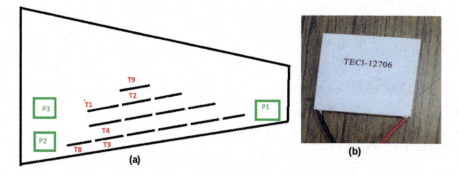

Fig. 105.4 **a** Schematic of temperature sensor and peltier module positions **b** Peltier module

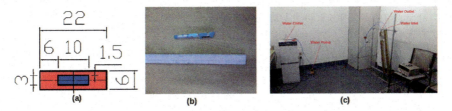

Fig. 105.5 **a** Cross section of the aluminum profile **b** aluminum profile used for water cooling **c** cooling setup at VECC

with the insertion of the 10 mm thick aluminum plate in detector acceptance. So it was decided to reduce material budget in detector acceptance.

Accordingly, we designed a cooling setup which uses less material budget. An extruded aluminum profile was chosen as shown in Fig. 105.5a with a rectangular hollow channel inside shown in blue color, through which water can flow.

In Fig. 105.5c, it can be seen that the "U" shaped cooling profile and one dummy detector are clamped to an aluminum support structure. The dummy FEBs are placed on the aluminum cooling profile. The blue color pu tubes coming out of the cooling channel are for water inlet and outlet. One DC power supply is used to drive current through the heating elements of dummy FEBs. Two temperature sensors are placed on the cooling profile from which the temperature values are logged and monitored. A water pump drives chilled water from the water chiller to the aluminum cooling channel and loops back to the chiller. This setup can maintain the temperature of each FEB well below 25 °C.

105.5   Conclusion

We have fabricated and tested a 10 mm aluminum cooling plate for MUCH electronics cooling with satisfactory performance. In search of alternative cooling which uses no liquid, we have demonstrated an air cooling setup, with few iterations of channelized air flow. A Peltier cooling setup is developed and demonstrated, but due to localized and uneven cooling and also heat extraction issue, this setup did not give satisfactory performance. A rectangular water channel cooling setup is developed at VECC Workshop for mini CBM experiment with low material budget.

References

1. B. Friman, C. Hohne, J. Knoll, S. Leupold, J. Randrup, R. Rapp, P. Senger, The CBM physics book: Compressed baryonic matter in laboratory experiments. Lect. Notes Phys. **814**, 1–980 (2011)

2. Technical Design Report for the CBM, Muon Chambers (November, 2014). https://repository.gsi.de/record/161297
3. C. Ghosh et al., DAE Symposium on Nuclear Physics, vol. 62 (2017)
4. A. Kumar et al., DAE DAE Symposium on Nuclear Physics, vol. 62 (2017)

Chapter 106
Transient Current Technique (TCT) Measurements and Simulations at the University of Delhi

Geetika Jain, Chakresh Jain, Ashutosh Bhardwaj, and Kirti Ranjan

Abstract The foreseen high fluence environment in future nuclear and particle physics experiments requires corresponding radiation hard silicon sensors. However, these sensors must be tested for their charge collection performance and long-term radiation sustainability prior to their usage. The Transient Current Technique (TCT) is a useful characterization technique for investigating the radiation damage effects. This work reports measurements on silicon pad detectors carried out using the TCT setup installed at the University of Delhi. Measurement results are complemented with TCAD simulation, which is useful to get an insight of the silicon detector.

106.1 Introduction

The silicon sensors, deployed in various Nuclear and High Energy Physics (HEP) experiments, are exposed to extensive radiation fluences. High luminosities correspond to large particle fluxes which distort the silicon structure both on the surface and in the bulk by displacing the atom from its lattice site. This leads to three major deterioration effects on the sensor performance (1) increase in the leakage current, (2) change in full depletion voltage (Vfd), and (3) decrease in charge collection (CC) efficiency. The Transient Current Technique (TCT), developed by the Ioffe Institute in St. Petersburg, Russia and the Brookhaven National Laboratory, USA [1], is a technique used to investigate the CC behaviour of the sensors. It is based on the generation of free charge carriers within the Device Under Test (DUT), when it is illuminated by a laser light. These charge carriers then drift towards the respective electrodes, inducing current at the electrodes by Ramo-Schokley Theorem, which states that charge induced (q) on the electrodes by the generated charge carriers (Q) is proportional to the displacement (x) of the carriers, where the proportionality con-

G. Jain (✉) · C. Jain · A. Bhardwaj · K. Ranjan
Centre for Detector and Related Software Technology, Department of Physics
and Astrophysics, University of Delhi, Delhi, India
e-mail: geetikajain.hep@gmail.com

© Springer Nature Singapore Pte Ltd. 2021   767
P. K. Behera et al. (eds.), *XXIII DAE High Energy Physics Symposium*,
Springer Proceedings in Physics 261,
https://doi.org/10.1007/978-981-33-4408-2_106

stant is the DUT thickness (w) [2, 3]. The induced current pulse signal is then fed
to the electronics for amplification and pulse shaping. The output TCT signal can be
used to extract various sensor properties like Vfd, CC efficiency, rise time and sensor
profiles like charge carrier mobility, electric field, etc.

106.2   TCT Measurement Setup at Delhi

The University of Delhi has developed a programmable TCT setup, as shown in
Fig. 106.1, in collaboration with CERN PH Detector Lab within the framework of
RD50 collaboration. The DUT, mounted on the PCB, is placed inside the electromagnetic shielded metal box and is illuminated by red laser (the trigger rate is set by
the function generator). While a Source Measuring Unit (SMU) is used to reverse
bias the DUT, a bias tee filters the transient TCT signal, which is then amplified by
an amplifier.

106.3   TCT Simulations at Delhi

Silvaco TCAD tool [4] is being used at the University of Delhi to simulate the
electrical and optical behaviour of the DUT. The schematic of the simulation Mixed-
Mode TCT circuit is shown in Fig. 106.2. The values of the circuit elements used in
the simulations are used as implemented in the measurement setup. C_{stray} and L_{stray}
represent the capacitance and inductance of the cables and connectors in the circuit.

Fig. 106.1 The red laser TCT measurement setup at University of Delhi

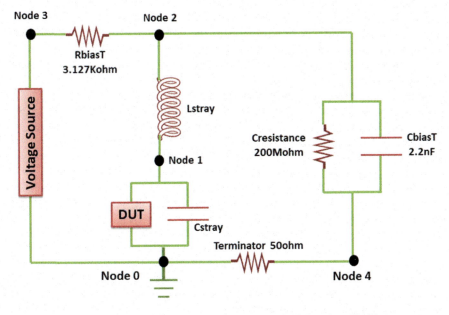

Fig. 106.2 The mixedMode TCT simulation circuit diagram at University of Delhi

106.4   Results of Measured and Simulated TCT Signals

In this work, TCT measurements were carried out on two test diode structures, viz. N3 and L11; from the same wafer. A comparison of their TCT signals is shown in Fig. 106.3. The overlap of the TCT signals from two diodes validate the working of the TCT setup at the University of Delhi.

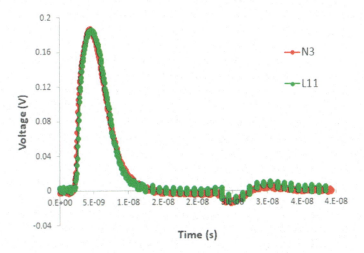

Fig. 106.3 The TCT signal (Voltage versus Time) for 2 diodes at 200 V bias

Figure 106.4 shows the electron signal at different bias voltages from 0 to 200 V, when the p^+-n-n^+ diode is illuminated from the front side. It can be seen from the plot that as the bias voltage is increased, more and more charge pairs are collected. This means that at higher biases, the recombination and trapping effects reduce significantly.

Both C_{stray} and L_{stray} have been tuned to 14 pF and 15 nH (see Fig. 106.5), respectively, such that there is a good agreement of the simulated TCT signal with the measured TCT signal.

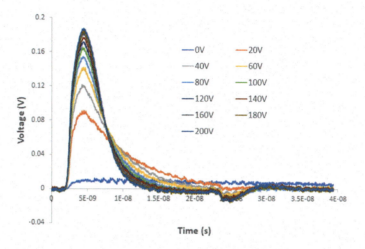

Fig. 106.4 The electron TCT signal at various bias voltages for diode N3

Fig. 106.5 Variation of C_{stray} and L_{stray} values

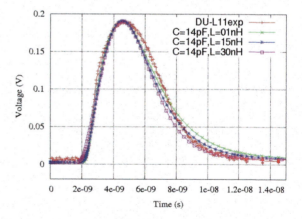

106.5 Summary

The TCT setup using red laser is installed and commissioned at University of Delhi. TCT measurements have been carried out on two diode structures. To complement the measurements, Mixed-Mode simulations were performed and the results are found to be consistent.

Acknowledgements Authors are thankful to UGC-JRF, DST, R&D grant DU and ISJRP for financial research support and to PH-DT Lab at CERN to set up the TCT system.

References

1. H.W. Kraner, Z. Li, E. Fretwurst, The use of the signal current pulse shape to study the internal electric field profile and trapping effects in neutron damaged silicon detectors. NIM A **326**, 350–356 (1993)
2. S. Ramo, Current induced by electron motion. Proc. I.R.E. **27**, 584 (1939)
3. W. Schokley, Currents to conductors induced by a moving point charge. J. Appl. Phys. **9**, 635 (1938)
4. ATLAS Silvaco, Version 5.15.32.R Nov 2009, Users manual, Silvaco webpage. http://www.silvaco.com

Chapter 107
Progress of the Charged Pion Semi-inclusive Neutrino Charged Current Cross Section in NOvA

Jyoti Tripathi

Abstract The NOvA experiment is a long baseline neutrino oscillation experiment designed to measure the rates of electron neutrino appearance and muon neutrino disappearance. It uses the NuMI beam that has recently been upgraded to 700 kW as the neutrino beam source. The NOvA Near Detector (ND) is placed onsite at Fermilab, 800 m from the NuMI target while the far detector is placed 810 km away at a site near Ash River, Minnesota, both the detectors being functionally identical and 14.6 mrad off-axis w.r.t the NuMI beam. The primary physics goal of the experiment include precise measurement of θ_{23}, CP violating phase δ_{CP} and the neutrino mass hierarchy. In addition to oscillation physics, the NOvA experiment provides an excellent opportunity to study neutrino-nucleus cross-sections at the ND as it is situated close to the neutrino source and observes an intense rate of neutrino interactions. This work is a proceeding to a talk showing the status of the charged pion production in muon neutrino introduced charged current interactions at the NOvA ND based on the exposure equivalent to 8.85×10^{20} for the neutrino beam using convolutional neural networks (CNN).

107.1 Introduction

Neutrino oscillation physics has entered into a precision era. The accelerator based neutrino experiments need reduction of systematic uncertainties to a level of few percent for precise measurement of the neutrino oscillation parameters. Neutrino-nucleus cross-sections are an important source of systematic uncertainty as they are known with a precision not exceeding 20% in the hundred-MeV to a few GeV neutrino energy region. This energy region is dominated by several cross-sectional

J. Tripathi (✉)
Department of Physics, Panjab University, Chandigarh 160014, India
e-mail: jyoti.tripathi571@gmail.com

© Springer Nature Singapore Pte Ltd. 2021
P. K. Behera et al. (eds.), *XXIII DAE High Energy Physics Symposium*,
Springer Proceedings in Physics 261,
https://doi.org/10.1007/978-981-33-4408-2_107

channels that are not well measured: charged and neutral current quasielastic, single pion production, multipion resonant production and collective nuclear responses like the 2p2h. The peak neutrino energy for the NOvA experiment lies between 1 and 3 GeV where there is an overlap between the charged current quasielastic interactions (CCQE) and the resonant pion production. The pion production processes are of great interest as they act as a source of background and systematic uncertainties because pions can mimic an electron in the final state. Also, a single charged pion produced in the interaction can make the event mimic the CCQE topology. Estimating the rate of a single charged pion production in the charged current (CC) interactions is important for the correct estimation of the incoming neutrino energy.

107.2 Single Charged Pion in ν_μ CC Interactions

This analysis looks at one muon and a single charged pion in the final state where the kinetic energy of the charged pion lies between 250 and 900 MeV.

$$\nu_\mu + N \rightarrow \mu^\mp + \pi^\pm + X$$

where N is the nucleus in the detector and X is the recoil nucleus plus any other particle. Figure 107.1 shows the feynman diagram of a ν_μ CC interaction producing a charged pion and a neutron through Δ^+ resonance along with the event display of a simulated interaction producing a muon and a charged pion in the final state.

The NOvA ND Monte Carlo sample has been used for the selection and identification of the signal events.

107.3 Event Selection and Classification

To reject events from the neutrino interactions in the rock surrounding the detector, the reconstructed vertex of the interaction is required to be within a well-defined fiducial volume. The events that reconstruct outside the detector location or failing

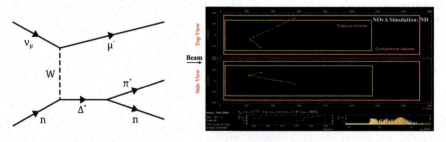

Fig. 107.1 A single charged pion produced in CC interaction (left). Simulation of a NOvA Near Detector event showing a muon and a π^+ in the final state (right)

to get reconstructed well are rejected. Also, only two 3D prong events are considered where *prong* is defined as a cluster of hits with a starting point and a specific direction. These cuts applied together are referred to as *Preselections*. After the candidate neutrino interactions are selected, each event is examined to determine if it belongs to the signal sample.

107.3.1 Convolutional Neural Network

Convolutional Neural network (CNNs) is widely used in the field of computer vision for image recognition. NOvA has pioneered the use of CNNs for particle classification in neutrino physics. NOvA has developed its own CNN, called CVN (convolutional visual network) for identifying the neutrino interaction [3]. Each interaction topology is treated as an image with cells as pixels and charge as color value. The convolutional layers optimally extract features from the images providing good separation between the different interaction modes. Figure 107.2 shows extraction of features by different convolutional layers from a simulated ν_μ CC interaction producing muon and charged pion in the final state. Another implementation of CVN in NOvA is of particle classification using *Prong CVN* as particle identification is important for the cross-sectional analyses looking for specific final states. It takes complete image of the event and the image of the individual prong in each view as an input and assigns 5 different scores to each prong in the event under a specific particle hypotheses (MuonID, ElectronID, ProtonID, Charged PionID and PhotonID) [5]. Currently, Prong CVN has been used for this analysis for identifying a muon and a charged pion in the final state.

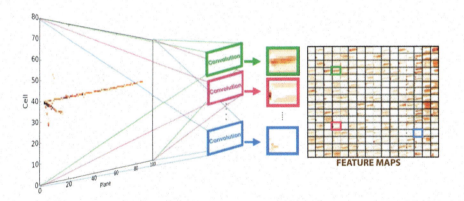

Fig. 107.2 Simulated ν_μ CC event. Some track activity is visible in one of the feature maps [4]

107.3.2 Muon and Pion Selection

The selected event sample consists of a two prong events. For muon selection, out of all the prongs that are identified as muon (based on the CVN score), the one with the highest CVN muon score is identified as the muon. Figure 107.3 shows the distribution of CVN MuonID (left) and the highest CVN MuonID (right) for each true particle type. This cross-sectional measurement will be limited by systematic uncertainties, and therefore the Figure of Merit (FoM) for the event selection is based on the minimization of fractional uncertainty on the total cross section [6].

$$\frac{\delta\sigma}{\sigma} = \sqrt{\frac{(\delta N_{bkg}^{syst})^2}{(N_{sel} - N_{bkg})^2} + \left(\frac{\delta\epsilon}{\epsilon}\right)^2}$$

where N_{sel} and N_{bkg} are the number of selected and background events, $\delta\epsilon$ is the fractional uncertainty in the signal efficiency, and δN_{bkg}^{syst} is the systematic uncertainty in the background.

The second non-muon prong is considered as the potential pion candidate. Pion selection is based on the optimized sum of the pion and muon score since the pion and the muon look alike a lot in the detector and the muon has already been separated out in the event.

The sum of the pion and the muon score where the $\delta\sigma/\sigma$ distribution is minimized is considered as the optimum value for the pion selection. Figure 107.4 shows the minimum in the distribution where the MuonID + PionID is 0.8 [7].

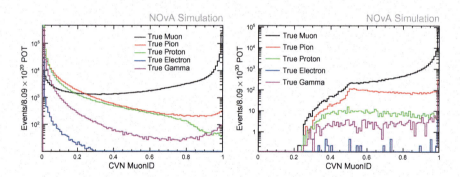

Fig. 107.3 Distribution of the CVN MuonID for all the true particle prongs (left) and the true particle prongs that are most likely to be the muons as per the muon CVN score (right)

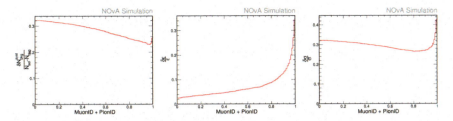

Fig. 107.4 The fractional systematic uncertainty on the background (top left) and the fractional uncertainty on the signal (top right). The fractional uncertainty on the total cross section (bottom) shows a minimum at 0.8

107.4 Efficiency and Purity of Event Selection

The efficiency is defined as the fraction of signal events getting selected where as the purity is the fraction of selected events which is signal. Table 107.1 shows the efficiency and purity of the selected sample after each selection cut. The selected signal has 72.78% events from resonance, 7.5% from coherent and 19.6% from DIS, respectively. Figure 107.5 is showing the purity and efficiency as a function of true kinetic energy of the charged pion after each selection cut is applied. The selection efficiency of the signal sample is 21% and 1.5% with respect to the preselection and the total, respectively, where as the purity of the selected sample is 93%.

107.5 Summary and Future Prospects

Charged current single charged pion measurement has a potential to look at the interactions with low W (invariant hadronic mass) leading to the improved model of these processes. The analysis presented here uses deep learning-based technique for particle classification. The charged pion selection criteria is optimized based on the

Table 107.1 Purity, Efficiency, and different background fractions as each cut is applied sequentially

Selection	Selected	Efficiency (%)	Purity (%)	Background			
				CC Resonance (%)	CC DIS (%)	NC (%)	Coherent (%)
Total	7.16e+07	100	2.7	19.6	64	9.5	0.58
Preselections	5.27e+05	6.9	27.25	36	16.4	16.9	1.5
Muon Selection	3.56e+05	5.6	32.48	48.4	14.1	2.1	1.27
Pion Selection	43,927	1.5	70.3	15.7	10.78	1.2	1.9

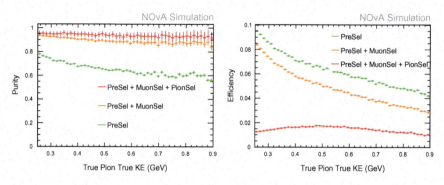

Fig. 107.5 Purity (left) and Efficiency (right) of the event selection after applying all the selection cuts

minimization of the systematic uncertainties on the total cross section. The analysis further looks to the energy estimation of the charged pion so that a differential cross-sectional measurement w.r.t to the charged pion kinematics could be made.

References

1. NOvA Technical Design Report, FERMILAB-DESIGN-2007-01
2. D.A. Harris, The State of the Art of Neutrino Cross Section Measurements (2015), arXiv:1506.02748 [hep-ex]
3. A. Aurisano et al., A Convolutional Neural Network Neutrino Event Classifier (2016), arXiv:1604.01444 [hep-ex]
4. Fernanda Psihas and NOvA Collaboration, J. Phys. Conf. Ser. **888**, 012063 (2017)
5. F. Psihas et al., Particle Identification with a Context-Enriched Convolutional Network for Neutrino Events, NOvA-doc-34747-v1 (2018)
6. L. Aliaga et al., Measurement of the ν_μ CC inclusive cross section in the NOvA near detector, NOVA-doc-32688-v2 (2019)
7. J. Tripathi, Efficiency and Purity Studies for Charged Pion Semi-Inclusive Muon Neutrino Charged Current Cross-Section in NOvA, NOvA Internal Document, DocDB-34660-v1 (2018)

Chapter 108
Electronics and DAQ for the Magnetized mini-ICAL Detector at IICHEP

M. N. Saraf, Pathaleswar, P. Nagaraj, K. C. Ravindran, B. Satyanarayana,
R. R. Shinde, P. K. Kaur, V. B. Chandratre, M. Sukhwani, H. Kolla,
M. Thomas, Umesh Shas, D. Sil, S. S. Upadhya, A. Lokapure, U. Gokhale,
Rajkumar Bharathi, E. Yuvaraj, A. Behere, S. Moitra, N. Ayyagiri, S. Sikder,
S. R. Joshi, A. Manna, A. Padmini, M. Punna, Gobinda Majumder,
and V. M. Datar

Abstract Magnetized iron calorimeter (ICAL) detector built using Resistive Plate Chambers (RPCs), which are interleaved between the iron plates, is proposed by India-based Neutrino Observatory (INO), to study atmospheric neutrinos [1]. Ten layers of RPCs are deployed in the mini-ICAL which is an 85-ton magnetized detector stack with 20 RPCs in 10 layers. The mini-ICAL is being used for prototyping data acquisition electronics, its integration with the magnetized detector, back-end software, etc. The analog front-end Amplifier-Discriminator, digital front-end RPC-DAQ and bipolar HV unit to bias RPC are embedded with the RPC. The RPC-DAQ mainly generates pre-triggers, acquires event data on a final trigger and monitors the health of RPC. A central trigger system generates the final trigger by processing the pre-trigger signals. The relative arrival time offsets of the final trigger at each of the RPC-DAQs are measured and event timer stamp clocks are synchronized by the Calibration and Auxiliary Unit (CAU) housed next to the trigger system. All the RPC-DAQ nodes are connected to back-end servers over LAN for control, data collection and monitoring. The back-end server manages the overall run control and detector health monitoring. The data concentrator server collects data from all the digital front-ends (DFEs) and the event builder server builds events based on event time stamps. The overall design is motivated by modularity, flexibility, reduced cost and power, and the use of open-source software. This paper will highlight the design details and performance of prototype electronics.

M. N. Saraf (✉) · Pathaleswar · P. Nagaraj · K. C. Ravindran · B. Satyanarayana · R. R. Shinde ·
P. K. Kaur · U. Shas · D. Sil · S. S. Upadhya · A. Lokapure · U. Gokhale · R. Bharathi · E. Yuvaraj
· S. R. Joshi · G. Majumder · V. M. Datar
Department of High Energy Physics, Tata Institute of Fundamental Research, Homi Bhabha Road,
Navy Nagar, Colaba 400005, Mumbai, India
e-mail: mandar@tifr.res.in

V. B. Chandratre · M. Sukhwani · H. Kolla · M. Thomas · A. Behere · S. Moitra · N. Ayyagiri ·
S. Sikder · A. Manna · A. Padmini · M. Punna
Electronics Division, Bhabha Atomic Research Centre, Trombay 400085, Mumbai, India

© Springer Nature Singapore Pte Ltd. 2021   779
P. K. Behera et al. (eds.), *XXIII DAE High Energy Physics Symposium*,
Springer Proceedings in Physics 261,
https://doi.org/10.1007/978-981-33-4408-2_108

108.1 Introduction

The ICAL detector will study atmospheric neutrino oscillation parameters and mass ordering by tracking charged particles in the magnetized ICAL detector, produced during neutrino interactions [1]. The mini-ICAL is a stack of 10 layers of RPCs interleaved between magnetized iron layers. On a physics event, each participating RPC-DAQ records event time stamp, interaction x-y co-ordinates and Time Of Flight (TOF) relative to the final trigger arrival, and event data packets are pushed over the LAN using the TCP/IP protocol. These packets are collected by the back-end Data Concentrator (DC) servers, and events are built by collating the event time stamps. In addition to event data recording, the DAQ system sets high voltage to the RPCs and monitors the high voltage applied, RPC current and RPC noise rates periodically in the background as the RPC health monitoring process. The mini-ICAL is a scaled-down version of the ICAL made of a magnetized detector stack and is currently being used for prototyping the ICAL electronics and the DAQ system.

The architecture of the ICAL's electronics and DAQ system (Fig. 108.2) is based on designating the RPC as the minimum standalone unit of the detector. Every RPC tray assembly (Fig. 108.1) consists in addition to the RPC, the analog and digital front-end processing electronics as well as a high voltage power supply module for biasing the RPC.

Fig. 108.1 RPC tray

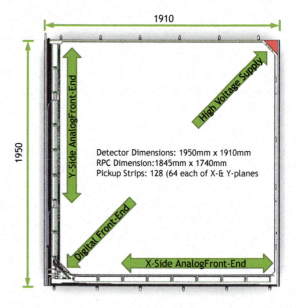

Detector Dimensions: 1950mm x 1910mm
RPC Dimension:1845mm x 1740mm
Pickup Strips: 128 (64 each of X-& Y-planes

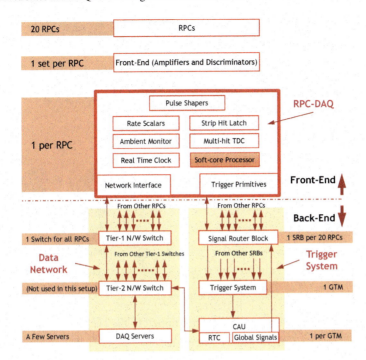

Fig. 108.2 DAQ architecture

108.2   Front-End Electronics

The analog front-end (AFE) consists of boards (Fig. 108.3) made using NINO ASIC which is an ultra-fast, low power, 8-channel pre-amplifier discriminator chip developed at CERN [1]. The AFE boards are mounted along the two orthogonal sides of the RPC corresponding to the X and Y planes. The single-ended signals from the RPC are converted into the differential type and fed to the differential inputs of the NINO chip. The output of the AFE is in the form of pseudo-LVDS signals and feeds the digital front-end. Meanwhile, AFE boards using indigeneously designed 4-channel voltage amplifier and 8-channel leading edge discriminator ASICs are being extensively tested and getting ready to be mass-produced.

The RPC signal processing electronics called the digital front-end (DFE) is located at one corner of the RPC tray as shown in Fig. 108.1. The DFE module comprises several functional blocks such as a Time-to-Digital Converter (TDC), Strip-hit latch, Rate monitor, Pre-trigger generator, ambient parameter monitor and analog front-end (AFE) control [2]. A softcore processor takes care of all the data acquisition (DAQ) needs, configuration of the front-end components as well as data transfer between the RPC unit and the back-end servers. A considerable part of the DFE module's hardware, including the soft-processor, is implemented inside a Field Programmable

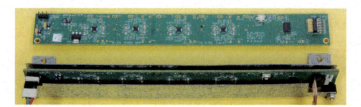

Fig. 108.3 NINO AFE

Fig. 108.4 DFE

Gate Array (FPGA). Digitized data is transmitted to the back-end using the DFE's network interface. Figure 108.4 shows a DFE module with its dismantled power supply board.

108.3 Trigger System

The trigger criteria for the ICAL electronics have been decided considering the characteristic hit patterns of various physics events that will occur in the ICAL detector. The trigger system will generate a global trigger signal if the event topology satisfies such a trigger criteria set by the user. The trigger criteria thus is solely based on the event topology and is defined as $M \times N / P$, i.e. a trigger is generated when M consecutive strips have simultaneous signals in at least N layers in a group of P consecutive layers.

The design of the trigger system follows a distributed and hierarchical architecture as shown in Fig. 108.5. The pre-trigger signals from DFEs are routed to the Signal Router Boards (SRBs) and sorted M fold signals are sent to the Trigger Logic Boards (TLBs). The second-level trigger logic is implemented in the TLBs and the boundary coincidences are resolved by the Global Trigger Logic Boards (GTLBs). The entire control of the trigger system and monitoring of various signal rates, etc. is handled by the Trigger Control and Monitor (TCAM) module. Further, the Control and Monitoring (CAM) module pushes a trigger packet on every final trigger, to back-end DC servers where it is used as a reference time stamp while collating the RPC event data packets.

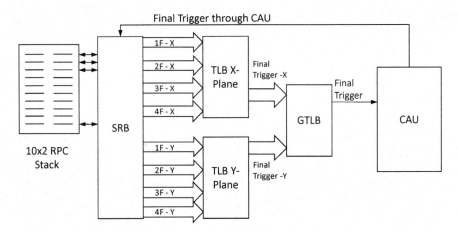

Fig. 108.5 The Trigger System for mini-ICAL

The CAU is a common interface between all the ICAL subsystems for control, calibration and time synchronization. The CAU performs the functions of distribution of global clocks and trigger signals as well as the measurement of global trigger time offset at each DFE, which are present due to disparate signal path lengths from the trigger system to the DFEs, by measuring the round path delay using calibration signals. The local TOF in each DFE is then translated to a common timing reference by adding the respective offsets for reconstructing the particle trajectories with direction. The CAU unit was tested extensively on the RPC stacks and was found to provide offset corrections of better than 100 ps. The CAU also pre-loads the Real-Time Clocks (RTCs) of all the DFEs with epoch time and keeps them synchronized up to a tenth of a microsecond using a Pulse Per Second (PPS) signal and a global clock. These RTC time stamps enable the back-end to built events.

108.4   Back-End Data Acquisition

All RPC nodes, Trigger node and CAU node are configured as a network element along with back-end server nodes, in Ethernet-based LAN using network switches. Thus, the entire mini-ICAL detector is a small Ethernet LAN between the front- and back-end processing nodes for control, data acquisition and monitoring [3].

The back-end Data Concentrator (DC) server receives an event and monitors data on different TCP/IP ports from the DFE modules. The DC also receives a trigger packet from the TCAM containing a trigger number and trigger time stamp. The DC adds a common event marker to all the RPC data packets by collating the event time stamp with the trigger packet time stamp within a window of $20\,\mu s$ and forwards this data to the event builder. This event data is then collated by the event builder using the common event marker. Finally, the back-end system performs various quick quality checks on the data in addition to providing user interfaces, slow control and monitoring, event display, data archival and so on.

108.5 Power Supplies

RPCs require a variable High Voltage (HV) supply of up to 12 kV for the generation of the electric field across their detector gas medium. In order to eliminate the requirement of bulky HV cables and connectors, it was decided to generate the HV bias locally on each RPC. This significantly reduces the space usage in the cable trays as well as the overall cost. However, the modules have to be very compact and should not generate electromagnetic interference to the RPC or to other electronics [5].

The supply for RPC biasing is designed as a $\pm$ 0–6 kV split supply, so as to minimize the HV leak problem on the chamber surface. The HV outputs are generated by two low cost, low noise HV DC-DC converters. The converter topology, shown in Fig. 108.6, is based on a current-fed resonant Royer circuit, which minimizes harmonic generation and RFI due to quasi-sinusoidal power delivery of the inverter. The stepped-up secondary output is further raised by a multi-stage Cockcroft–Walton multiplier. In addition to the programming of output voltages, the supply has provision for remote on/off control, setting adjustable HV ramp rates and output voltage/load current readback facility. Figure 108.7 shows a fully assembled HV power supply.

The low voltage power supplies required for the analog and digital front-end boards as well as the HV DC-DC module on-board the RPC unit are individually supplied, controlled and monitored through a commercial and centrally located low-voltage power supply distribution and monitoring sub-system.

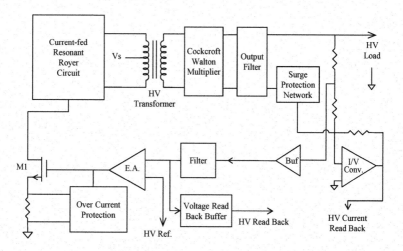

Fig. 108.6 HV generation schematic

Fig. 108.7 HV module

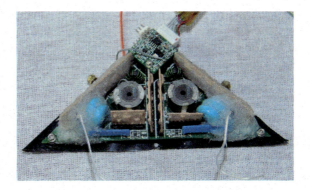

108.6   Status

The mini-ICAL electronics and DAQ has been installed at Madurai, Tamilnadu, with 10 RPCs and is performing very well. More RPC units are being added to the mini-ICAL stack. The integration of the front-end and back-end electronics and the back-end data acquisition software has been done successfully. The integration of electronics, especially the analog and digital front-end and the HV module on-board the RPC detector module, posed a big challenge.

Meanwhile, extensive cosmic muon data has been taken with the detector with and without the magnetic field to study the detector parameters like RPC multiplicity, efficiency and noise rate stability. Figure 108.8 shows a glimpse of muon tracks captured by the DAQ; slight bending can be seen in the tracks because of the magnetic field. RPC multiplicity and region-wise normalized efficiency can be seen in the figure. The yellow spots are RPC button locations.

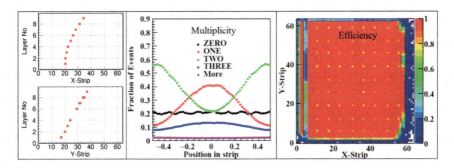

Fig. 108.8   Results showing a bent track, RPC multiplicity and region-wise efficiency

References

1. INO-Collaboration, India-based neutrino observatory, Tech. Rep. INO/2006/01, Tata Institute of Fundamental Research (January 2006). http://www.ino.tifr.res.in/ino/OpenReports/INOReport.pdf
2. P.K. Kaur et al., Development of Fast, Low Power 8-Channel Amplifier-Discriminator Board for the RPCs. In: *XXII DAE High Energy Physics Symposium*. Springer Proceedings in Physics, vol. 203 (Springer, Cham, 2018). https://doi.org/10.1007/978-3-319-73171-1_134
3. M.N. Saraf et al., Soft-core processor based data acquisition module for ICAL RPCs with network interface. In: *XXI DAE-BRNS High Energy Physics Symposium*. Springer Proceedings in Physics, vol. 174 (Springer, Cham, 2016) https://doi.org/10.1007/978-3-319-25619-1_86
4. P. Nagaraj et al., Ethernet scheme for command and data acquisition for the INO ICAL detector. In: *XXII DAE High Energy Physics Symposium*. Springer Proceedings in Physics, vol. 203 (Springer, Cham, 2018) https://doi.org/10.1007/978-3-319-73171-1_210
5. A. Manna et al., Development of a ±6 kV Bias Supply for ICAL Detectors of INO. In: M. Naimuddin (eds) *XXII DAE High Energy Physics Symposium*. Springer Proceedings in Physics, vol. 203. (Springer, Cham, 2018) https://doi.org/10.1007/978-3-319-73171-1_206

Chapter 109
Deep Learning-Based Energy Reconstruction of Cosmic Muons in mini-ICAL Detector

Deepak Samuel, Manjunath Omana Kuttan, Amrutha Samalan, and Lakshmi P Murgod

Abstract The mini-ICAL is a magnetized prototype of the proposed iron calorimeter (ICAL) detector of the upcoming India-Based Neutrino Observatory (INO). The mini-ICAL is designed to study the performance of Resistive Plate Chambers (RPCs) in a magnetic field and the efficiency of reconstruction algorithms. In this study, we investigate the possibility of using a deep learning-based algorithm for muon energy reconstruction in the detector. Deep learning models were developed using the data generated from a simplified geometry of mini-ICAL simulated using the Geant4 package. The models are evaluated for their accuracy in predictions and reconstruction time.

109.1 Introduction

The INO-ICAL aims to study atmospheric neutrinos using a 50 kton iron calorimeter (ICAL) detector [1]. The mini-ICAL is a miniature prototype of this detector built at IICHEP, Madurai, for the R&D activities of the ICAL experiment. The mini-ICAL detector is a stack of 10 layers of RPCs of dimension 2 m × 2 m interleaved between iron plates of thickness 56 mm. The current-carrying copper coils wound through the detector can magnetize the iron plates to a field of up to 1.5 T in the central region. A muon traversing the stack produces signals in the RPCs which are picked up by orthogonal pickup panels having a strip pitch of 3 cm, placed over

D. Samuel (✉) · M. Omana Kuttan · A. Samalan · L. P. Murgod
Department of Physics, Central University of Karnataka, Kalaburagi 585367, India
e-mail: deepaksamuel@cuk.ac.in

M. Omana Kuttan
e-mail: okmanjunath660@gmail.com

A. Samalan
e-mail: amruthasyamalan@gmail.com

L. P. Murgod
e-mail: lakshmipmurgod@gmail.com

© Springer Nature Singapore Pte Ltd. 2021   787
P. K. Behera et al. (eds.), *XXIII DAE High Energy Physics Symposium*,
Springer Proceedings in Physics 261,
https://doi.org/10.1007/978-981-33-4408-2_109

and below each RPC [2]. The RPCs also offer a time resolution of about 1 ns. The position and timing information are used to reconstruct the properties such as charge, momentum and direction of the incoming muon. A Kalman filter-based algorithm is presently employed for the purpose. In this study, we examine the possibility of using a deep learning algorithm for reconstructing the energy of cosmic muons reaching the detector. Simplified geometry of mini-ICAL is simulated using the Geant4 package with the aim of benchmarking the code, and this data is used to develop machine learning models based on XGBoost [3] and Artificial Neural Networks (ANN). The performance of these models is then studied based on the accuracy of their predictions and the time taken to reconstruct the energy of the muon.

109.2 Simulation Setup and Datasets

The geometry of the mini-ICAL detector was simulated using an in-house developed software based on Geant4. The simulated geometry consists of 10 glass plates of dimension 4 m $\times$ 4 m $\times$ 6mm interleaved between iron plates of dimension 4 m $\times$ 4 m $\times$ 56mm. The glass plates record the position information of muons crossing it. A rectangular slab of vacuum material of dimensions 4 m $\times$ 4 m $\times$ 1mm placed above the module is used to record the energy of the incoming muons. The fine details of the geometry such as copper coils, gas gap and pickup panels were not included in the simulations. Two datasets (Dataset 1, Dataset 2) containing 30000 μ^- events each with a uniform energy spectrum of 200 MeV–2000 MeV were generated from a point source above the detector without angular smearing. Dataset 1 was generated without any magnetic field while Dataset 2 was generated using a magnetic field of 1.5 T along the plane in the iron plates.

109.3 Training the Models

XGBoost and ANN algorithms were used to develop 4 models (XGB1, ANN1, XGB2 and ANN2) using 75% the events in the datasets (Dataset 1, Dataset 2) while the remaining events were used to test the models. The details of the models are tabulated in Table 109.1. Each event was represented as a sparse matrix where the location of hits was marked 1 leaving the other cells as 0. The hit positions in a detector were digitized to a strip number between 0 and 133 so that the strip pitch is ~3 cm. Therefore, an event in a stack of 10 layers can be represented by a matrix of dimension 20 $\times$ 134 where the first 10 rows represent the x strip hits while the next 10 rows represent y strip hits. The use of such a large feature set for training is computationally inefficient and consumes a lot of memory. This is overcome by using a matrix of smaller dimension. Figure 109.1 shows the sparse matrix representations of two different events having similar features but passing through different locations in the detector. As shown in the figure, if we translate event 2 by certain values, we don't

Table 109.1 Table summarizing the models developed in this study. In each dataset, 75% of the events were used for training and remaining 25% for testing the model

Model	Dataset used	Algorithm used
XGB1	Dataset 1	XGBoost
XGB2	Dataset 2	XGBoost
ANN1	Dataset 1	ANN
ANN2	Dataset 2	ANN

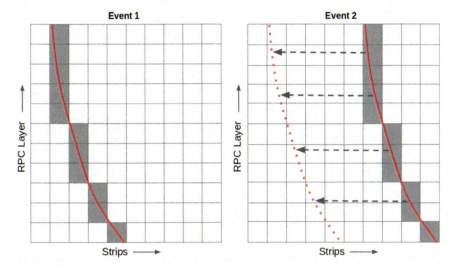

Fig. 109.1 Sparse matrix visualization of two events. The solid red line is the actual track of the muon and the gray cells are the strips fired. The dashed red line is the track after translating the event

lose any features, and the size of the matrix required to store the event reduces. As an alternative to translating the events, we have chosen a point source of particles so that the representation of all the events is a uniform 20×30 matrix.

109.4   Results and Discussions

All the models were tested using the remaining 25% of the events in their corresponding dataset. Figure 109.2 shows the correlation plot for 4 different models. Both XGBoost and ANN algorithms (XGB1 and ANN1) gave comparable results with small error bars for energies up to around 750 MeV (region 1) and large error bars for higher energies (region 2). The presence of a magnetic field (XGB2 and ANN2) made the results slightly better in region 2 while keeping the same trend. On a system running on an Intel i3 processor with 8 GB physical memory and Ubuntu

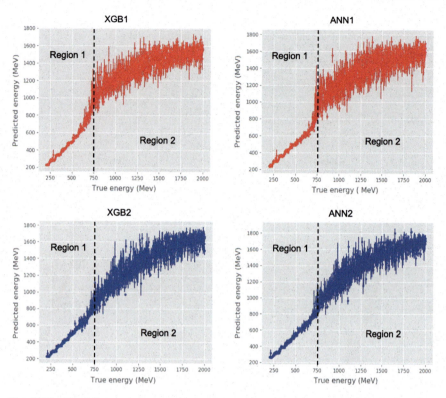

Fig. 109.2 Correlation plots of different models. All plots have two distinct regions: region 1 having small error bars where the true energy is less than 750 MeV and region 2 with large error bars where the true energy is greater than 750 MeV

16.04 Operating System, the XGBoost models predicted an event in ∼1 mS while the ANN models were much faster with a prediction time of ∼30 μS for a single event.

Both XGBoost and ANN models were comparable in terms of accuracy in the predictions but the ANN models were found to be faster in predicting. For the simulated geometry, the energy loss of muons [4] is such that the events become partially contained for energies beyond ∼750 MeV. Thus, the large error band in region 2 is due to the inability of the model to accurately predict the energy of partially contained events compared to the fully contained events. In this study, we have shown the first results of an energy reconstruction algorithm using machine learning for the INO-ICAL experiment. Further fine-tuning of the algorithm and addition of finer details of geometry in the simulation will be the next step of this study. The first results show that the existing Kalman filter technique can be replaced by the simpler and faster machine learning models to use, for example, in triggerless systems.

References

1. A.M. Sajjad et al., India-based neutrino observatory: project report, volume I. No. INO-2006-01 (2006)
2. S. Bheesette, Design and characterisation studies of resistive plate chambers (2009)
3. T. Chen, C. Guestrin, Xgboost: A scalable tree boosting system. In: *Proceedings of the 22nd ACM sigkdd International Conference on Knowledge Discovery and Data Mining* (ACM, 2016), pp. 785–794
4. D.E. Groom, N.V. Mokhov, S.I. Striganov, Muon stopping power and range tables 10 MeV–100 TeV. Atomic Data Nucl. Data Table **78**(2), 183–356 (2001)

Chapter 110
Simulation Studies for a Shallow Depth ICAL

N. Panchal, Gobinda Majumder, and V. M. Datar

Abstract The Iron CALorimeter (ICAL) is an upcoming neutrino detector at India-based Neutrino Observatory (INO). The observatory is planned to be located at Bodi West Hills, Madurai, under a mountain cover of $\sim$1 km. The rock overburden reduces the cosmic muon background by a factor of 10^6 and also casts aside the hadronic component of cosmic ray background. It is possible to remove the hadronic component almost completely by a shallow depth (100 m) of the rock where the cosmic muon flux also reduces by a factor of 100. For achieving the same background reduction of cosmic muons at a shallow depth of 100 m, a veto detector with veto efficiency of 99.99% is required which appears to be feasible Panchal et al. (J Instrum 12:T11002, 2017 [1]). The neutral particles (neutrons and K_L^0) produced by muon-nuclear interaction can pass through the veto detector undetected which can mimic neutrino events in the ICAL detector. With this motivation, Geant4 based simulation studies have been performed to estimate such false positive event rate in the ICAL detector and the results are presented in this paper.

110.1 Introduction

One of the prime goals of the proposed ICAL at INO is to determine the mass hierarchy in the neutrino sector [2]. ICAL will be a 51 kTon detector consisting of three 17 kton modules, each having 150 layers of iron interleaved with Resistive Plate Chamber (RPC) detectors [3]. INO is planned to be situated under a mountain cover

N. Panchal (✉)
Homi Bhabha National Institute, Anushaktinagar, Mumbai 400094, India
e-mail: neha.dl0525@gmail.com

N. Panchal · G. Majumder · V. M. Datar
Tata Institute of Fundamental Research, Homi Bhabha Road, Colaba, Mumbai 400005, India
e-mail: majumder.gobinda@gmail.com

V. M. Datar
e-mail: vivek.datar@gmail.com

© Springer Nature Singapore Pte Ltd. 2021   793
P. K. Behera et al. (eds.), *XXIII DAE High Energy Physics Symposium*,
Springer Proceedings in Physics 261,
https://doi.org/10.1007/978-981-33-4408-2_110

of about ~ 1 km from all the sides at Pottipuram in Bodi West Hills of Theni, India. The rock overburden above ICAL significantly reduces the background arising due to cosmic ray muons (by a factor of 10^6). Building an efficient cosmic muon veto detector (CMVD) can lead to achieving a similar background reduction at shallow depth say ~ 100 m. The reduction in cosmic muon flux at a depth of 100 m is about a factor of 100. Therefore, an ICAL detector placed at a shallow depth along with a CMVD with veto efficiency 99.99 % could result in similar muon flux as that at the Theni site. Such a detector is referred to as a shallow depth ICAL (SICAL). At shallow depths, the most significant background is due to cosmic ray muons which can be vetoed with a high veto efficiency. However, the secondaries generated due to muon-nucleus interaction with the rock material can be a serious concern which needs to be carefully investigated.

In this paper, we present results of a GEANT4-based study to estimate the contribution of this background and compare it with the event rate expected from atmospheric neutrinos in the SICAL detector. We first present the details of the simulation framework to characterize muon-nuclear interactions in the rock in Sect. 110.2. In Sect. 110.3, we discuss the estimation of false-positive event rate by propagating the neutrals through the ICAL detector using the INO-ICAL simulation code. The summary of the major results of the simulation is presented in Sect. 110.4.

110.2 Simulation of Neutral Particles Following Cosmic Muon-Rock Interaction

A full simulation of the neutrino-like events in ICAL would involve propagating the cosmic muons at the surface in the intervening rock corresponding to the chosen location. This would include keeping track of the secondary particles produced in muon-nucleus interactions anywhere along their path toward the ICAL detector at a specified depth. We performed the simulation for a depth of 100 m and the results are presented in this paper. The propagation of low-energy muons and secondary hadrons, produced in high-energy muon interactions in the upper part of the 100 m rock overburden, increases the computation time as these will not survive the remaining rock thickness. Therefore, to decrease the computation time, the simulation was done in two parts with the corresponding geometry using GEANT4. In the first part of the simulation, we study the energy loss of muons in the rock to parametrize it. In this part, the geometry was defined to be a cube of dimension 100 m × 100 m × 100 m with SiO_2 as the material. Muons were incident from the center of the top surface and propagated in the positive z-direction (downward going) with energies uniformly distributed between 0.01 and 500 GeV. The energy of the muon (E_μ') after traversing 100 m of the rock is plotted as a function of incident muon energy (E_μ) as shown in Fig. 110.1 (left). The full range of E_μ was divided into energy bins of width 1 GeV and the corresponding E_μ' distribution was fit to a Crystal Ball function [4]. The typical E_μ' distribution for $E_\mu = 80$ GeV and $E_\mu = 300$ GeV are shown in Fig. 110.1.

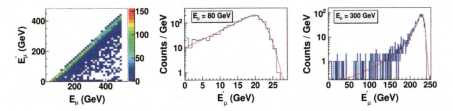

Fig. 110.1 Two-dimensional plot of incident muon energy E_μ and muon energy after 100 m rock E_μ' (left) and E_μ' distribution for 80 GeV (center) and 300 GeV (right) incident muon energy fitted to Crystal Ball fit function

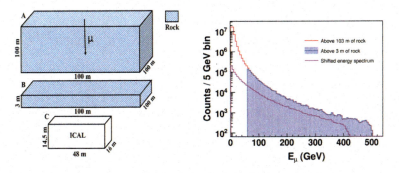

Fig. 110.2 (Left) Schematic showing the geometry and placement of the blocks of rock and ICAL detector used in the simulation. (Right) Primary cosmic muon energy spectrum generated from CORSIKA at position A (Red), muon energy spectrum at position B, i.e., after traversing 100 m of the rock (Blue) and the shifted muon energy spectrum as described in Sect. 110.2

In the second part of the simulation, CORSIKA software with SIBYLL [5] model was used to generate the primary cosmic muon spectrum. For any particular incident muon energy, E_μ' was chosen generating a random number from the Crystal Ball fit function as shown in Fig. 110.1 (center and right). The muon energy spectrum was then shifted according to the two-dimensional plot in Fig. 110.1 (left). The simulated E_μ' and the shifted energy spectrum were given as an input to the GEANT4 simulation with geometry as a cuboid of dimensions $100\,\text{m} \times 100\,\text{m} \times 3\,\text{m}$ (thickness) and SiO_2 as rock material. The neutrals produced in muon-nuclear interactions, mostly from the last part of the 3 m depth of the rock, could exit the rock. The choice of 3 m was guided by the hadronic interaction length for the rock which is 36 cm [6], i.e., 10 times smaller.% This was verified by performing the simulation for 5 m and 10 m of the rock which produced, within error, the same number of outgoing particles as with the 3 m rock.

We have used the Kokoulin model [7] to simulate muon-nuclear interactions in the material. The hadronic interactions of the secondaries were also considered. In the simulation, the primary muons, along with all the secondaries in the hadronic cascade, were propagated to the bottom of the block B shown in Fig. 110.2 (left) and the associated information about the particles were stored. The secondaries produced

Table 110.1 Fraction of secondaries produced in muon-nuclear interaction

n	π^0	K_L^0	Σ^0	η^0	η^1	λ^0	Σ^-	Σ^+	ρ^0	Ξ^0	Ξ^-
93.8	4.9	0.65	0.21	0.16	0.05	0.04	0.03	0.03	0.02	0.001	0.001

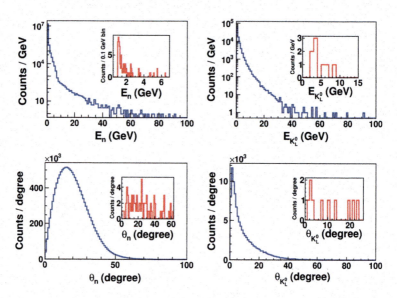

Fig. 110.3 Energy and theta spectra of neutrons (left top and left bottom) and K_L^0s (right top and right bottom). The blue histogram shows the total neutrons (or K_L^0s) coming out of the rock and the red histogram shows neutrons (or K_L^0s) coming out without any accompanied charged particle

in muon-nuclear interactions with their fractions are listed in Table 110.1. The two most populous and relevant long-lived neutrals are neutrons and K_L^0, and the typical energy and θ distributions of both the particles from this simulation are shown in Fig. 110.3. The π^0s are not considered here as they have a very short lifetime ($\sim 10^{-16}$ s) and decay into 2 γ-rays leading to an electromagnetic shower which can be vetoed out.

In the simulation for total 10^{10} incident muons, there are total $\sim 4.5 \times 10^7$ muon-nuclear interactions. The total secondaries produced were about 3.0×10^8 out of which 1.3×10^7 come out of the rock and the number of events in which no charged particle is present is about a hundred. Out of a total of $\sim 6.8 \times 10^6$ events, in roughly 8 % of them, neutrons and K_L^0s having an energy more than 1 GeV come out of the rock. The number of events with neutrals unaccompanied by any charged particle that are relevant for this study should satisfy the following conditions:

(a) the charged particle coming out of the rock along with the neutral particle should have energy more than 50 MeV, as below this energy the charged particle may not give any signal in the veto detector,

(b) the neutral particle should have energy greater than 1 GeV which is required to produce a charged particle in nuclear interaction, which can pass through 5 layers of RPC detector in ICAL to mimic a muon in a ν_μ charged current interaction.

110.3 False-Positive Event Rate in ICAL Due to Muon-Induced Neutrals

Any event that is generated as a result of the interaction of particles different from a neutrino but is very likely to be classified as a neutrino induced event in the ICAL detector is called a false-positive signal. All the secondaries resulting from the muon-nuclear interactions, as listed in Table 110.1, were propagated in the ICAL simulation code [8] with their respective (E, θ) distributions taken from previous simulation (as shown in Fig. 110.3). The incident particle was chosen in accordance with the fraction by which it was produced. A trajectory of a charged particle due to false-positive signal is considered to be track-like, if it gives hits in minimum 5 layers of ICAL detector and $\chi^2/ndf < 10$. In the simulation, the fraction in which a track-like signal F_{trk} was obtained is 2×10^{-3}.

Due to the almost 100 % efficiency of the veto detector, a large fraction of all the secondaries coming out of the rock would be vetoed. However, due to the small inefficiency, the rest will traverse the veto detector undetected. From an earlier measurement with a small Cosmic Muon Veto Detector [1], the veto efficiency achieved was 99.978 % which is equivalent to a reduction in muon flux, resulting in reducing false-positive events, by about 10^4. So, the number of secondaries coming out of the rock which could lead to false-positive events will also effectively reduce by a factor of 10^4. The fraction of false positives due to neutrals in ICAL is given as

$$F_{FP} = N_{out} \times F_{trk} \times (\epsilon_{veto})^{n_q} \qquad (110.1)$$

where n_q is the number of charged particles coming out of the rock along with at least one neutral particle, N_{out} (E > 1 GeV) is the number of events in which a neutron or K_L^0 comes out of the rock with E > 1 GeV and ϵ_{veto} is the veto inefficiency. While the probability of not detecting only one charged particle is more than the probability of not detecting at least one out of two, for the sake of completeness the event-wise breakup is done for estimating the F_{FP}. Table 110.2 shows the distribution of events with a different number of charged and neutral particles coming out of the rock.

The total number of incident muons in the simulation at 100 m depth was 10^{10}. As the surface flux reduces after traversing the 100 m rock by a factor of ~ 100, the effective number of muons at the surface is 10^{12}. From Eq. 3.1, the false-positive signal rate comes out to be 0.2×10^{-12}. The primary cosmic ray muon flux at sea level is $70 \, \text{m}^{-2} \, \text{sec}^{-1} \, \text{sr}^{-1}$ from [6]. The dimensions of each module of ICAL are $16 \, \text{m} \times 16 \, \text{m} \times 14.5 \, \text{m}$ and there will be three such modules resulting in a total surface area of $48 \, \text{m} \times 16 \, \text{m}$. For an overground ICAL, the muon event rate will be $\sim 10^{10}$ /day and the false-positive event rate due to neutrals is 0.002 /day. If $\sigma_{CC} = 10^{-38} \, \text{cm}^2$ [6]

Table 110.2 Breakup of events with a specified number of charged particles (q) and neutral particles (n) for 10^{10} muons interacting with the last 3 m rock just above ICAL

Configuration	N_{out} (E > 1 GeV)	F_{FP}
0q & 1n	120	2.4×10^{-1}
0q & 2n	1	2×10^{-3}
0q & 3n	0	0
1q & 1n	119381	2.3×10^{-2}
1q & 2n	0	0
1q & 3n	0	0
2q & 1n	70430	1.4×10^{-6}
2q & 2n	4636	9.3×10^{-8}
2q & 3n	0	0
Others	329818	0
Total	524386	2×10^{-1}

is the ν_μ and $\bar{\nu}_\mu$ inclusive of scattering cross section (per nucleon), $\rho = 7.8$ gm-cm^{-3} is the density of iron (Fe), A = 56 is the atomic weight (in units of amu) of Fe and λ is the thickness of the iron plate in the ICAL detector, and N_A is the Avogadro number, then from $\frac{N_{ev}}{N_{inc}} = \frac{\sigma_{\mu N} \cdot \rho \cdot \lambda \cdot N_A}{A}$, $\frac{N_{ev}}{N_{inc}} \sim 4 \times 10^{-14}$ for 150 layers of iron in ICAL. The primary cosmic ray flux from [9] is 10^3 m^{-2} sec^{-1} sr^{-1} which gives the neutrino event rate for ICAL to be ~ 3 /day. The signal to false positive for SICAL is therefore, about 1000 which makes SICAL a feasible proposition.

110.4 Summary

In this paper, the possibility of locating the INO-ICAL detector at a shallow location with a rock overburden of ~ 100 m along with a CMVD of veto efficiency 99.99 % was explored. The muon-induced neutral background is a vital background which needs to be considered in such a case. Therefore, a GEANT4-based simulation for estimating the neutral particles produced considering only the vertical going muon passing through a block of 103 m of the rock was carried out. These neutral particles are further propagated in the ICAL simulation code for estimating the neutrino-like events produced by such particles. A background of ~ 0.1% was obtained from the simulation which makes SICAL a viable option to be persuaded. However, the validation of the simulation is required on a proof-of-principle detector built at a shallow depth of about 30 m rock overburden, together with the Cosmic Muon Veto Detector.

Acknowledgements   We would like to thank Naba K. Mondal, D. Indumathi, Paschal Coyle, John G. Learned, Morihiro Honda, Thomas K. Gaisser, Shashi R. Dugad, Deepak Samuel and Satyajit Saha for useful comments and suggestions and Apoorva Bhatt and P. Nagaraj for help in simulation related issues.

References

1. N. Panchal et al., A compact cosmic muon veto detector and possible use with the Iron Calorimeter detector for neutrinos. J. Instrum. **12**, T11002 (2017)
2. A. Ghosh et al., Determining the neutrino mass hierarchy with INO, T2K, NOvA and reactor experiments. J. High Energy Phys. **2013**, 9 (2013)
3. M.S. Athar et al., INO Collaboration: Project Report **I**, (2006)
4. J. E. Gaiser, *Appendix-F Charmonium Spectroscopy from Radiative Decays of the J/Psi and Psi-Prime*, Ph.D. Thesis, SLAC-R-**255** (1982) 178
5. R.S. Fletcher, T.K. Gaisser, Lipari et al., sibyll: An event generator for simulation of high energy cosmic ray cascades. Phys. Rev. D, **50**, 5710–5731 (1994)
6. C. Patrignani et al., Particle data group. Chin. Phys. C **40**, 100001 (2016)
7. A.G. Bogdanov et al., IEEE Trans. on Nuc. Sci., **53**(2) (2006)
8. Apoorva Bhat, G. Majumder, priv. comm.
9. M. Honda et al., Atmospheric neutrino flux calculation using the NRLMSISE-00 atmospheric model. Phys. Rev. D **92**, 023004 (2015)

Chapter 111
ALICE Inner Tracking System Upgrade at the LHC

Nirbhay Kumar Behera

Abstract The design objective of ALICE is to study the properties of the Quark–Gluon plasma using the pp, p–Pb, and Pb–Pb collisions at the LHC. Using the excellent tracking and particle identification capability, ALICE has achieved several milestones in understanding the hot and dense nuclear matter produced in these collisions. During the long shutdown period in 2019–20, ALICE will undergo a major upgrade to improve its precision of the present physics measurements. As part of this upgrade plan, the ALICE Inner Tracking system (ITS) will significantly be upgraded to enhance its capabilities to measure the rare probes with greater precision. The key goal of the ITS upgrade is the construction of a new detector with high resolution, low material budget, and high read-out rate. This upgrade will enable to collect data at the rate of 400 kHz and 50 kHz in pp and Pb–Pb collisions, respectively, during the LHC high luminosity program in Run 3. The new ITS detector has 7 concentric layers of CMOS pixel detector based on Monolithic Active Pixel Sensor (MAPS) technology and is designed to achieve all these requirements. It is one of the first applications of MAPS technology in a high-energy physics experiment. This MAPS-based sensor, with other stringent mechanical design of the detector support material, will reduce the material budget to 0.3% X_0 for the inner layers and 1% X_0 of the outer layers. In this presentation, we will discuss the design goal, layout of the new detector, its performance during the research and development phase, and the production status.

111.1 Introduction

The design goal of A Large Ion Collider Experiment (ALICE) at the LHC is to study the properties of Quark–Gluon Plasma (QGP) [1]. ALICE has collected the pp and Pb–Pb collision data during the Run 1 and Run 2 periods of LHC and helped to

Nirbhay Kumar Behera—for the ALICE collaboration.

N. K. Behera (✉)
Department of Physics, Inha University, 402-751 Incheon, Republic of Korea
e-mail: nbehera@cern.ch

© Springer Nature Singapore Pte Ltd. 2021
P. K. Behera et al. (eds.), *XXIII DAE High Energy Physics Symposium*,
Springer Proceedings in Physics 261,
https://doi.org/10.1007/978-981-33-4408-2_111

unravel many interesting physics results. It has enriched our understanding of the particle production mechanisms, thermal properties of QGP, and the parton interaction with the QGP medium. However, some of the physics results from rare probe measurements need further improvement. Current results of rare probes face challenges due to small event statistics and limited coverage of transverse momentum (p_T) range, which is due to the limitations of the detectors. After Run 2 data taking finished in December 2018, LHC is now in the second long shutdown (LS2) up to 2020. During LS2, ALICE plans for a detector upgrade to meet the following physics goals [2, 3]:

- measurements of the elliptic flow of charm and beauty hadrons to study the thermalization and hadronization of heavy quarks;
- nuclear modification factor (R_{AA}) of D and B mesons for a wide range of transverse momentum (p_T) to understand the in-medium energy loss of heavy quarks.

These measurements demand high event statistics and precise measurement of the primary and secondary vertices. After LS2, LHC is planning for the high luminosity run of Pb–Pb collisions at 50 kHz interaction rate. Therefore, to meet the requirements of the LHC high luminosity run, ALICE plans for major upgrades. The objectives of the upgrades are (i) new beam pipe with smaller radius, (ii) upgrade of Time Projection Chamber (TPC) with Gas Electron Multiplier (GEM) detectors and new read-out electronics, (iii) upgrade of the read-out electronics of TRD, (iv) upgrade of the forward trigger detectors, (v) upgrade of online–offline reconstruction framework, and (vi) upgrade of Inner Tracking System (ITS) with high-resolution and low-material-budget detectors.

In this report, the upgrade plans of the ALICE ITS are discussed.

111.2 ALICE ITS Upgrade Plan

The current ITS detector consists of six layers. The two innermost layers are made of Silicon Pixel Detectors (SPD), the two middle layers are made of Silicon Drift Detectors (SDD), and the two outer layers are made of double-sided Silicon Strip Detectors (SSD). Although the current ITS is used for high precision measurements of rare probes in a broad range of p_T, it has the following limitations:

- low read-out rate: the SDD has a read-out rate of about 1 kHz. Therefore, ALICE can only record a small fraction of the Pb–Pb collisions delivered at the rate of 8 kHz by the LHC;
- high material budget: current ITS has about 1.1 % X_0 per layer;
- low spatial resolution: current ITS has the spatial resolution about 110–120 μm at $p_T \sim 0.5$ GeV/c. The path length of Λ_c baryon is 60 μm. Therefore, it limits the measurement of Λ_c baryon in Pb–Pb collisions.

To overcome these limitations and to meet the requirements of the Run 3 physics program of LHC, the ITS detector will go for a major upgrade. The new upgraded

Table 111.1 The upgrade objectives of the ALICE ITS detector

	Current ITS	Upgraded ITS
Number of layers	6	7
Distance to IP	39 mm	23 mm
Pointing resolution in z (at p_T 0.5 GeV/c)	120 μm	40 μm
Material budget	1.1 % X_0	0.3 % X_0
Pixel size	50$\times$425 μm^2	O(30$\times$30 μm^3)
Read-out rate	1 kHz	100 kHz (Pb–Pb)

ITS detector will be made up of seven concentric cylindrical layers of silicon pixel detectors based on Monolithic Active Pixel Sensor (MAPS) technology of 180 nm CMOS technology of TowerJazz [3]. The new upgraded ITS detector has the following objectives tabulated in Table 111.1 [4, 5]:

111.3   The Upgraded ITS Detector

111.3.1   Detector Layout

The new ITS detector has seven layers and is grouped into two barrels: Inner Barrel (IB) and Outer Barrel (OB). The IB consists of three layers, and OB consists of four layers. The schematic layout of the upgraded ITS detector is shown in Fig. 111.1. It will have a pseudorapidity coverage of $|\eta| < 1.3$. All layers are azimuthally segmented into mechanically independent units called Staves. Each stave consists of a space frame, cold plate, and hybrid integrated circuit (HIC). The staves of OB are further divided into two halves called half-staves. There is a total of 48 staves in the IB, and each stave consists of one module. The two middle layers have 54 staves and eight modules per staves. Similarly, the two outer layers have 90 staves and each stave consists of 14 modules. The modules are made up of a HIC, which is glued onto a carbon plate. Each module has eight sensors and 14 sensors for IB and OB staves, respectively. So there will be total 24×10^3 silicon chips making 10 m^2 active silicon area. This will act as a 12.5 G-Pixels 3D camera. The main active component is the silicon sensor, which is called the ALICE Pixel Detector (ALPIDE). Some of the salient features of the ALPIDE chip are discussed in the next section.

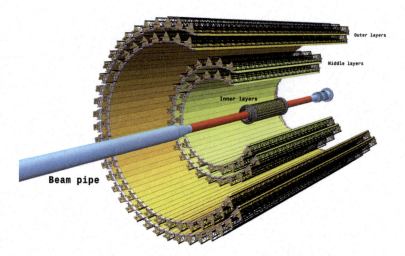

Fig. 111.1 (Color online) The schematic layout of the upgraded ALICE ITS detector

111.3.2 ALPIDE Chip

The ALPIDE chip uses the 180nm CMOS technology by TowerJazz. The dimension of each pixel is $27 \times 29\ \mu m^2$. The size of each ALPIDE chip is $30 \times 15\ mm^2$ with epitaxial layer thickness 50 μm (100 μm) for IB (OB). Each ALPIDE chip has 1024×512 pixels. Some of the highlights of the ALPIDE chips are as follows: (i) the n-well collection diode has 2 μm diameter, which leads to smaller capacitance ($\sim$ fF); (ii) the deep p-well shields the n-well of PMOS transistor allowing full CMOS circuitry within the active volume; (iii) the analog signal is amplified and digitized at the pixel level leading to low power consumption ($<$300 nW); (iv) reverse bias voltage can be applied to the substrate between $-6V$ and $0V$; (v) high resistivity (>1 kΩcm) p-type epitaxial layer on p-type substrate leads to a larger signal to noise ratio; (vi) the detection efficiency is higher than 99 %, and fake hit rate is less than 10^{-10} pixels per event; and (vii) high radiation hardness (2.7 Mrad for IB, 100 Krad for OB (TID), including a safety factor of ten, and, 1.7×10^{12} 1 MeV n_{eq}/cm^2 (NIEL)).

111.4 ITS Construction

The chip production starts with raw wafers. First, the CMOS wafer production is done by TowerJazz. After wafer probe testing is done at CERN, they are sent for thinning and dicing by FUREX in South Korea. The chips are picked and placed in the tray and sent to the mass chip test sites (50 μm chips to CERN, 100 μm chips to Pusan-Inha, Yonsei University in South Korea). After the series test, the chips are sent to different HIC assembly sites. In the HIC assembly sites, the HICs are produced by aligning the chips and glued with the pre-verified Flexible Printed Circuit (FPC) using the customized (ALICIA) machine. The electrical connection between the chips and FPC are established by wire bonding. The assembled HICs then go through the qualification test. The qualified HICs are then sent to stave assembly sites where the HICs are glued to the cold plates. The power bus is connected with cross-cables of HIC by soldering. Then the staves are glued on the carbon space frame. All assemblies are done with high precision. All the assembled staves will go through a number of qualification tests before the assembly of the detector. The qualification test ensures the electrical and mechanical functionality of different components.

111.5   Summary and Outlook

In summary, the ALICE ITS detector is going for a major upgrade to fulfill the requirements of the ALICE physics goal and the LHC high luminosity physics program. The new ALICE ITS detector will be equipped with a MAPS technology-based CMOS pixel sensor with high resolution, low material budget, and high radiation tolerance. The mass chip test, HIC assembly, and stave production are finished. Meanwhile, the quality assurance test of the staves is ongoing. The surface commissioning will take place in 2019. In 2020, the final detector will be installed in the ALICE pit. The upgraded ITS detector will take part in data taking in Run 3 and Run 4, which will help to study the rare probe with greater precision at the LHC.

Acknowledgements This work was supported by National Research Foundation of Korea (NRF), Basic Science Research Program through the National Research Foundation of Korea funded by Ministry of Education, Science and Technology (Grant number: NRF-2014R1A1A1008246).

References

1. K. Aamodt et al., ALICE collaboration. JINST **3**, S08002 (2008)
2. ALICE collaboration, conceptual design report for the upgrade of the ALICE ITS, CERN LHCC 2012, 005 (2012)
3. B. Abelev et al., ALICE collaboration. J. Phys. G **41**, 087002 (2014)
4. F. Reidt, ALICE collaboration, JINST **11**(12), C12038 (2016)
5. M. Mager, ALICE collaboration. Nucl. Instrum. Meth. A **824**, 434 (2016)

Chapter 112
Azimuthal Dependence of Cosmic Muon Flux by 2 m × 2 m RPC Stack at IICHEP-Madurai and Comparison with CORSIKA and HONDA Flux

S. Pethuraj, Gobinda Majumder, V. M. Datar, N. K. Mondal, S. Mondal, P. Nagaraj, Pathaleswar, K. C. Ravindran, M. N. Saraf, B. Satyanarayana, R. R. Shinde, Dipankar Sil, S. S. Upadhya, P. Verma, and E. Yuvaraj

Abstract The INO-ICAL experiment is a proposed underground particle physics experiment to study the neutrino oscillation parameters. Twelve layers of 2 m × 2 m RPC stack was built to study the performance of RPC detectors which are made in the Indian industry and test the indigenously made electronics and DAQ. The RPC Stack is used to study the azimuthal dependence of the cosmic ray muons in different zenith angular bins. The observed results are compared with CORSIKA and HONDA predictions.

112.1 Introduction

The primary cosmic rays coming from outer space mostly consist of protons (90 %), helium (9 %), and a small fraction of heavy elements like carbon, nitrogen, oxygen, etc. The interaction of primary elements with the earth' atmosphere creates a shower of secondary particles (mostly pions (π^+, π^- and π^0). The neutral pions mostly decay to 2 γ and the charged pions mainly decay to muon and neutrino. Muons are the most abundant charged particles at sea level from cosmic ray shower. The flux of primaries is modulated by the earth' magnetic field, which will cause azimuthal dependency in arrival directions. This effect is called the "east-west" asymmetry of cosmic rays. The same effect can be observable in the ground-based experiments.

S. Pethuraj (✉)
Homi Bhabha National Institute, Mumbai 400094, India
e-mail: spethuraj135@gmail.com

G. Majumder · V. M. Datar · S. Mondal · P. Nagaraj · Pathaleswar · K. C. Ravindran · M. N. Saraf · B. Satyanarayana · R. R. Shinde · D. Sil · S. S. Upadhya · P. Verma · E. Yuvaraj
Tata Institute of Fundamental Research, Mumbai 400005, India

N. K. Mondal
Saha Institute of Nuclear Physics, Kolkata 700064, India

© Springer Nature Singapore Pte Ltd. 2021
P. K. Behera et al. (eds.), *XXIII DAE High Energy Physics Symposium*,
Springer Proceedings in Physics 261,
https://doi.org/10.1007/978-981-33-4408-2_112

The east-west asymmetry depends on the latitude, longitude, altitude, and momentum cutoff. Many experiments studied the east-west asymmetry in a different locations on earth [1–4].

112.2 Experimental Setup and Data Analysis

The INO-ICAL experiment will be built using 50 k-tonne iron plates and RPC detector of 2 m × 2 m in the area chosen as an active detector element. The main aim of the INO-ICAL experiment is to measure the sign of $\Delta m_{32}^2 (m_3^2 - m_2^2)$ through earth matter effects [5]. A prototype of INO-ICAL is built using 12 layers of 2 m × 2 m RPC detectors [6]. A 3D view of the RPC stack is shown in Fig. 112.1. RPC detectors are made of two thin glass plates of 3 mm thickness, separated by a gap of 2 mm. In order to establish high potential, both sides of the chamber are coated by a thin layer of graphite. The suitable gas mixtures of $C_2H_2F_4$ (95.2 %), iC_4H_{10} (4.5 %), and SF_6 (0.3 %) are chosen to operate RPCs in avalanche mode in the differential bias voltage of ± 5 kV. The passage of muon through the gas gap ionises the gas mixtures and produces an avalanche and induces the signal in the pickup strips placed on both sides of the chamber orthogonally to get a localised position. The X-plane has 60 pickup strips and Y-plane has 63 pickup strips of width 2.8 cm and the inter-strip gap between the strips are 0.2 cm. The induced signal is amplified and discriminated by a front-end board (NINO [7] (charge sensitive) front-end board is used for layer 0 to layer 10, ANUSPARSH (voltage sensitive) is used for layer 11). The LVDS output from the front-end passes through FPGA-based RPC-DAQ digital front-end, the event data is latched by the arrival of the trigger from the Trigger system. The current study is done from the data taken with a coincidence of signals within 100 ns from trigger layers of 4, 5, 6, and 7 (either X-or Y-plane). The data collected from the experiment has two sets of information: 1. strip hit information and 2. time of muon arrival in each layer. The detailed description of the selection criteria and method to estimate detector efficiency, position residues, noise, and strip multiplicity are discussed in [8, 9]. For the sake of completeness, this section discusses a few points about the data analysis of muon data. The present analysis uses the hit information at most three consecutive hits per layer. The hit position in selected layers is fitted using a straight-line equation (112.1) in both XZ- and YZ-planes.

$$x(/y) = \alpha \times z + \beta \tag{112.1}$$

where x or y is the hit position from the X- or Y-plane, respectively, for Zth layer, α is the slope which is $\tan\theta\cos\phi$ ($\tan\theta\sin\phi$) for XZ (YZ) plane, and β is the intercept. The θ and ϕ of the muon is estimated using the fitted parameters.

Fig. 112.1 12-layer detector
stack of 2 m × 2 m RPCs

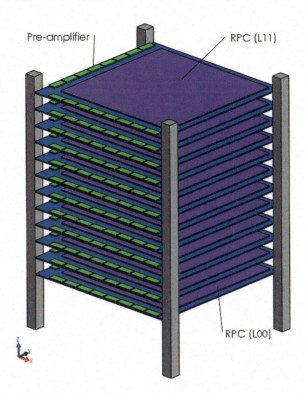

112.3   Monte Carlo Generation

The Geant4 toolkit [10] is used for the detailed simulation. The detector geometry
along with all support materials, detector building hall, and nearby building around
the experimental hall is also incorporated in the Geant4 geometry description. The
CORSIKA [11] software is used to generate 20 million primary protons and 2 million
primary helium. The generated primary energy is compared with cut-off rigidity at
different zenith-azimuth bins. The primary particle having energy more than a cut-off
rigidity will be allowed to progress, otherwise, the new primary will be generated.
The vertex, px, py, pz, and arrival time of secondary particles at an experimental
site are stored in the ROOT file. The vertex of the secondary particles is digitised
in the detector area. The position and momentum of the secondary particle are generated on the top trigger layer and extrapolated in to the bottom trigger layer to
check the acceptance condition. The event vertex on top of the roof is calculated for
accepted events and events are generated above the roof. The detector parameters
like inefficiency, trigger efficiency, position-dependent strip multiplicity, and noise
are calculated from data and incorporated in the digitisation process. The simulated
data is analysed by the same analysis code is used for experimental data.

112.4 Estimation of Muon Flux at Different (θ, ϕ) Bins

The reconstructed muons which have χ^2/ndf less than 8 and more than 5-layer muon hits used to estimate the muon flux at different (θ, ϕ) bins the using Eq. 112.2,

$$I_{\theta,\phi} = \frac{I_{\text{data}}}{\epsilon_{\text{trig}} \times \epsilon_{\text{selec}} \times \epsilon_{\text{daq}} \times T_{\text{tot}} \times \omega} \tag{112.2}$$

where I_{data} is the number of reconstructed muons at a (θ, ϕ) bin, ϵ_{trig} is the trigger efficiency in that (θ, ϕ) bin, ϵ_{selec} is the event selection efficiency in that (θ, ϕ) bin, ϵ_{daq} is the efficiency due to dead time in the data acquisition system, T_{tot} is the total time taken to record the data (in seconds) including DAQ's dead time (0.5 ms/event), and ω is the accepted solid angle times the surface area, which is further defined as

$$\omega = \frac{AN}{N'} \int_{\theta_1}^{\theta_2} \sin\theta d\theta \times \int_{\phi_1}^{\phi_2} d\phi \tag{112.3}$$

where A is the surface area of the RPC on the top triggered layer, N is the number of events accepted at a (θ, ϕ) bin when the generated position on the top and bottom trigger layers is inside the detector, and N' is the Number of events generated on the top trigger layer at (θ, ϕ) bin.

112.5 Systematic Studies

The systematic variation of muon flux at different (θ, ϕ) bins are studied. The parameters changed to study the systematic are given as follows: 1. To account the uncertainties in the inefficiency and trigger efficiency of the detector, the inefficiency and trigger efficiency are increased and decreased by $\pm 1 \sigma$; 2. increase and decrease the noise by 10% during the digitisation process; 3. To account uncertainty in the material description of the detector, the thickness of the roof and density of the detector materials are changed in geometry description in Geant4; 4. The selection criteria on reconstructed muons are changed, and the muon flux is calculated with muons having more than 4 layer hits; 5. The data set is split in to odd numbered and even numbered events, and the muon flux is calculated for two data sets; 6. The input interaction model is changed in CORSIKA to generate cosmic ray shower.

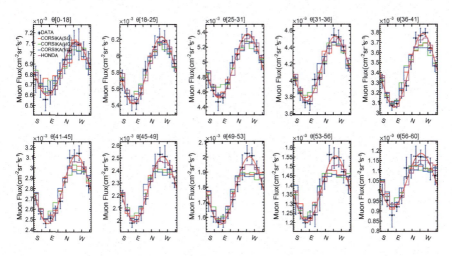

Fig. 112.2 Comparison of the azimuthal muon flux with CORSIKA and HONDA predictions

112.6   Comparison of Data with CORSIKA and HONDA Predictions

The shape of the observed muon flux is compared with CORSIKA and HONDA predictions. The observed muon flux, CORSIKA with different interaction models (namely SIBYLL-GEISHA (SG), VENUS-GEISHA (VG), and HDPM-GEISHA (VG)), and HONDA predictions are shown in Fig. 112.2.

The data and predictions are fitted using Eq. 112.4,

$$f(\phi) = P_0(1 + A sin(-\phi + \phi_0)),\tag{112.4}$$

here parameter A from the fit will give the amplitude of asymmetry of the fit function. The fitted asymmetry parameters, A and ϕ_0, are shown in Fig. 112.3a and b, respectively.

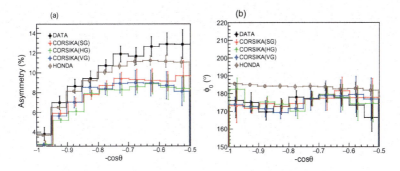

Fig. 112.3   a Asymmetry parameter for data, CORSIKA and HONDA, **b** ϕ_0 parameter for data, CORSIKA and HONDA

112.7 Conclusion

The intensity of cosmic ray muons is calculated for different (θ, ϕ) bins. The detailed systematic studies are done to estimate the variation of muon flux by varying different detector parameters in Geant4 and by changing the input model in CORSIKA. The azimuthal asymmetry of the data is compared with different CORSIKA and HONDA predictions. The asymmetry from data is better matching with the predictions for lower zenith angle. The discrepancy between the data and predictions at a higher zenith angle will be a better input neutrino event generator to estimate neutrino flux at the INO experimental site.

Acknowledgements We would like to thank Dr. P. K. Mohanty and Mr. Hariharan for providing the direction dependent rigidity map for the experimental site. We would like to acknowledge the crucial contributions by A. Bhatt, S. R. Joshi, Darshana Koli, S. Chavan, N. Sivaramakrishnan, B. Rajeswaran, and Rajkumar Bharathi in setting up the detector, electronics, and the DAQ systems.

References

1. D.S.R. Murty, Proc. Indian Acad. Sci. **37**, 317 (1953). https://doi.org/10.1007/BF03052714
2. P.N. Diep et al., Measurement of the east-west asymmetry of the cosmic muon flux in Hanoi. Nucl. Phys. **B678**, 3–15 (2004)
3. T.H. Johnson et al., The east-west asymmetry of the cosmic radiation in high latitudes and the excess of positive mesotron. Phys. Rev. **59**(1), 11 (1941)
4. S. Tsuji et al., Measurements of muons at sea level. J. Phys. G Nucl. Part. Phys. **24**1805–1822 (1998)
5. I.C.A.L. Collaboration, Physics Potential of the ICAL detector at the India-based Neutrino Observatory (INO). Pramana J. Phys. **88**, 79 (2017)
6. R. Santonica, R. Cardarelli, Development of resistive plate counters. Nucl. Instrum. Meth. **187**, 377–380 (1981). https://doi.org/10.1016/0029-554X(81)90363-3
7. F. Anghinolfi et al., NINO: an ultra-fast and low-poer front-end amplifier/discriminator ASIC designed for the multigap resistive plate chamber. Nucl. Instrum. Meth. A **533**, 183–187 (2004)
8. S. Pethuraj et al., Measurement of cosmic muon angular distribution and vertical integrated flux by 2 m × 2 m RPC stack at IICHEP-Madurai JCAP09, 021 (2017). https://doi.org/10.1088/1475-7516/2017/09/021
9. S. Pal, B. Acharya, G. Majumder, N. Mondal, D. Samuel, B. Satyanarayana, Measurement of integrated flux of cosmic ray muons at sea level using the INO-ICAL prototype detector. J. Cosmol. Astropart. Phys.**07**, 033 (2012). https://doi.org/10.1088/1475-7516/2012/07/033
10. GEANT4 Collaboration, S. Agostinelli et al., GEANT4: A Simulation toolkit. Nucl. Instrum. Meth. A**506**, 250 (2003)
11. D. Heck, J. Knapp, J.N. Capdevielle, G. Schatz, T. Thouw, Report FZKA 6019, Forschungszentrum Karlsruhe (1998). https://www.ikp.kit.edu/corsika/70.php

Chapter 113
Development of an Extended Air Shower Array at Darjeeling: An Update

S. Roy, S. Chakraborty, S. Chatterjee, S. Biswas, S. Das, S. K. Ghosh,
A. Maulik, and S. Raha

Abstract The only cosmic ray extended air shower (EAS) array in the eastern part of India, consisting of 7 active plastic scintillator detectors, has been commissioned at an altitude of about 2200 m above sea level in the Eastern Himalayas (Darjeeling) at the end of January 2018. Six detectors are kept at the vertices of a hexagon and one at the center of it. The distance between two consecutive detectors is 8 m. Each detector element is made up of four plastic scintillators of dimension 50 cm × 50 cm × 1 cm thereby forming a total active area of 1 m × 1 m. These scintillators are fabricated indigenously in the Cosmic Ray Laboratory (CRL), TIFR, Ooty, India. All four scintillators of a detector are coupled with a single Photo Multiplier Tube (PMT) using wavelength shifting (WLS) fibers. A custom-built module with seven inputs is used to generate a multifold trigger. Measurement of the number of cosmic ray air showers is going on since the end of January 2018. The secondary cosmic ray flux and its variation over time are also recorded at the laboratory in Darjeeling using a threefold coincidence technique with plastic scintillators. All the details of the experimental setup and techniques of measurement are reported earlier. The updates in the results are presented in this article.

113.1 Introduction

To study the properties of cosmic rays at high altitudes and to compare the results with measurements from other experiments at different parts of the world [1–6], a cosmic ray air shower array has been commissioned at Darjeeling Campus of Bose Institute in the beginning of 2018. This detector array is constructed for a better understanding of the dependence of cosmic ray fluxes on the geographical parameters such as latitudes, longitudes, and altitude. The method of fabrication of the detectors, experimental setup, and the preliminary results are discussed in detail

S. Roy (✉) · S. Chakraborty · S. Chatterjee · S. Biswas · S. Das · S. K. Ghosh · A. Maulik · S. Raha
Bose Institute, Department of Physics and CAPSS, EN-80, Sector V, Kolkata 700091, India
e-mail: shreyaroy2509@gmail.com

© Springer Nature Singapore Pte Ltd. 2021
P. K. Behera et al. (eds.), *XXIII DAE High Energy Physics Symposium*,
Springer Proceedings in Physics 261,
https://doi.org/10.1007/978-981-33-4408-2_113

in [8, 9]. We have carried out the measurements for a longer time duration to increase the statistics of the result in our earlier work. The updated result is presented in this paper.

113.2   Description of the Air Shower Array

A hexagonal array consisting of 7 plastic scintillator detectors is built to study cosmic ray extended air showers at an altitude of about 2200 m above sea level in the Eastern Himalayas (Darjeeling, 27° 3' N 88° 16' E) as described in Ref. [9]. Six detectors are placed at the vertices of the hexagon and one at the center. The arms of the hexagon are 8 m and the array covers an area of $168 \, m^2$. Each detector element consists of four plastic scintillators of dimension $50 \, cm \times 50 \, cm \times 1 \, cm$ making the total active area of $1 \, m \times 1 \, m$. These plastic scintillators are fabricated indigenously at the Cosmic Ray Laboratory (CRL), TIFR, Ooty, India [5–7]. The details of the fabrication method and experimental setup are mentioned in Ref. [8]. The schematic of the cosmic ray extended air shower detector array at Darjeeling is shown in Fig. 113.1.

All detectors are biased with $-1740 \, V$ from a single high voltage (HV) power supply using an external HV distribution network. The signals from the detectors are passed through a leading edge discriminator (LED) which sets a common threshold of $-20 \, mV$ for all signals to eliminate the noise. A custom-built logic module with seven inputs is used to generate a multifold trigger. Seven individual signals from the discriminator are fed to the trigger module and the shower trigger is generated when the central detector and any two detectors give signal simultaneously. The NIM output from the trigger module is counted using a scaler module also fabricated at TIFR. The trigger output is counted for 60 min to get each data point. The counts are

Fig. 113.1 Schematic of the hexagonal cosmic ray extended air shower detector array at Darjeeling

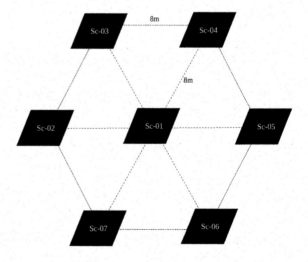

made so far manually. It has been observed that in 60 min, a significant number of counts are accumulated.

113.3   Observation and Results

The data for the number of cosmic ray air shower, i.e. the output of the trigger module, is being taken using a NIM scaler since the end of January 2018. Each data is taken for a duration of 60 min. The shower rate as a function of date and time is shown in Fig. 113.2. The distribution of the shower rate is fitted with a Landau function as

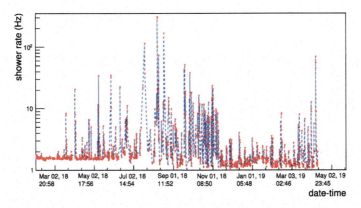

Fig. 113.2 Cosmic ray air shower rate as a function of date and time. Data from February 2018 to May 2019 is presented. Each data point is the average count rate measured in 60 min duration

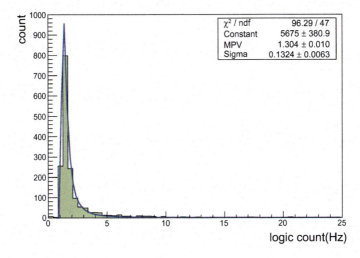

Fig. 113.3 Distribution of shower rate

shown in Fig. 113.3. It has been found that the most probable value of the air shower rate is ~ 1.3 Hz with a sigma of 0.13. This value could be compared with that obtained from a large array at a similar altitude but a different geographical location [14].

113.4 Summary

An array of seven plastic scintillator detectors is operational at an altitude of about 2200 m above sea level in the Himalayas at the Centre for Astroparticle Physics and Space Sciences, Darjeeling campus of Bose Institute, for the detection of cosmic ray air showers since the end of January 2018. From this array, it has been found that at an altitude of about 2200 m, the average air shower rate is ~ 1.3 Hz with a sigma of 0.13.

In the shower rate plot (Fig. 113.2), it can be observed that on some days the shower rate is significantly high. It is found that there was either some solar flares or proton fluence, electron fluence, and increase of K_p index on these particular days [11–13]. So far we have counted the number of showers manually using a NIM scaler as mentioned earlier during daytime only (Morning 7 a.m. to Evening 6 p.m. Indian Standard Time (IST)). Most of the jumps are observed during 12:00–15:00 h.

This small array is a pilot project and will be extended to an array of 64 such detectors, at the Darjeeling campus of Bose Institute.

Since the installation is located at a relatively high altitude (2200 m), the detector can also be made more sensitive to the primary radiation with respect to the similar installation at sea level. We are also planning an estimate of the energy of the primary cosmic ray that initiated the shower that we are measuring with the detector.

Acknowledgements The authors would like to thank the scientists of Cosmic Ray Laboratory (CRL), Tata Institute of Fundamental Research (TIFR), Ooty, India for the fabrication of these scintillators indigenously. We would also like to thank the Bose Institute workshop for all the mechanical work. We are thankful to Mrs. Sumana Singh for her assistance in the assembly of the detector modules. We are also thankful to Mr. Deb Kumar Rai, Mrs. Yashodhara Yadav, Mr. Sabyasachi Majee, and Mr. Vivek Gurung for their help in the course of this work. Thanks to Mrs. Sharmili Rudra of Seacom Engineering College for some important measurements after building the scintillator detectors at Darjeeling. Sincere thanks to the efforts made by Ms. Debonita Saha of St. Xaviers College for formatting the raw data from the detectors. Finally, we acknowledge the IRHPA Phase-II project (IR/S2/PF-01/2011 Dated 26/6/2012) of Department of Science and Technology, Government of India.

References

1. M. Aglietta et al., Nucl. Phys. B **16**, 493 (1990)
2. M. Aglietta et al., Europhys. Lett. **15**(81), 1 (1991)
3. KASCADE Collaboration, Astropart. Phys. **14**, 245 (2001)
4. W.D. Apel et al., Nucl. Instrum. Meth. A **620**, 202 (2010)

5. S.K. Gupta et al., Nucl. Phys. B Proc. Suppl. **196**, 153M (2009)
6. S.K. Gupta et al., Nucl. Instrum. Meth. A **540**, 311 (2005)
7. P.K. Mohanty et al., Astropart. Phys. **31**, 24 (2009)
8. S. Biswas et al., JINST **12**, C06026 (2017)
9. S. Roy et al., Nucl. Instrum. Meth. A **936**, 249 (2019)
10. S. Roy et al., *Proceedings of ADNHEAP 2017*, Springer Proceedings in Physics vol. 201 (2017), ISBN: 978-981-10-7664-0
11. https://www.spaceweatherlive.com/en/solar-activity/top-50-solar-flares/year/ 2018
12. ftp://ftp.swpc.noaa.gov/pub/forecasts/SGAS/
13. http://tesis.lebedev.ru/en/forecast$_$activity.html?m=3&d=19&y=2018
14. M. Zuberi et al., PoS (ICRC2017) 302

Chapter 114
Experimental and Numerical Studies of the Efficiency of Gaseous Detectors

Promita Roy, Supratik Mukhopadhyay, Sandip Sarkar, and Nayana Majumdar

Abstract We have estimated the gain and efficiency of resistive plate chambers experimentally, as well as numerically. The experimental setup has a stack of three plastic scintillators and a detector. We have calculated threefold and 4-fold efficiency of RPC with respect to plastic scintillators by designing a FPGA-based DAQ (Hu in Nucl Instr Methods Phys Res A, 2011). Garfield++ has been used in various possible modes to numerically simulate the detector response of RPCs. The results have been compared and attempts have been made to interpret their nature.

114.1 Introduction

Gaseous detectors have played an important role in the discovery of various particles. These radiation detectors are based on the effects produced by a charged particle while traversing the gas volume and provide us with a range of information regarding the detected particles that allows us to study their properties and understand the intricacies of the world at a microscopic scale. However, it is not always possible to have access to a detector or to predict what to expect from a particular setup. This is where the exciting world of simulation comes and to know whether a detector is worthy enough, we need to calculate the gain and efficiency of a detector. We tried calculating efficiency using a Field Programmable Gate Array (FPGA)-based Data

P. Roy (✉) · S. Mukhopadhyay · S. Sarkar · N. Majumdar
Saha Institute of Nuclear Physics, HBNI, 1/AF, Bidhannagar, Kolkata 700064, India
e-mail: promita.roy@saha.ac.in

S. Mukhopadhyay
e-mail: supratik.mukhopadhyay@saha.ac.in

S. Sarkar
e-mail: sandip.sarkar@saha.ac.in

N. Majumdar
e-mail: nayana.majumdar@saha.ac.in

© Springer Nature Singapore Pte Ltd. 2021
P. K. Behera et al. (eds.), *XXIII DAE High Energy Physics Symposium*,
Springer Proceedings in Physics 261,
https://doi.org/10.1007/978-981-33-4408-2_114

Acquisition (DAQ). In simulations, we first tried the approximated RKF (Runge–Kutta–Fehlberg) method of charge transport and then tried a detailed microscopic charge transport method.

114.2 Experimental Setup and Results

The experimental setup has a stack of three scintillators and a detector, here it is Resistive Plate Chambers (RPC). An FPGA-based DAQ has been used for the readout. FPGA programming has been done using Labview software. Similar attempts have become increasingly popular in recent times [1].

The RPC used in our lab is a bakelite one, which has a gap of 2 mm, and was operated at voltages ranging from 5500 to 6500 V.

114.2.1 Field Programmable Gate Arrays

FPGA is a semiconductor device on which a function can be designed even after manufacturing. It enables one to reconfigure hardware for specific applications even after the product has been installed in the field, hence the name field programmable. It has a large number of channels and gates which are not permanently connected. One can always erase the previous configuration and reprogram it using software to design some new hardware. Also, it is very cost-effective compared to the customized circuit design (Fig. 114.1).

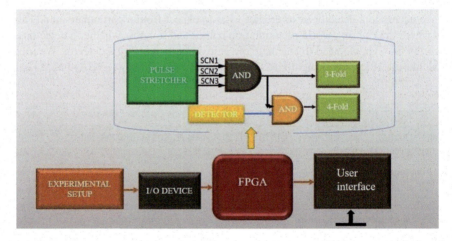

Fig. 114.1 Block diagram for coincidence detection with FPGA

The experimental setup has a pulse stretcher, two logical AND units, a 64-bit timer, and a counter. This whole setup has been interfaced with FPGA card through an i/o device from NI instruments.

114.2.2 Experimental Results

Using the experimental setup, the efficiency of RPC has been obtained. The gas mixture used is Freon_isobutane_SF6 in the ratio of 95.5:4.3:0.2. The detector is operated at different voltages starting from 5500 to 6500 V.

3-fold and 4-fold coincidence readings have been recorded. Efficiency has been calculated by taking the ratio of a 4-fold coincidence with 3-fold coincidence rates. 3-fold coincidence has the rates from three scintillators and 4-fold coincidence has rates from three scintillators and the detector (Fig. 114.2).

From the figure, we see that efficiency increases with the applied field and finally saturates.

114.3 Simulation Details and Results

Simulation of gaseous detectors is done using a package named GARFIELD++ [2]. It is an object-oriented toolkit for the detailed simulation of particle detectors that uses a gas mixture or a semiconductor material as a sensitive medium. Using programs

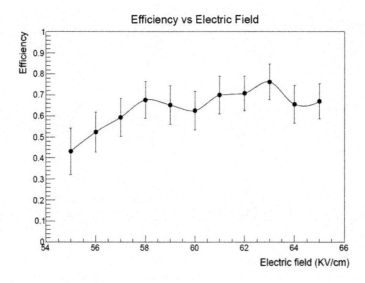

Fig. 114.2 Variation of efficiency of the detector with the applied field using FPGA DAQ

a) Freon:Isobutane:SF6= 95.5:4.3:0.2   b) Argon:CO$_2$:N$_2$ = 5:65:30

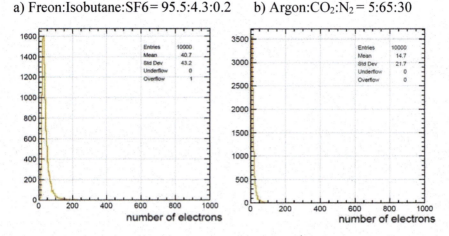

Fig. 114.3 Distribution of primaries

like Magboltz and HEED, transport properties of gas mixture and distribution of primary electrons and their energy deposition are calculated, respectively. Following which the electron multiplication and the required characteristics of the gas mixture have been calculated.

114.3.1 Primary Ionization

Generally, a gas mixture consists of three components:
    An ionizing gas, a photon quencher, and an electron quencher [3]. HEED is used to calculate the number of primaries generated.
    Two mixtures have been analyzed.

(a)   Freon:Isobutane:SF6 = 95.5:4.3:0.2
(b)   Argon:CO$_2$:N$_2$ = 5:65:30

From Fig. 114.3, we see that number of primaries in the freon mixture is more than that in argon–nitrogen–CO$_2$ mixture.

114.3.2 Transport Properties

Different gas properties like Townsend coefficient, attachment coefficient, and drift velocities have been calculated using MAGBOLTZ [4] (Figs. 114.4, 114.5, 114.6 and 114.7).

(a)   Freon: Isobutane: SF6 = 95.5: 4.3: 0.2

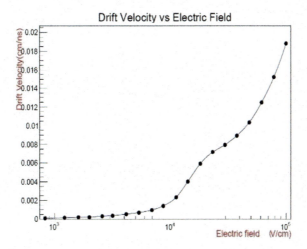

Fig. 114.4 Variation of drift velocity

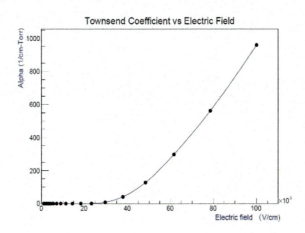

Fig. 114.5 Variation of Townsend coefficient

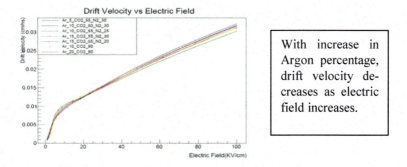

With increase in Argon percentage, drift velocity decreases as electric field increases.

Fig. 114.6 Variation of drift velocity for combinations of Ar, CO_2, and N_2

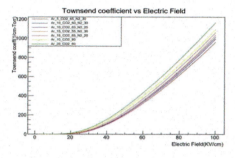

Townsend coefficient increases with increase in argon percentage with increasing applied field.

Fig. 114.7 Variation of Townsend coefficient for combinations of Ar, CO_2, and N_2

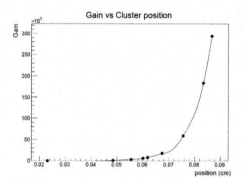

From this figure, we see how RKF gain varies exponentially with the electron cluster position following the equation $G = e^{\alpha x}$.

Fig. 114.8 Variation of RKF gain with electron cluster position

(b) Argon: CO_2: N_2 mixture.

114.3.3 Charge Transport

114.3.3.1 RKF Method

It is a modified form of RK4 method. It was developed by the German mathematician Erwin Fehlberg. It does not take into account diffusion and attachment of electrons (Fig. 114.8).

114.3.3.2 Microscopic Method of Charge Transport

In Microscopic tracking, all the microscopic parameters are considered while calculating the track of the particle, position of the electron clusters, and finally the gain of the detector (Fig. 114.9).

A more detailed approach has been adopted which includes

Fig. 114.9 Drift lines of
particle showing the number
of clusters and their positions

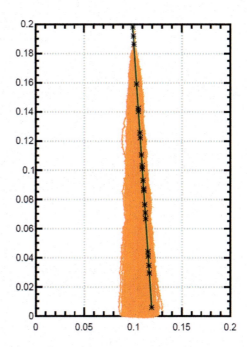

- gas number density,
- scattering cross-sections,
- instantaneous velocity, and
- fractional energy loss.

114.4   Comparison of Results

The efficiency obtained from RKF method, microscopic tracking method, and experimental setup has been compared in Fig. 114.10. Although the trends of both the curves are the same, the values are not in good agreement. We have traced the reason for improper area correction in the experimental estimates which is now being repeated. The numerical simulation is also being improved by including the effects of additional experimental details. However, it is clear from the agreement in the trend that the FPGA-based DAQ is working as designed.

Summary

- We have successfully designed a coincidence unit using FPGA and validated the results by measuring the efficiency of RPC.
- We tried to explore different mathematical models of charge transport within the detector volume and calculate the efficiency.

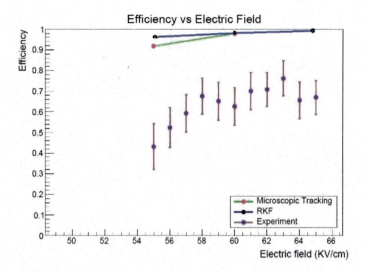

Fig. 114.10 Efficiency of RPC obtained from various methods

- RKF45 method gives a good estimate of gain but it does not take into account several physical processes such as attachment and diffusion.
- Microscopic Tracking gives the realistic picture of what actually happens in the detector, but it lacks computational efficiency.

References

1. W. Hu et al., Free-running ADC-and FPGA-based signal processing method for brain PET using GAPD arrays. Nucl. Instr. Methods Phys. Res. A (2011). https://doi.org/10.1016/j.nima.2011.05.053
2. Documentation (User Guide, Doxygen, FAQ). https://garfieldpp.web.cern.ch/garfieldpp
3. G.F. Knoll, Radiation detection and measurement. 4th Edition
4. S.F. Biagi, Magboltz 8. https://magboltz.web.cern.ch/magboltz

Chapter 115
^{32}Si and ^{32}P Background Estimate in CDMS II Silicon Detectors

Rik Bhattacharyya

115.1 Introduction

Several observational evidence indicate that almost 26.8% of the mass-energy budget of the Universe is composed of 'dark matter', which does not seem to interact with normal baryonic matter except through gravitational force [1]. The Super Cryogenic Dark Matter Search (SuperCDMS) identifies nuclear recoils from a dark matter-nucleus elastic scattering. The next iteration of SuperCDMS will take place in SNOLAB (2 km underground, $\sim$ 6010 m.w.e.), Canada, where we will employ germanium and silicon detectors at cryogenic temperatures. These detectors measure charge and phonon signals that result from the interaction of dark matter with the detectors. Due to the very low dark matter-nucleon interaction cross-section (upper bound is of the order of 10^{-45} cm^2) and tiny nuclear recoil energy ($\sim$ few keV) [2], identifying dark matter interaction becomes very challenging. To detect this rare and weak signal, a rigorous understanding of backgrounds is essential. ^{32}Si is an isotope that exists in the Si detectors right from the time of its fabrication. The provenance of Si materials is very difficult to control throughout the commercial production [3]. ^{32}Si undergoes beta emission. These beta particles act as a source of background as they interact with the electrons of the detector materials and create a charge signal. DAMIC has reported ^{32}Si activity for their Si-based CCD detectors at 80 $^{+110}_{-65}$ decays/kg-day [4] and more recently at 11.5 $\pm$ 2.4 decays/kg-day at 95% CL [5]. The large variation in the reported activity could be due to different ingots used to fabricate detectors. Different ingots may have different levels of ^{32}Si contamination.

On behalf of the SuperCDMS collaboration.

R. Bhattacharyya (✉)
School of Physical Sciences, National Institute of Science Education and Research, HBNI, Jatni 752050, India
e-mail: rikbhattacharyya22@gmail.com

© Springer Nature Singapore Pte Ltd. 2021
P. K. Behera et al. (eds.), *XXIII DAE High Energy Physics Symposium*,
Springer Proceedings in Physics 261,
https://doi.org/10.1007/978-981-33-4408-2_115

For our ^{32}Si analysis, CDMS II data is used [6, 7]. To estimate this background, both beta decay model spectrum and charge energy spectrum from CDMS II data is required. Since CDMS II uses Si detectors, it allows us to set an upper limit on ^{32}Si contamination level for the upcoming SuperCDMS SNOLAB experiment.

115.2 ^{32}Si and ^{32}P Beta Decay Model Spectrum

When a dark matter particle or neutron interacts with the detector, it scatters off the nucleus, creating a nuclear recoil (NR). Similarly, when a gamma or beta particle interacts with the detector, it scatters off the electron and produces an electron recoil (ER). ERs or NRs coming from any sources other than dark matter are considered to be backgrounds. In SuperCDMS SNOLAB, we will be using detectors operated at High Voltage (HV), which cannot discriminate between these two types of recoils unlike detectors used in CDMS II. Hence, we are unable to reject this ^{32}Si background for HV detectors [2].

^{32}Si decays into daughter radioactive ^{32}P, which decays into stable ^{32}S via beta emission. The endpoint energies of β decay are 227.2 keV and 1710.6 keV for ^{32}Si and ^{32}P, respectively. The half-lives of these ^{32}Si and ^{32}P are 153 years and 14.27 days, respectively. Fermi's theory of beta decay predicts the distribution of beta particles ($N(T_e)$) with kinetic energy (T_e) [8–10]

$$N(T_e) = C\sqrt{T_e^2 + 2T_e m_e}(Q - T_e)^2(T_e + m_e)F(Z, T_e) \,, \qquad (115.1)$$

where Q is the end-point energy of beta particles, Z and m_e are the atomic number of daughter nuclei and mass of beta, respectively, C is the normalization constant and $F(Z, T_e)$ is the Fermi function to account for the Coulomb interaction between the charged daughter nuclei and emitted beta particle after the decay.

In the relativistic regime when kinetic energy is greater than the rest energy of beta particle, such as in the case for the ^{32}P decay, which occurs with roughly three times the electron rest energy, the form of the Fermi function becomes a bit complicated [8, 11, 12]. We will use the Bethe–Bacher formalism for the relativistic Fermi function given as [11, 12]:

$$F(Z, T_e) = F_{NR}(Z, T_e)\left\{T_e^2(1 + 4\gamma^2) - 1\right\}^S \,, \qquad (115.2)$$

with

- Non-relativistic Fermi function $F_{NR}(Z, T_e)$ given by the numerical solution of electron wave function to the Schrödinger equation [8, 13].

$$F_{NR}(Z, T_e) = \frac{2\pi\eta}{1 - e^{-2\pi\eta}} \quad \text{with} \quad \eta = \frac{\alpha Z E}{pc}, \qquad (115.3)$$

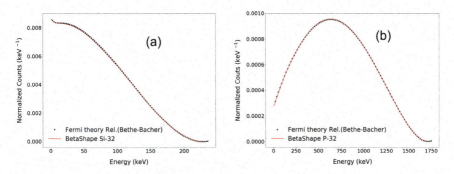

Fig. 115.1 [Color online] A comparison between the beta spectrum from the relativistic Fermi model (Bethe–Bacher) and BetaShape are shown for (a) ³²Si and (b) ³²P

where

- α is the fine-structure constant ($\sim 1/137$) ;
- E, p and c are the total energy, momentum of emitted beta particles and speed of light in vacuum, respectively.

• γ and S are defined by

$$\gamma = \alpha Z \quad ; \quad S = (1 - \gamma^2)^{1/2} - 1.$$

At low energy, the theoretical Fermi curve does not fit with the experimental data. Some correction factors (called shape correction factor or spectral correction factor) are needed to account for the exchange and screening effects [14, 15]. We will use BetaShape by Laboratoire National Henri Becquerel (LNHB) [16]. It uses experimental data from Decay Data Evaluation Project (DDEP).

The spectrum obtained from BetaShape in Fig. 115.1 deviates at most 2% from the relativistic Fermi theory. Thus, the spectrum from BetaShape will be used as beta decay model for ³²Si and ³²P.

115.3 CDMS II Data and Experimental Setup

During the years 2003–2008, CDMS II took data at Soudan Underground Laboratory (SUL, 780 m underground), Minnesota, U.S.A. It used a total of 30 Z-sensitive Ionization and Phonon (ZIP) detectors: 19 Ge ($\sim$ 239 g each), 11 Si ($\sim$ 106 g each), operated at <50 mK. ERs and NRs create electron-hole (e/h) pairs in ZIP detectors, which are then drifted to the respective electrodes due to the applied bias voltage. This gives the ionization signals. In addition, it also measures nonequilibrium phonon signals as a result of that interaction. Detectors can measure these ionization and phonon energies [7, 17, 18]. CDMS II deployed 11 Si detectors during its run. For

Fig. 115.2 [Color online]
Summed charge distribution
from silicon detectors. The
red and black dashed lines
are the end-point energies of
^{32}Si and ^{32}P, respectively

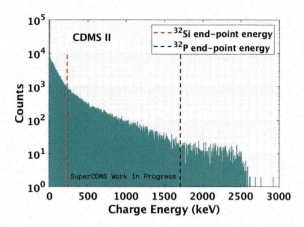

this analysis, we consider the 8 detectors that were used for the CDMS II WIMP
search as these detectors were the best understood [6].

As a beta particle from ^{32}Si and ^{32}P creates ER, they will produce both charge
and phonon energies. For some detectors, phonon channels get saturated at lower
energies. Hence, charge energy has been chosen as the main observable for this
analysis. We can measure ^{32}Si content from the charge energy spectrum. Glitches,
pulses with baseline fluctuations, events not producing proper amplification, time
periods with some improper hardware settings, time periods with DAQ issues, etc.
were excluded from the data.

We summed up the energy of each of the 8 detectors on an event-by-event basis
because, for the energies of decays and weak stopping power of Si, events might not
be entirely contained within a single detector. The total charge energy spectrum for
the 8 Si detectors is obtained as shown in Fig. 115.2. The red and black dashed lines
in Fig. 115.2 are the end-point energies of ^{32}Si and ^{32}P beta decays (227.2 keV and
1710.6 keV, respectively).

We will further optimize event selection. We will put an upper limit using the
optimum interval method [19] and the likelihood method on the ^{32}Si contamination
level for SuperCDMS detectors.

115.4 Summary

We have presented the status of the analysis related to ^{32}Si background. Currently,
we have obtained the total charge energy spectrum using CDMS II data. We have
evaluated models of the Si beta spectrum and determined that BetaShape is appropriate as it has a basis in theory but incorporates data-driven corrections. ^{32}Si will be
a dominant background for future silicon-based direct dark matter detection experiments like SuperCDMS SNOLAB. This analysis will also be helpful in putting better
limits on SuperCDMS dark matter search results.

References

1. P. A. Ade *et. al.* (PLANCK collaboration), Astronomy & Astrophysics **594** (2016)
2. R. Agnese et al., SuperCDMS collaboration. Phys. Rev. D **97**, 022002 (2018)
3. J.L. Orrell, I.J. Arnquist, M. Bliss, R. Bunker, Z.S. Finch, Astroparticle Physics **99**, 9 (2018)
4. A. Aguilar-Arevalo *et. al.* (DAMIC collaboration), J. Instrum. **10(08)** (2015)
5. G. Rich (DAMIC collaboration), *Status and plans for the DAMIC experiment at SNOLAB and Modane* in IDM 2018, Brown University. https://indico.cern.ch/event/699961/contributions/ 3043431/
6. R. Agnese et al., SuperCDMS collaboration. Phys. Rev. Lett. **111**, 251301 (2013)
7. R. Agnese et al., SuperCDMS collaboration. Phys. Rev. D. **92**, 072003 (2015)
8. E. Fermi, Z. Phys. **88**, 161 (1934)
9. K.S. Krane, D. Halliday, *Introductory nuclear physics* (Wiley, New York, 1988)
10. R. A. Dunlap, *An introduction to the physics of nuclei and particles* (Thomson/Brooks-Cole,2003)
11. P. Venkataramaiah, K. Gopala, A. Basavaraju, S. Suryanarayana, H. Sanjeeviah, J. Phys. G: Nucl. Phys. **11**(3), 359 (1985)
12. H.A. Bethe, R.F. Bacher, Rev. Mod. Phys. **8**, 82 (1936)
13. F.L. Wilson, Am. J. Phys. **36**(12), 1150 (1968)
14. X. Mougeot, C. Bisch, M. e. a. Loidl., Nuclear Data Sheets, **120** (2014) p.129
15. M. Morita, Prog. Theor. Phys. Suppl. **26**, 1 (1963)
16. X. Mougeot, Phys. Rev. C **91(5)** (2015) p.055504.; Phys. Rev. C **92** (2015) p.059902
17. M. D. Pepin (SuperCDMS collaboration), Ph.D. thesis, University of Minnesota (2016)
18. Z. Ahmed (SuperCDMS collaboration), Ph.D. thesis, Caltech (2012)
19. S. Yellin, Phys. Rev. D **66**(3), 032005 (2002)

Chapter 116
Prototype Test for Electromagnetic Calorimeter, FOCAL, at CERN-SPS Using Large Dynamic Range Readout Electronics

Sanjib Muhuri, Sinjini Chandra, Sourav Mukhopadhyay, Jogender Saini, V. B. Chandratre, R. N. Singaraju, T. K. Nayak, Ton van den Brink, and Subhasis Chattopadhyay

Abstract A silicon–tungsten prototype calorimeter, for the proposed ALICE LS3-upgarde, was fabricated and tested at the CERN-SPS beamline facility in the year 2017. The calorimeter was designed with the help of GEANT4 simulation to perform

S. Muhuri · S. Chandra (✉) · J. Saini · R. N. Singaraju · T. K. Nayak · S. Chattopadhyay
Variable Energy Cyclotron Centre, Kolkata, India
e-mail: s.chandra@vecc.gov.in

S. Muhuri
e-mail: sanjibmuhuri@gmail.com

J. Saini
e-mail: jogendra.saini@gmail.com

R. N. Singaraju
e-mail: rncingaraj@gmail.com

T. K. Nayak
e-mail: tapankumarnayak@gmail.com

S. Chattopadhyay
e-mail: sub.chattopadhyay@gmail.com

S. Chandra
Homi Bhabha National Institute, Mumbai, India

S. Mukhopadhyay · V. B. Chandratre
Bhabha Atomic Research Centre, Mumbai, India
e-mail: onlysourav@gmail.com

V. B. Chandratre
e-mail: v.b.chandratre@gmail.com

T. K. Nayak
CERN, Geneva, Switzerland

T. van den Brink
Utrecht University, 3844 CC Utrecht, The Netherlands
e-mail: a.vandenbrink@uu.nl

© Springer Nature Singapore Pte Ltd. 2021
P. K. Behera et al. (eds.), *XXIII DAE High Energy Physics Symposium*,
Springer Proceedings in Physics 261,
https://doi.org/10.1007/978-981-33-4408-2_116

in a high multiplicity density environment with optimised energy and position resolutions for incident energy up to 200 GeV. There were several tests conducted with different prototype configurations both at laboratory and CERN beamline facilities. The latest prototype in the series consists of 20 layers, each layer consisting of a 6 * 6 array of 1cm^2 silicon pad detector, fabricated on a single 300 um thin wafer, and 1 radiation length (X_R) thick, 10 cm * 10 cm pure tungsten plates as absorbers. Data was collected for pion, muon, and electron over a wide range of incident energies (20–180 GeV). Because of large energy deposition within the calorimeter by the electromagnetic showers, the readout electronics need to have a large dynamic range. Here, we will discuss the performances of the prototype, experimented at CERN-SPS, using a newly developed readout electronics (ANUINDRA) with a large dynamic range (up to 2.6 pC) and compare the improvements in this regard.

116.1   Introduction

A new sampling type electromagnetic calorimeter, FOrward CALorimeter (FOCAL), has been proposed as an upgrade for the ALICE experiment at CERN to extend its physics capabilities in the forward pseudorapidity region ($3.5 \leq \eta \leq 5.5$). This can help to probe the Parton distribution functions in the gluon-dominated region which is still unexplored. A detailed GEANT-4 simulation led to an optimised geometry for the calorimeter, with depth 20 X_R and transverse size with inner and outer radii of 6 cm and 60 cm, respectively, at 7 m from the interaction point. Using tungsten as the absorber/convertor and silicon pad detectors for the calorimeter would confine and tighten the electromagnetic showers within the longitudinal and transverse directions giving desired energy and position resolutions in the incident energy range of 1–200 GeV [1–3].

A series of prototype calorimeters were fabricated and tested both with a radioactive source and test beam to check the feasibility and functionality of the calorimeter. Initially, a mini-prototype was constructed and tested at the CERN-PS [4]. This was followed by the development of a full-depth (20 X_R) prototype calorimeter, which was tested at CERN-SPS beamline [5]. Both the prototypes were tested with pions to study MIP-response and electrons (EM-shower) of different incident energies from 1 to 60 GeV. The analysis showed the performance of the prototype getting affected at higher incident energies of particles (saturation in ADC distribution was observed) because of the limited dynamic range (up to 600 fC) of the ASICs used. This led to the requirement and fabrication of a new ASIC, named ANUINDRA with a large dynamic range (from 0 to 2.6 pC), which was used in the prototype calorimeter tested at CERN-SPS beamline facility in July 2017 [6].

116.2 FOCAL Prototype SetUp at CERN-SPS

Figure 116.1 shows the prototype calorimeter arrangement at CERN-SPS. It consists of 20 tungsten layers sandwiched between 21 layers of silicon pad sensors and readout electronics. Each detector layer is a 6 * 6 array of 1 cm * 1 cm Si-sensors fabricated on a single 4-inch wafer of thickness 300 um. The tungsten layers are 3.5 mm thick each, with a transverse dimension of 10 cm * 10 cm. A mechanical frame made of stainless steel has been used for keeping the tungsten plates, silicon detector layers and the readout electronics in place. The readout electronics are taken out either from the top or from the side of the mechanical frame using different backplane printed circuit boards (BP-PCBs) for MANAS and ANUINDRA ASICs. Five detector layers (8th–12th) around the shower–max were readout using ANUINDRA ASICs, whereas MANAS ASICs (with linear dynamic range from −300 fC to 500 fC) have been used to read out signals from the other layers as shown in Fig. 116.2. The prototype has been tested at the T4-H8 beamline at the CERN-SPS facility for a wide range of incident energies for different types of particles (20–150 GeV electrons, 120 GeV pions and 180 GeV muons). The trigger for positioning and selecting beam-type during the test beam is provided by three scintillators, Presence (P), Horizontal (H) and Vertical (V) along with a Cherenkov counter upstream. The Cherenkov detector helps to improve the purity of electron beams. The trigger unit helps to select the position of the incoming beam and restrict it to the centre of the detector or tungsten layers.

Prototype test for electromagnetic calorimeter, FOCAL, at CERN-SPS using large dynamic range readout electronics.

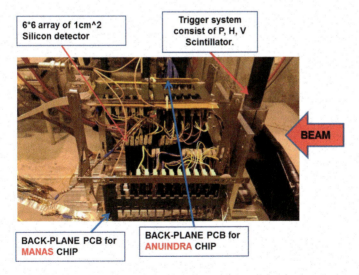

Fig. 116.1 The experimental setup at CERN-SPS in July 2017, along with ASIC ANUINDRA (linear dynamic range up to 2.6 pC)

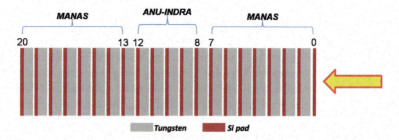

Fig. 116.2 Schematic diagram showing the different ASICs used to readout signals from different layers of the prototype

116.3   Results and Discussions

Analysis of the test beam data shows satisfactory calorimetric performances for the prototype. Also, the newly developed large dynamic range ASIC, ANUINDRA is found to resolve the saturation effect up to 80 GeV, which was seen with MANAS ASIC for the higher incident energies (beyond 20 GeV) in earlier experiments. The results have been discussed in detail in the next sections.

116.3.1   Response to Pion Beam

The prototype calorimeter was exposed to a pion beam of energy 120 GeV, to understand the behaviour of the prototype to minimum ionising particles (MIPs). Pions behave like MIPs within the EM-calorimeter depth under consideration and are unlikely to produce a shower within the longitudinal depth of the prototype calorimeter used in the experiment. The energy deposited by the pion beam in a single silicon pad detector (that lies directly in the path of the beam) is plotted in Fig. 116.3 in units of ADC. The energy spectrum is fitted with a Landau distribution which gives a good fit matching with theoretical predictions. The most probable value (MPV) obtained is 17.81 ± 0.06 ADC which matches with earlier calculations and experimental results [4].

116.3.2   Response to Electron Beam

For studying the response of the prototype calorimeter to EM showers, the experimental setup was exposed to electron beams of different incident energies in the range from 20 to 150 GeV. Due to the segmentation in longitudinal as well as in the transverse directions, it has been possible to track the propagation of the shower within the depth of the prototype calorimeter. The total energy deposited in the prototype

Fig. 116.3 Response of Si-pad detector to 120 GeV pion beam

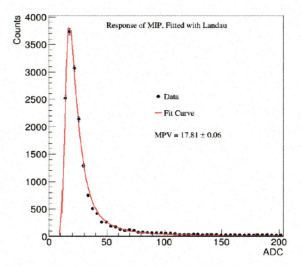

by electrons in the incident energy range 20–110 GeV has been plotted in Fig. 116.4. The distributions are well- separated from each other and have distinct peaks, which shows a reasonably good discrimination capability of the calorimeter over a wide incident energy range. However, beyond 100 GeV, the distributions started to get distorted which may be because of limitation in a dynamic range of the readout electronics or contamination of electron beam with hadrons.

The measured energy is found to increase monotonically with the incident energy as shown in Fig. 116.5, which proves that the prototype behaves linearly. A deviation from linearity has been observed beyond 90 GeV and is considered as an important input for further development. Prototype test for electromagnetic calorimeter, FOCAL, at CERN-SPS using large dynamic range readout electronics.

Fig. 116.4 Total measured energy reconstructed for electrons of different incident energies

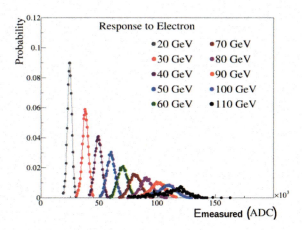

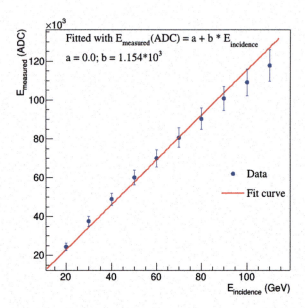

Fig. 116.5 Calibration of measured energy with respect to incident energy shows the prototype is working in the linear region

To find the energy resolution of the prototype calorimeter, the data has been fitted with a two-parameter fit function as shown in Fig. 116.6, which includes contributions from the compactness of the detector (a_f) and the fluctuation in shower development (b_f). The prototype is found to have a good energy resolution of 22.1% as obtained from the equation $\sigma/E_{dep}(\%) = 5.6 + (22.1/\sqrt{E_i})$. This leaves the opportunity to

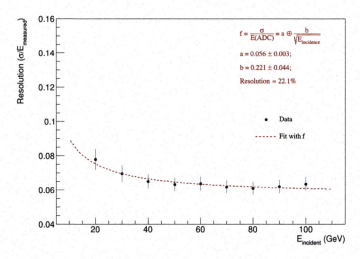

Fig. 116.6 Energy resolution plot of the prototype calorimeter

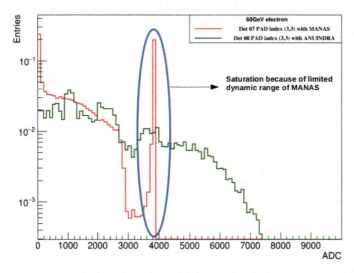

Fig. 116.7 Response of MANAS and ANUINDRA to 60 GeV electrons. With MANAS, saturation is seen due to its limited dynamic range, whereas no saturation is seen with ANUINDRA

improve the calorimeter with better performances and taken up in the next step of development.

The ADC distribution of a single pad detector has been found to get saturated for higher incident energies (20 GeV e^- onwards) when readout using MANAS due to its limited dynamic range. However, such saturation has been successfully removed while using ANUINDRA as the readout electronics, as shown in Fig. 116.7.

References

1. R. Sandhir, S. Muhuri, T.K. Nayak, NIMA, 681(2012) 34
2. T. Peitzmann et al. (ALICE FoCal collaboration). https://arxiv.org/abs/1308.2585 [physics.ins-det]
3. https://indico.cern.ch/event/102718/contributions/14229/attachments/9309/13663/focal-loi-0-1.pdf
4. S. Muhuri et al, NIMA 764 (2014) 24
5. Conference Proceedings of the DAE Symp on Nucl Phys 61 (2016) (Contribution No. - E8)
6. Conference Proceedings of the DAE Symp on Nucl Phys 62 (2017) (Contribution No. - E35)

Chapter 117
Charge Threshold Study of a Glass RPC in Avalanche Mode

Anup Kumar Sikdar and Prafulla Kumar Behera

Abstract India-based Neutrino Observatory (INO) is proposed to detect atmospheric neutrinos and measure the earth's matter effect to address the mass hierarchy problem in neutrino physics. INO is going to host the world's largest 50 kton iron calorimeter (ICAL) using 28,800 Resistive Plate Chambers (RPC) as an active element. These glass RPCs are very efficient for the energy of coming muons on the GeV scale. The performance of the RPC depends on how we are separating the signal from the background. We are studying the performance of a glass RPC of dimension $30 \times 30 \, \text{cm}^2$ by measuring the efficiency of RPC and calculating charge and timing of signal with varying charge threshold at flow rate 10 standard cubic centimeters per minute (SCCM) of gas mixture $C_2H_2F_4/iC_4H_{10}/SF_6$ at 10.4 kV and 21 °C with relative humidity (RH) $\sim$53%.

117.1 Introduction

The India-based Neutrino Observatory (INO) is a non-accelerator-based high-energy physics project. The primary goal of INO is to precisely measure the oscillation parameter by studying atmospheric neutrinos. According to the Standard Model (SM) of particle physics, neutrinos are massless and they come in three different flavors associated with electron, muon and tauon. However, a recent experiment from Super-Kamiokande shows these neutral fundamental particles have small but finite mass. The determination of neutrino masses and mixing parameters for oscillation is one of the most important open problems in physics. INO aims to determine the neutrino mass hierarchy and improve the precision bounds on θ_{23} and Δm^2_{32} from the Earth's matter effects on the atmospheric neutrinos.

A. K. Sikdar (✉) · P. K. Behera
Indian Institute of Technology Madras, Chennai 600036, India
e-mail: akssikdar@gmail.com

P. K. Behera
e-mail: prafulla.behera@gmail.com

© Springer Nature Singapore Pte Ltd. 2021
P. K. Behera et al. (eds.), *XXIII DAE High Energy Physics Symposium*,
Springer Proceedings in Physics 261,
https://doi.org/10.1007/978-981-33-4408-2_117

INO is proposed to build an underground laboratory, which will host 50,000 tons of Iron CALorimeter (ICAL) in three stakes of 151 layers with dimension $16\,m \times 16\,m \times 14.5\,m$ [1]. Each layer consists of 5.6 cm thick iron plates and 4 cm air gap to place $2\,m \times 2\,m$ RPC as an active detector element. RPCs are used in most of the high energy experiments as tracking detectors because of their excellent efficiency, position and time resolutions [2].

RPC is a gas ionization chamber which consists of two parallel plates of glass separated by a gas gap. These plates are painted with graphite coating and very high voltage ($\sim$10.4 kV) is applied to produce a uniform magnetic field across the gap. We operated RPC in the avalanche mode and a mixture of three different gases continuously flows (10 SCCM) through the gap. The gas mixture contains R134a ($C_2H_2F_4$, 95%), Isobutane (iC_4H_{10}, 4.5%) and Sulfur hexafluoride (SF_6, 0.5%). When an incoming particle passes through the medium, it ionizes gases and produces e^--ion pair [3]. These ions propagate through the medium and further give rise to charge multiplication and are later collected by the copper read-out strips as a signal. How the variation of different thresholds on these signals affects the charge collection, time resolution and efficiency of the RPC was studied and presented here.

117.2   Experimental Setup

We have used two Saint-Gobain glass plates of 3 mm thick and $30\,cm \times 30\,cm$ in dimension. These plates were pasted with conductive tape (T-9149) and separated by 2 mm gap to form the RPC. The RPC was sandwiched between two pick-up panels consisting of 10 copper strips each of 28 mm separated by 2 mm gap. The middle strip was used to read out the signal.

Figure 117.1 shows the schematic setup of RPC with scintillators. Three plastic scintillator paddles P_1, P_2 and P_3 were arranged vertically one above the other to get a threefold coincidence and the RPC was placed between P_1 and P_2. The dimensions

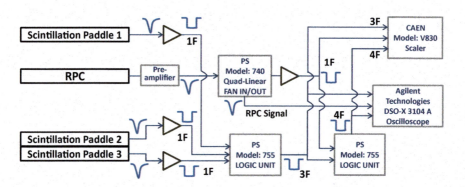

Fig. 117.1 Schematic of experimental setup

of P_1, P_2 and P_3 in length $\times$ width $\times$ thickness are $30 \times 2 \times 1\,\text{cm}^3$, $30 \times 3 \times 1\,\text{cm}^3$ and $30 \times 5 \times 1\,\text{cm}^3$, respectively. The central read-out strip of RPC was vertically aligned with three paddles. Each scintillator was attached to a photo multiplier tube followed by the discriminator to convert the analog signal into a digital one. Three logic signals from scintillators are ANDed using a logic unit. In the avalanche mode, the total charge produced by the muon is $\sim$pC. A preamplifier of gain 70 is used to amplify the signal coming from the RPC and is fed into the Data Acquisition System (DAQ). The amplified analog signal from RPC was connected to a linear FAN IN/FAN OUT (FIFO) to get two buffered output signals. One output signal was connected to an oscilloscope to measure the charge and time of the signal, and the other was converted to a logic signal by feeding it to a discriminator with a varying threshold voltage.

117.2.1   Calibration of the Gas System

A gas mixing system with multi-channel distribution, developed by the INO team [4], was used to supply a different proportion of gases to the RPC. We used Mass Flow Controllers (MFCs) to control the flow rates of different gases. The actual flow rate which is different from the set value in the MFC system is calibrated using the water downward displacement method. The calibration plots of MFCs corresponding to Freon, Isobutane and SF_6 are shown in Fig. 117.2. The required proportion (95.5% : 4.2% : 0.3%) of three (Freon:Isobutane:SF_6) individual gases in the mixture was determined, and the flow rates were adjusted in the corresponding MFCs according to the calibration.

117.3   Effect of Threshold Variation on the Performance of RPCs

We have used a mixture of R134a ($C_2H_2F_4$, 95.5%), Isobutane (iC_4H_{10}, 4.2%) and Sulfur hexafluoride (SF_6, 0.3%). Freon acts as a secondary electron quencher

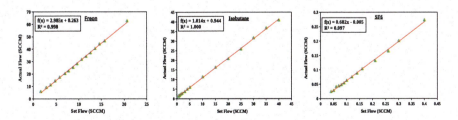

Fig. 117.2 Calibration plots of MFCs corresponding to Freon, Isobutane and SF_6

with a high probability for primary ionization and Isobutane acts as an absorber of UV photons. SF_6 is required to control the excess number of electrons in the avalanche mode [4]. When atmospheric muons pass through the RPC, ionization takes place which induces signals in the copper strips. The threshold voltage was applied to the discriminator to separate these signals from background noise. The effects of the variation on threshold voltage to the discriminator are studied through the performance of RPC by measuring charge distribution and time resolution from the 1000 signals for each threshold.

117.3.1 Charge Distribution

We have applied four different threshold voltages 13.20, 19.66, 26.90 and 40.86 mV. The effects of the variation of these voltages are shown in Fig. 117.3 and fitted with Gaussian distribution function.

The Mean and Sigma of charge distribution are shown in Fig. 117.4 (left plot) and Efficiency of the RPC at different thresholds is shown in Fig. 117.4 (right plot).

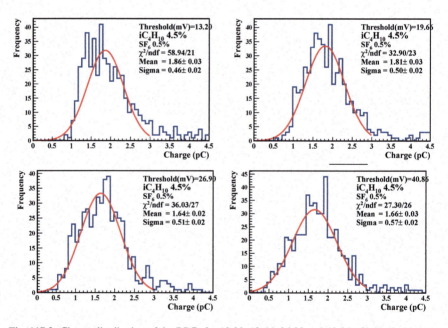

Fig. 117.3 Charge distribution of the RPCs for 13.20, 19.66, 26.90 and 40.86 mV threshold

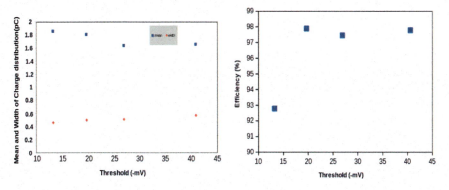

Fig. 117.4   Mean and sigma of charge distribution and efficiency of the RPC

117.3.2   Time Resolution

The time distribution of the signals collected with the RPC at various thresholds is shown in Fig. 117.5. The time resolution of the RPC is found to be 1.3 ns approximately.

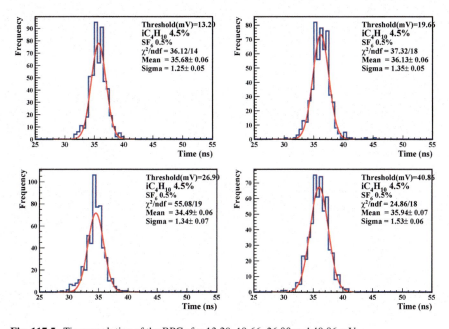

Fig. 117.5   Time resolution of the RPCs for 13.20, 19.66, 26.90 and 40.86 mV

117.4 Conclusion

The total charge collection decreases with increasing threshold because of removing more and more background as well as signals. At the lowest threshold $-13.20\,\text{mV}$, the efficiency is quite small. It increases with threshold and gets flat at $\sim98\%$ because at a low threshold, the noise rate is high. Time resolution is not affected significantly by the threshold but becomes poorer at a higher value of the threshold. Larger statistical data can provide more accurate and precise results presented here.

References

1. S. Ahmed et al., Invited review: physics potential of the ICAL detector at the India-based Neutrino Observatory (INO). Pramana **88**, 79 (2017). https://doi.org/10.1007/s12043-017-1373-4
2. R. Santonico, R. Cardarelli, Development of resistive plate counters. Nucl. Instrum. Methods A **187**, 377–380 (1981). https://doi.org/10.1016/0029-554X(81)90363-3
3. M. Abbrescia et al., The simulation of resistive plate chambers in avalanche mode: charge spectra and efficiency. Nucl. Instrum. Methods A **431**, 413–427 (1999). https://doi.org/10.1016/S0168-9002(99)00374-5
4. S.D. Kalmani et al., On-line gas mixing and multi-channel distribution system. Nucl. Instrum. Methods A **602**, 845–849 (2009). https://doi.org/10.1016/j.nima.2008.12.153

Chapter 118
Effect of FCNC-Mediated Z Boson in Semileptonic Decay $B_s \to \varphi\mu^+\mu^-$

P. Nayek and S. Sahoo

Abstract Rare B meson decays induced by flavour-changing neutral current (FCNC) transition play the most promising role to probe the flavour sector of the standard model (SM). Basically, FCNC processes are forbidden at the tree level in SM and will arise from loop diagrams which are generally suppressed in comparison to tree diagrams. This provides an excellent testing ground for NP. On the basis of various experimental studies, it is found that the FCNC processes having quark-level transition $b \to s$ are challenging because of their small branching ratio $(\mathcal{O}(10^{-6})$. So we would like to study such a type of semileptonic rare B decay mode $B_s \to \varphi\mu^+\mu^-$ involving the quark-level transition $b \to sl^+l^- (l = \mu)$. Here, we analyse the effect of non-universal Z boson in the differential decay rate of the decay mode $B_s \to \varphi\mu^+\mu^-$. The non-universal Z model is a simple model beyond the SM with an extended matter sector due to an additional vector-like down quark, as a consequence of which it allows the CP-violating Z-mediated FCNC process at the tree level. We find a significant deviation of the differential decay rate for this decay $B_s \to \varphi\mu^+\mu^-$ from the SM value because of non-universal $Z - bs$ coupling.

118.1 Introduction

In recent years, semileptonic decays of bottom hadrons are in the focus of many theoretical and experimental studies due to increasing experimental evidences of new physics (NP). Rare B meson decays which are induced by FCNC transition $b \to s$ play one of the most important roles in the research area of the particle physics, especially in the flavour sector of the standard model (SM). These FCNC transitions arise at the loop level and are suppressed in the SM due to the dependency on the weak mixing angles of the CKM matrix V_{CKM} [1, 2]. At the loop level, they are induced by the GIM mechanism [3]. This provides an excellent testing ground for NP. Here, we are interested to study $B_s \to \varphi\mu^+\mu^-$ in the non-universal Z model.

P. Nayek (✉) · S. Sahoo
Department of Physics, National Institute of Technology, Durgapur 713209, West Bengal, India
e-mail: mom.nayek@gmail.com

© Springer Nature Singapore Pte Ltd. 2021   847
P. K. Behera et al. (eds.), *XXIII DAE High Energy Physics Symposium*,
Springer Proceedings in Physics 261,
https://doi.org/10.1007/978-981-33-4408-2_118

FCNC coupling of this Z boson can be generated at the tree level in various exotic scenarios, e.g. (i) by including an extra $U(1)$ symmetry in the gauge group of the SM [4] and (ii) by adding the non-sequential generation of quarks [5].

118.2 Standard Model Contribution

The semileptonic B meson decay channel $B_s \rightarrow \varphi \mu^+ \mu^-$ is governed by $b \rightarrow s l^+ l^-$ quark-level transition for which the effective Hamiltonian can be written as [6–9]

$$H_{eff} = -\frac{4G_F}{\sqrt{2}} V_{tb} V_{ts}^* \sum_{i=1}^{10} C_i(\mu) O_i(\mu), \tag{118.1}$$

where $O_i(\mu)(i = 1, \ldots \ldots 6)$ are the four-quark operators, $i = 7, 8$ are dipole operators and $i = 9, 10$ are semileptonic electroweak operators given in. $C_i(\mu)$ are the corresponding Wilson coefficients at the energy scale $\mu = m_b$. In our study, we use the light-cone sum rules for calculating these form factors related to the matrix element of the decay $B_s \rightarrow \varphi l^+ l^-$. Now the free quark decay amplitude for this transition $b \rightarrow s l^+ l^-$ can be written as

$$\mathcal{M} = \frac{G_F \alpha}{\sqrt{2}\pi} V_{tb} V_{ts}^*$$
$$\left\{ -2C_7^{eff} \frac{m_b}{q^2} \left(\bar{s} i\sigma_{\mu\nu} q^\nu P_R b \right)\left(\bar{l} \gamma^\mu l \right) + C_9^{eff} \left(\bar{s} \gamma_\mu P_L b \right)\left(\bar{l} \gamma^\mu l \right) + C_{10} \left(\bar{s} \gamma_\mu P_L b \right)\left(\bar{l} \gamma^\mu \gamma_5 l \right) \right\}. \tag{118.2}$$

The Wilson coefficient C_9^{eff} corresponding to the operator O_{10} has three parts and can be written as

$$C_9^{eff} = C_9^{SM}(\mu) + Y_{SD}(z, s') + Y_{LD}(z, s'), \tag{118.3}$$

where z and s' are denoted as $z = \frac{m_c}{m_b}$ and $s' = \frac{q^2}{m_b^2}$. The function $Y_{SD}(z, s')$ defines the short-distance perturbative part that involves the indirect contributions from the matrix element of the four quark operators $\sum_{i=1}^{10} \langle l^+l^- s | O_i | b \rangle$ and lies at the region far away from the $c\bar{c}$ resonance regions. The long distribution contributions $Y_{LD}(z, s')$ come from four quark operators near the $c\bar{c}$ resonance and cannot be obtained from the first principle of QCD. In our study, we have excluded this long-distance resonance part because this is far away from the part of the interest and experimental analysis also ignore this [6, 10, 11]. The expression of the differential decay rate of the decay can be written as [12, 13]

$$\frac{d\Gamma(B_s \to \varphi l^+ l^-)}{d\hat{s}} = \frac{G_F^2 m_B^5 \alpha^2}{1536\pi^5} |V_{tb} V_{ts}^*|^2 \lambda^{\frac{1}{2}}(1, \hat{s}, \hat{r}) \sqrt{1 - \frac{4\widehat{m}}{\hat{s}}} \left[\left(1 + \frac{2\widehat{m}}{\hat{s}}\right) \beta_V + 12\widehat{m}\delta_V \right]. \quad (118.4)$$

The functions β_V, δ_V, G, F, H_+ and R are taken from [12, 13]. The form factors are taken from [14, 15].

118.3   Contribution of Z Boson on Decay Mode $B_s \to \varphi \mu^+ \mu^-$

We consider a model beyond the SM with an extended matter sector due to the addition of extra vector-like down quark D_4. In this model, the corresponding effective Hamiltonian in this model is given as [16–19]

$$\mathcal{H}_{eff} = \frac{G_F}{\sqrt{2}} U_{sb} \left[\bar{s} \, \gamma^\mu (1 - \gamma_5) b \right] \left[\bar{l} \, (C_V^l \gamma_\mu - C_A^l \gamma_\mu \gamma_5) l \right], \quad (118.6)$$

where C_V^l and C_A^l are the vector and axial-vector couplings of the leptons with the Z boson, i.e. $Z l^+ l^-$. The total contributions on two Wilson coefficients C_9 and C_{10} can be written as

$$C_9^{Total} = C_9^{eff} + \frac{2\pi}{\alpha} \frac{U_{sb} C_V^l}{V_{tb} V_{ts}^*}, \quad (118.7)$$

$$C_{10}^{Total} = C_{10} - \frac{2\pi}{\alpha} \frac{U_{sb} C_A^l}{V_{tb} V_{ts}^*}. \quad (118.8)$$

The coupling U_{sb} representing the Z-b-s strength is in a general complex quantity and can be parameterized as $U_{sb} = |U_{sb}| e^{i\varphi_{sb}}$, and it induces the weak phase difference φ_{sb} between the SM and NP contributions. The value of U_{sb} can be obtained from $B_s^0 - \bar{B}_s^0$ mixing parameters. It indicates that for $|U_{sb}| \leq 0.00048$, the total range of φ_{sb} is allowed, i.e. from 0 to 2π. The value of the coupling parameter is consistent with the value obtained from the branching ratio of decay mode $B \to X_s l^+ l^-$.

118.4   Numerical Analysis

We have taken all input parameters from [20]. We have calculated $|U_{sb}| = 0.00048$ in the previous section and the weak phase difference can be considered as 0 for destructive interference and π for constructive interference between SM and NP amplitude. Considering Eqs. (118.7) and (118.8), we show the variation of differential branching ratio graphically for $B_s \to \varphi \mu^+ \mu^-$ decay channel with the coupling parameter and weak phase and also with q^2 in this section. In Fig. 118.1a, we have

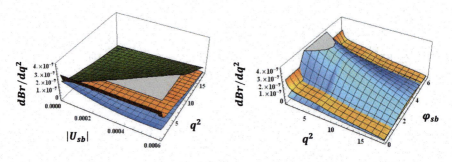

Fig. 118.1 The variations of differential branching ratio $\frac{dBr}{dq^2}$ (DBR) are shown. In Fig. 118.1a, blue plate is for S_1, green plate is for S_2 and orange plate represents SM value

considered the new weak phase $\varphi_{sb} = 0, 180$ degree and shown the variation of DBR with the coupling parameter and q^2. For S_2, DBR initially crosses the SM value at a small value of coupling parameter and q^2 then increases sharply. The contribution of scenario 2 data on DBR is more from the SM than scenario 1 for this decay channels. Therefore, we can conclude that with the higher values of the coupling parameter and momentum, the differential branching ratio decreases. In Fig. 118.1b, we have fixed the value of the coupling parameter as $|U_{sb}| = 0.00048$ and vary DBR with the variation of q^2 as well as the new weak phase φ_{sb}. Here, we can see that the enhancement of DBR from SM value through NP is significant only in the low q^2 region but in the high q^2 region, this enhancement is quite small. This deviation of DBR from the SM value provides a clear conjecture for NP.

118.5 Summary and Conclusions

A set of intriguing anomalies are found in the measurements of branching fractions, ratios of branching fractions and angular distributions in those rare decays having the quark-level transition $b \rightarrow sl^+l^-$. So in this situation, the theoretical analyses of physical observables for $b \rightarrow sl^+l^-$ transition are required [21–23]. In this paper, we have studied differential branching fraction for $b \rightarrow sl^+l^-$-mediated decays $B_s \rightarrow \varphi\mu^+\mu^-$ in SM and the non-universal Z model. We have also predicted the value of the branching fraction for this muon decay channel as 7.30×10^{-7} (for S_1) and 5.78×10^{-6} (for S_2) over the whole kinematic region. Recently, the LHCb Collaboration [24] has found the branching fraction of the decay mode $B_s \rightarrow \varphi\mu^+\mu^-$ as $\mathcal{B}(B_s \rightarrow \varphi\mu^+\mu^-) = \left(7.07^{+0.64}_{-0.59} \pm 0.17 \pm 0.71\right) \times 10^{-7}$ over the full q^2 range. From the significant enhancements of DBR for the decay process $B_s \rightarrow \varphi\mu^+\mu^-$ in the non-universal Z model, we can conclude that FCNC-mediated Z boson modifies the SM picture and gives a signal for NP beyond the SM. The other parameter R_φ is also more interesting. The future measurements of these decays which are sensitive to lepton flavour non-universality establish the lepton flavour non-universal NP.

Acknowledgements P. Nayek and S. Sahoo would like to thank SERB, DST, Govt. of India, for the financial support through grant no. EMR/2015/000817.

References

1. N. Cabibbo, Phys. Rev. Lett. **10**, 531 (1963)
2. M. Kobayashi, K. Maskawa, Prog. Theor. Phys. **49**, 652 (1973)
3. S.L. Glashow, J. Iliopoulos, L. Maiani, Phys. Rev. D **2**, 1285 (1970)
4. P. Langacker, M. Plümacher, Phys. Rev. D **62**, 013006 (2000)
5. Y. Nir, D. Silverman, Phys. Rev. D **42**, 1477 (1990); D. Silverman, Phys. Rev. D **45**, 1800 (1992); Int. J. Mod. Phys. A **11**, 2253 (1996); Y. Nir, R. Rattazzi, in Heavy Flavours II, edited by A. J. Buras and M. Linder (World Scientific, Singapore, 1998), p. 755
6. I. Ahmed, M.J. Aslam, M.A. Paracha, Phys. Rev. D **88**, 014019 (2013)
7. A.J. Buras, M. Münz, Phys. Rev. D **52**, 186 (1995); M. Misiak, Nucl. Phys. B **393**, 23 (1993); ibid. **439**, 461 (E) (1995)
8. W. Altmannshofer, P. Ball, A. Bharucha, A.J. Buras, D.M. Straub, M. Wick, JHEP **0901**, 019 (2009). ([arXiv: 0811.1214])
9. K.G. Chetyrkin, M. Misiak, M. Münz, Phys. Lett. B **400**, 206 (1997); ibid. **425**, 414 (1998) [arXiv: hep-ph/9612313]
10. Y. Li, J. Hua, Eur. Phys. J. C **71**, 1764 (2011)
11. M. Beneke, G. Buchalla, M. Neubert, C.T. Sachrajda, Eur. Phys. J. C **61**, 439 (2009)
12. S. R. Singh, B. Mawlong, . D'cruz, Int. J. Mod. Phys. A **33**, 1850125 (2018)
13. D. Melikov, N. Nikitin, S. Simula, Phys. Rev. D **57**, 6814 (1998)
14. A. Ali, P. Ball, L.T. Handoko, G. Hiller, Phys. Rev. D **61**, 074024 (2000)
15. P. Ball, R. Zwicky, Phys. Rev. D **71**, 014029 (2005)
16. M. Bando, T. Kugo, Prog. Theor. Phys. **101**, 1313 (1999); M. Bando, T. Kugo and K. Yoshioka, Prog. Theor. Phys. **104**, 211 (2000)
17. R. Mohanta, Phys. Rev. D **71**, 114013 (2005)
18. A.K. Giri, R. Mohanta, Eur. Phys. J. C **45**, 151 (2006)
19. R. Mohanta, Eur. Phys. J. C **71**, 1625 (2011)
20. T. Tanabashi et al., Particle Data Group. Phys. Rev. D **98**, 030001 (2018)
21. T. Hurth, F. Mahmoudi, D.M. Santos, S. Ncshatpour, Phys. Rev. D **96**, 095034 (2017). ([arXiv: 1705.06274 [hep-ph]])
22. A. Arbey, T. Hurth, F. Mahmoudi, S. Ncshatpour, Phys. Rev. D **98**, 095027 (2018). ([arXiv: 1806.02791 [hep-ph]])
23. A. Arbey, T. Hurth, F. Mahmoudi, D. M. Santos, S. Ncshatpour, [arXiv: 1904.08399 [hep-ph]]
24. R. Aaij et al. (LHCb Collaboration), JHEP **1307**, 084 (2013) [arXiv: 1305.2168[hep-ex]]

Chapter 119
Testing Lepton Nonuniversality in $b \to c\tau\bar{\nu}_l$ Decay Modes

Suchismita Sahoo, Rukmani Mohanta, and Anjan K. Giri

Abstract We present a model-independent analysis of $B \to D^{(*)}\tau\bar{\nu}_l$ and $B_c \to (\eta_c, J/\psi)\tau\bar{\nu}_l$ processes involving $b \to c\tau\bar{\nu}_l$ quark-level transitions by considering the most general effective Lagrangian in the presence of new physics. We perform a global fit to various sets of new coefficients, including the measurement on $R_{D^{(*)}}$, $R_{J/\psi}$ and the upper limit on Br($B_c \to \tau\nu_l$). We then show the implication of constrained new couplings on the lepton nonuniversality ratios of these decay modes in bins of q^2.

119.1 Introduction

Though the study of rare semileptonic B channels with tau-lepton in the final state is experimentally quite challenging, however, they are quite interesting from the theoretical point of view, due to the presence of several observables besides branching fractions, such as decay distributions and tau polarizations, which are quite sensitive to new physics (NP) beyond the Stadard Model (SM). Furthermore, the measurements on lepton nonuniversality (LNU) ratios associated $\bar{B} \to \bar{D}^{(*)}\tau\bar{\nu}_\tau$ processes [1]

$$R_D^{\text{Expt}} = 0.340 \pm 0.027 \pm 0.013\,, \quad R_{D*}^{\text{Expt}} = 0.295 \pm 0.011 \pm 0.008\,, \quad (119.1)$$

S. Sahoo (✉)
Physical Research Laboratory, Navrangpura, Ahmedabad 380009, India
e-mail: suchismita8792@gmail.com

R. Mohanta
University of Hyderbad, Hyderabad, Telangana 500046, India
e-mail: rukmani98@gmail.com

A. K. Giri
IIT Hyderabad, Kandi, Telangana 502285, India
e-mail: giria@iith.ac.in

© Springer Nature Singapore Pte Ltd. 2021   853
P. K. Behera et al. (eds.), *XXIII DAE High Energy Physics Symposium*,
Springer Proceedings in Physics 261,
https://doi.org/10.1007/978-981-33-4408-2_119

have 3.08σ disagreement with their corresponding SM predictions [2]

$$R_D^{\text{SM}} = 0.300 \pm 0.008, \qquad R_{D^*}^{\text{SM}} = 0.252 \pm 0.003. \qquad (119.2)$$

Besides these, the $R_{J/\psi}^{\text{Expt}} = 0.71 \pm 0.17 \pm 0.18$ [3] value measured by the LHCb Collaboration also shows a discrepancy of 1.7σ from its SM result, $R_{J/\psi}^{\text{SM}} = 0.289 \pm 0.01$ [4]. In this concern, we would like to perform model-independent analysis of $B \to D^{(*)}l\bar{\nu}_l$ and $B_c \to (\eta_c, J/\psi)l\bar{\nu}_l$ processes in different q^2 bins.

The paper is structured as follows. In Sect. 119.2, we discuss the effective Hamiltonian in the presence of NP and the global fit to new parameters. The numerical analysis of LNU parameters of $b \to cl\bar{\nu}_l$ decay modes are presented in Sect. 119.3 followed by the conclusion in Sect. 119.4.

119.2 General Effective Hamiltonian and the Numerical Fit

The effective Hamiltonian of $b \to c\tau\bar{\nu}_l$ processes by using only the left-handed neutrinos can be written as [5]

$$\mathcal{H}_{\text{eff}} = \frac{4G_F}{\sqrt{2}} V_{cb} \Big[(\delta_{l\tau} + V_L) \, \mathcal{O}_{V_L}^l + V_R \mathcal{O}_{V_R}^l + S_L \mathcal{O}_{S_L}^l + S_R \mathcal{O}_{S_R}^l + T \mathcal{O}_T^l \Big], \quad (119.3)$$

where G_F is the Fermi constant, V_{cb} is the CKM matrix element, $X (= V_{L,R}, S_{L,R}, T)$ are the Wilson coefficients and $\mathcal{O}_X^l$ are their respective effective operators

$$\mathcal{O}_{V_L}^l = (\bar{c}_L \gamma^\mu b_L) (\bar{\tau}_L \gamma_\mu \nu_{lL}), \ \mathcal{O}_{V_R}^l = (\bar{c}_R \gamma^\mu b_R) (\bar{\tau}_L \gamma_\mu \nu_{lL}), \ \mathcal{O}_{S_L}^l = (\bar{c}_L b_R) (\bar{\tau}_R \nu_{lL}),$$

$$\mathcal{O}_{S_R}^l = (\bar{c}_R b_L) (\bar{\tau}_R \nu_{lL}), \ \mathcal{O}_T^l = (\bar{c}_R \sigma^{\mu\nu} b_L) (\bar{\tau}_R \sigma_{\mu\nu} \nu_{lL}), \qquad (119.4)$$

with $q_{L(R)} = L(R)q$, $L(R) = (1 \mp \gamma_5)/2$ being the projection operators.

We first perform a χ^2 fit to know the disagreement of SM with the measured data, which is defined as

$$\chi^2(X) = \sum_i \frac{(\mathcal{O}_i^{\text{th}}(X) - \mathcal{O}_i^{\text{exp}})^2}{(\Delta \mathcal{O}_i^{\text{exp}})^2 + (\Delta \mathcal{O}_i^{\text{SM}})^2}. \qquad (119.5)$$

Here, $\mathcal{O}_i^{\text{exp}}$ (i are the no. of observables) represent the measured central values of the observables where $\Delta \mathcal{O}_i^{\text{exp}}$ stand for the respective error values. The theoretical predictions as a function of X are denoted by $\mathcal{O}_i^{\text{th}}$ with $\Delta \mathcal{O}_i^{\text{SM}}$ as the theoretical uncertainties arising due to input parameters. We include the measurements on $R_{D^{(*)}}$ and $R_{J/\psi}$ parameters and the upper limit on $\text{Br}(B_c^+ \to \tau^+ \nu_l)$ as 30% [6] for the evaluation of χ^2. Considering one new coefficient at a time, which is real, the computed best-fit values of all coefficients are given in Table 119.1. We also provide the $\chi_{\text{min,SM+NP}}^2/\text{d.o.f}$ and pull=$\sqrt{\chi_{\text{min,SM}}^2 - \chi_{\text{min,SM+NP}}^2}$ values of individual coefficients

Table 119.1 Predicted best-fit, $\chi^2_{min}/$d.o.f and pull values of new real Wilson coefficients

New Wilson coefficients	Best-fit values	$\chi^2_{min}/$d.o.f	Pull
V_L	−2.086	0.842	3.692
V_R	−0.0671	3.298	2.502
S_L	0.097	4.492	1.64
S_R	−1.443	2.695	2.841
T	−0.034	1.45	3.44

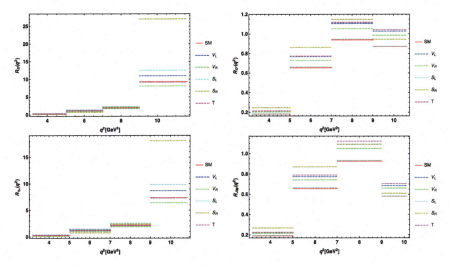

Fig. 119.1 The plots for R_D (top-left), R_{D^*} (top-right), R_{η_c} (bottom-left) and $R_{J/\psi}$ (bottom-right) ratios in four q^2 bins

in this table. One can see that the theory with only V_L coefficient fits the data very well.

119.3 Implication on $B_{(c)} \to (P, V)l\bar{\nu}_l$ Decay Modes

The effects of these real new coefficients on the $R_{P(V)}$ ratios of $B_{(c)} \to (P, V)l\bar{\nu}_l$ processes, defined as

$$R_{P(V)} = \frac{\mathrm{Br}(B \to P(V)\tau\bar{\nu}_\tau)}{\mathrm{Br}(B \to P(V)l\bar{\nu}_l)}, \quad l = e, \mu. \tag{119.6}$$

where $P = D, \eta_c$ are the pseudoscalar mesons and $V = D^*, J/\psi$ are the vector mesons, are discussed in this section. For numerical estimation, the particle masses and the lifetime of $B_{(c)}$ mesons are taken from [7] and the form factors from [8]. Using the best-fit values of $V_{L,R}$, $S_{L,R}$ and T coefficients from Table 119.1, the binwise graphical representation of R_D (top-left), R_{D^*} (top-right), R_{η_c} (bottom-left) and $R_{J/\psi}$ (bottom-right) ratios are shown in in Fig. 119.1. Here, the red solid lines represent the SM contribution and the blue, green, cyan, dark yellow and magenta color dashed lines stand for the contributions from the additional V_L, V_R, S_L, S_R and T coefficients, respectively. The presence of either V_L or V_R or S_L show profound deviation from the SM values of R_D and R_{η_c} parameters only in the last bin, whereas tensor coupling has a vanishing effect. The S_R coefficient has maximum impact on $R_{D(\eta_c)}$. None of the new couplings affect these ratios in the first three q^2 bins. All the coefficients excluding S_L have significant impact on the R_{D^*} and $R_{J/\psi}$ parameters.

119.4   Conclusion

We perform a binwise analysis of lepton nonuniversality ratios of $B \to D^{(*)}\tau\bar{\nu}_l$ and $B_c \to (\eta_c, J/\psi)\tau\bar{\nu}_l$ processes in a model-independent approach. We constrain the new vector, scalar and tensor couplings from the χ^2 fit to $b \to c\tau\bar{\nu}_l$ data. We found that all coefficients have larger impact on $R_{D(\eta_c)}$ except T in the last bin and on R_{D^*}, $R_{J/\psi}$ except S_L coefficient in the four q^2 bins.

References

1. Heavy Flavor Averaging Group (2019), https://hflav-eos.web.cern.ch/hflav-eos/semi/spring19/html/RDsDsstar/RDRDs.html
2. H. Na, C.M. Bouchard, G.P. Lepage, C. Monahan, J. Shigemitsu, Phys. Rev. D **92**, 054510 (2015), [Erratum: Phys. Rev.D **93**,no.11,119906(2016)]; S. Fajfer, J. F. Kamenik, I. Nisandzic, and J. Zupan, Phys. Rev. Lett. **109**, 161801 (2012)
3. R. Aaij et al. (LHCb), Phys. Rev. Lett. 120, 121801 (2018), http://orcid.org/10.1103/PhysRevLett.120.121801
4. W.-F. Wang, Y.-Y. Fan, Z.-J. Xiao, Chin. Phys. C37, 093102 (2013) http://orcid.org/10.1088/1674-37/37/9/093102
5. V. Cirigliano, J. Jenkins, M. Gonzalez-Alonso, Nucl. Phys. B **830**, 95 (2010), http://orcid.org/10.1016/j.nuclphysb.2009.12.020
6. A.G. Akeroyd, C.-H. Chen, Phys. Rev. D **96**, 075011 (2017)
7. M. Tanabashi et al., Particle data group. Phys. Rev. D **98**, 030001 (2018)
8. J. A. Bailey et al. (MILC), Phys. Rev. D **92**, 034506 (2015); I. Caprini, L. Lellouch, and M. Neubert, Nucl. Phys. B **530**, 153 (1998); J. A. Bailey et al. (Fermilab Lattice, MILC), Phys. Rev. D **89**, 114504 (2014); Y. Amhis et al. (HFAG) (2014);, 1412.7515 Z.-R. Huang, Y. Li, C.-D. Lu, M. A. Paracha, C. Wang, Phys. Rev. D **98**, 095018 (2018); R. Watanabe, Phys. Lett. B **776**, 5 (2018)

Chapter 120
Measurement of the CKM Angle ϕ_3 Using B→DK with Belle II

M. Kumar, K. Lalwani, and K. Trabelsi

Abstract We present a preliminary study of using the decay $B^{\pm} \rightarrow D_{CP}K^{\pm}$ to measure ϕ_3 at Belle II, where D_{CP} represents a D meson decay to a CP even eigenstate, i.e. K^+K^- and $\pi^+\pi^-$. We discuss the ϕ_3 measurement one may expect at Belle II with an integrated luminosity of 50 ab^{-1}. We also present the preliminary results on the reconstruction of B and D mesons from a Belle II data sample corresponding to an integrated luminosity of 472 pb^{-1}.

120.1 Introduction

The CKM angle ϕ_3 is one of the least well-constrained parameters of the Unitarity Triangle [1, 2]. The measurement of ϕ_3 from $B^{\pm} \rightarrow D^0K^{\pm}$ and $B^{\pm} \rightarrow \overline{D^0}K^{\pm}$ decays is theoretically clean as they occur at the tree level [3] as shown in Fig. 120.1. If the D^0 or $\overline{D^0}$ is reconstructed as a CP eigenstate, the $b \rightarrow c$ and $b \rightarrow u$ processes interfere. This interference may lead to direct CP violation. To measure D meson decays to such final states, a large number of B mesons is required since the branching fraction of these modes are only of the order 0.01% [5]. Then a large number of B decays are required to extract ϕ_3. To extract ϕ_3 using the GLW method [6], the observables sensitive to CP violation are

$$\mathcal{A}_{1,2} \equiv \frac{\mathcal{B}(B^- \rightarrow D_{1,2}K^-) - \mathcal{B}(B^+ \rightarrow D_{1,2}K^+)}{\mathcal{B}(B^- \rightarrow D_{1,2}K^-) + \mathcal{B}(B^+ \rightarrow D_{1,2}K^+)} \tag{120.1}$$

$$= \frac{2r\sin\delta'\sin\phi_3}{1 + r^2 + 2r\cos\delta'\cos\phi_3}, \tag{120.2}$$

On behalf of the Belle II Collaboration.

M. Kumar (✉) · K. Lalwani
Malaviya National Institute of Technology, Jaipur, India
e-mail: 2016rpy9052@mnit.ac.in

K. Trabelsi
Laboratoire De L'accelerateur Lineaire, Orsay, France

© Springer Nature Singapore Pte Ltd. 2021
P. K. Behera et al. (eds.), *XXIII DAE High Energy Physics Symposium*,
Springer Proceedings in Physics 261,
https://doi.org/10.1007/978-981-33-4408-2_120

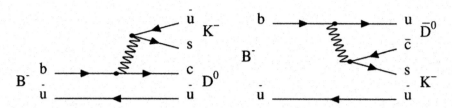

Fig. 120.1 Feynman diagram for $B^- \to D^0 K^-$ (left) and $B^- \to \overline{D}^0 K^-$ (right) [4]

and the double ratios

$$\mathcal{R}_{1,2} \equiv \frac{R^{D_{1,2}}}{R^{D^0}} = 1 + r^2 + 2r\cos\delta'\cos\phi_3 \qquad (120.3)$$

$$\delta' = \begin{cases} \delta & \text{for } D_1 \\ \delta + \pi & \text{for } D_2. \end{cases} \qquad (120.4)$$

The ratios $\mathcal{R}^{D_{1,2}}$ and $\mathcal{R}^{D^0}$ are defined as

$$\mathcal{R}^{D_{1,2}} = \frac{\mathcal{B}(B^- \to D_{1,2}K^-) + \mathcal{B}(B^+ \to D_{1,2}K^+)}{\mathcal{B}(B^- \to D_{1,2}\pi^-) + \mathcal{B}(B^+ \to D_{1,2}\pi^+)}$$

$$\mathcal{R}^{D^0} = \frac{\mathcal{B}(B^- \to D^0 K^-) + \mathcal{B}(B^+ \to \overline{D}^0 K^+)}{\mathcal{B}(B^- \to D^0\pi^-) + \mathcal{B}(B^+ \to \overline{D}^0\pi^+)},$$

where D_1 and D_2 are CP-even and CP-odd eigenstates, respectively. Here $r = |A(B^- \to \overline{D}^0 K^-)/A(B^- \to D^0 K^-)|$ is the ratio of the magnitude of the tree amplitudes and δ is their strong-phase difference. Note that the asymmetries $\mathcal{A}_1$ and $\mathcal{A}_2$ are of opposite sign.

There have been many efforts by BaBar, Belle and LHCb collaborations to measure the CKM angle ϕ_3, which are summarized in Table 120.1, but a measurement with a precision of $1°$ is desirable to compare to the indirect measurement. Therefore, the determinations of ϕ_3 with high statistics are required, as the measurement is dominated by the statistical uncertainty.

Table 120.1 Previous ϕ_3 measurements

Experiment	Measurement of ϕ_3
Belle	$\left(73^{+13}_{-15}\right)^\circ$ [7]
BaBar	$\left(69^{+17}_{-16}\right)^\circ$ [8]
LHCb	$\left(74^{+5.0}_{-5.8}\right)^\circ$ [9]

Fig. 120.2 Distribution between the expected ϕ_3 from $B^+ \to D(K_S^0\pi^+\pi^-)K^+$ uncertainty versus luminosity accumulated by Belle II [10]

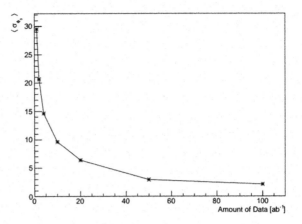

In this work, we present a preliminary Monte Carlo (MC) study of $B^\pm \to D^0K^\pm$ to extract ϕ_3 using the 50 ab^{-1} data to be accumulated by the Belle II detector. The Belle II [11] experiment at the SuperKEKB asymmetric e$^+$e$^-$ collider [12] will accumulate collision data at an unprecedented instantaneous luminosity of 8×10^{35}cm^{-2}sec^{-1}, which is 40 times larger than the Belle experiment. Figure 120.2 shows how the expected uncertainty on ϕ_3 scales with luminosity based on toy Monte Carlo studies for the mode $B^+ \to D(K_S^0\pi^+\pi^-)K^+$. It shows that the expected uncertainty is approximately 3° and the overall ϕ_3 projection is 1.6° after including GLW/ADS [13] and D^* modes with an integrated luminosity of 50 ab^{-1}. In this work, we present the study of $D^{*\pm} \to D^0(K^-\pi^+)\pi^\pm$ using the Phase II data of the Belle II experiment collected within the integrated luminosity of 472pb^{-1}. This decay is an important control channel for GLW and ADS analyses at Belle II. Here, Phase II data is incorporating single ladder per layer of the vertex detector, which is approximately $\frac{1}{8}$th of the complete vertex detector and all other subdetectors. We also show the study on $B^\pm \to D^0K^\pm$ with MC simulation.

120.2 Preliminary Results from Phase II Data

We reconstruct the decay of $D^{*\pm} \to D^0(K^-\pi^+)\pi^\pm$ using Phase II data, corresponding to an integrated luminosity of 472 pb^{-1}. To select $c\bar{c}$ events, the center-of-mass momentum of D^* is required to be greater than 2.5 GeV/c. The distribution of ΔM is shown in Fig. 120.3 (left), where ΔM is the difference between the invariant mass of $D^{*\pm}$ and D^0 meson. The invariant mass distribution of D^0 from $K^-\pi^+$ is shown in Fig. 120.3 (right).

Further, the reconstruction of B mesons is carried out with an MC data sample of 2×10^6 $B^\pm \to D^0(K^+K^-)K^\pm$ events. In order to select B mesons, two important variables, the energy difference, $\Delta E = \sum E_i - E_{\text{beam}}$ and the beam-constrained

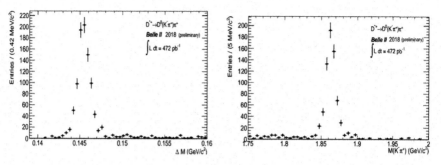

Fig. 120.3 ΔM (left) and M_D (right) distribution in Phase II data for $D^0 \to K^- \pi^+$ final state

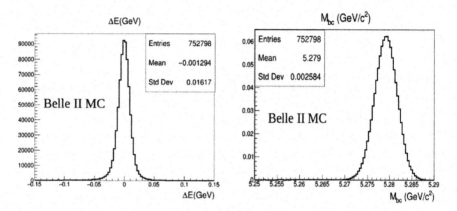

Fig. 120.4 ΔE (left) and M_{bc} (right) distribution from MC simulation

mass, $M_{bc} = \sqrt{(E_{beam})^2 - \sum (\vec{p_i})^2}$, are used. Where E_{beam} is the center-of-mass (CM) beam energy, E_i and p_i are the CM energies and momenta of B candidate's decay product. Figure 120.4 shows the M_{bc} and ΔE distributions reconstructed from the MC sample; work is in progress to reconstruct B mesons in the Phase II data.

120.3 Summary

The full 50 ab^{-1} data sample to be collected by Belle II at SuperKEKB will provide a substantial improvement in the precision measurement of ϕ_3. A clear signature of $D^{*\pm} \to D^0(K^- \pi^+)\pi^\pm$ is observed in Phase II data corresponding to an integrated luminosity of 472 pb^{-1}. Further, the reconstruction of B mesons using M_{bc} and ΔE is carried out with MC simulation and the same is in progress with data.

Acknowledgements The authors are thankful to the Belle II Collaboration for providing the opportunity to present this work. We are thankful to MNIT Jaipur and SERB, DST Delhi, for providing the funding support from the research project (EMR Project File No: EMR/2016/007519) to participate and present this work in this symposium held at IIT Madras.

References

1. M. Kobayashi, T. Maskawa, Prog. Theor. Phys. **49**, 652 (1973)
2. N. Cabibbo, Phys. Rev. Lett. **10**, 531 (1963)
3. J. Brod, J. Zupan, The ultimate theoretical error on gamma from $B \to DK$ decays, arXiv:1308.5663 [hep-ph]
4. J. Brodzicka et al., Physics achievements from the Belle experiment, **PTEP 2012, 04D001**
5. M. Tanabashi et al., (Particle Data Group), Phys. Rev. D **98**, 030001 (2018)
6. M. Gronau, D. London, Phys. Lett. B**253**, 483 (1991); M. Gronau and D. Wyler, Phys. Lett. B**265**, 172 (1991)
7. J. Libby, Direct CP violation in hadronic B decays, arXiv:1412.4269 [hep-ex]
8. J. P. Lees et al., (The BaBar Collaboration), Phys. Rev. D **87**, 052015 (2013)
9. The LHCb Collaboration, CERN-LHCb-CONF-2018-002
10. E. Kou et al., The Belle II Physics Book, PTEP 2018, arXiv:1808.10567 [hep-ex]
11. T. Abe et al., Belle II Technical Design Report (2010), arXiv:1011.0352 [physics.ins-det]
12. B. Golob, (The Belle II Collaboration), Super KEKB / Belle II Project, Nuovo Cim. C**033N5** 319-326 (2010)
13. D. Atwood, I. Dunietz, A. Soni, Enhanced CP Violation with $B \to KD^0(\overline{D}^0$ Modes and Extraction of the Cabibbo-Kobayashi-Maskawa Angle γ. Phys. Rev. Lett. **78**, 3257(1997)

Chapter 121
Projection Operators in Second Class Constrained Systems

A. S. Vytheeswaran

Abstract We look at a method developed to convert second class constraints into first class. The method involves a certain operator introduced at the classical level. For two second class constraints, this operator is a projection operator. For more than two second class constraints, more than one operator, each one a projection operator by itself, are possible. We analyse here the conditions under which combinations of such operators are also projection operators.

121.1 Introduction

In the theory of Constrained Dynamical systems, developed in the phase space formalism, constraints are classified as first class and second class. First class constraints are the generators of gauge transformations. Second class constraints are traditionally known to merely reduce the physical degrees of freedom, and Dirac brackets are introduced to eliminate them.

It is also possible to convert the second class constraints into first class. This introduces new gauge invariances in the theory, and enlarges existing gauge symmetries. One procedure [1] to carry out this conversion is developed within the phase space of the original second class constrained system. This is done by defining and constructing one or more operators, at the classical level.

These operators, when considered individually, are projection operators. However, they may or may not commute among themselves. Also, combinations of such operators may or may not immediately result in projection operators.

Here, we analyse these operators and present some results. We start with a system with only two second class constraints, and later generalise to four constraints.

A. S. Vytheeswaran (✉)
Department of Theoretical Physics, University of Madras, Guindy Campus,
Chennai 600025, India
e-mail: asvythee@gmail.com; asvythee@unom.ac.in

© Springer Nature Singapore Pte Ltd. 2021
P. K. Behera et al. (eds.), *XXIII DAE High Energy Physics Symposium*,
Springer Proceedings in Physics 261,
https://doi.org/10.1007/978-981-33-4408-2_121

121.2   The Case of Two Second Class Constraints

Consider a system with two second class constraints $\phi_1 \approx 0$, $\phi_2 \approx 0$, defining a surface $\sum_2$ in the corresponding phase space. Their second class nature is reflected in the invertibility, everywhere in the phase space, of the matrix

$$C = \begin{bmatrix} 0 & \{\phi_1, \phi_2\} \\ \{\phi_2, \phi_1\} & 0 \end{bmatrix} = \begin{pmatrix} 0 & c \\ -c & 0 \end{pmatrix}, \quad \text{with } c = \{\phi_1, \phi_2\}. \quad (121.1)$$

To construct a gauge theory [1], we first define $\chi = c^{-1}\phi_1$, $\psi = \phi_2$. The redefined constraint χ alone is retained, and the ψ is discarded as a constraint. The relevant constraint surface is now $\sum_1$, defined by only $\chi \approx 0$. On this $\sum_1$, the χ and ψ form a canonically conjugate pair.

The $\chi \approx 0$ is now like a first class constraint. For a proper gauge theory, with gauge transformations generated by the χ, physical observables must be gauge invariant; their Poisson brackets with χ must be zero, at least on $\sum_1$. For this, we first define the operator $\hat{\chi}$ by its action on a phase space function $A : \hat{\chi}(A) = \{\chi, A\}$. Using this, we construct the operator $\mathbf{P}$ as follows:

$$\mathbf{P} = \; : e^{-\psi\hat{\chi}} : \; = \sum_{n=0}^{\infty} \frac{(-1)^n}{n!} \, \psi^n \hat{\chi}^n, \qquad \hat{\chi}^n = \{\chi, \{\chi, \{\chi, \ldots\}\}\ldots\}$$

$$(121.2)$$

with the operator $\hat{\chi}^n$ representing the n-tuple Poisson bracket with χ appearing n times.

The $\mathbf{P}$ operator acting on any quantity A gives the gauge invariant quantity $\tilde{A}$,

$$\tilde{A} = \mathbf{P}(A) = \sum_{n=0}^{\infty} \frac{(-1)^n}{n!} \, \psi^n \hat{\chi}^n(A) = \sum_{n=0}^{\infty} \frac{(-1)^n}{n!} \, \psi^n \{\chi, \{\chi, \{\chi, \ldots, A\}\} \ldots\}\}$$

$$(121.3)$$

This $\tilde{A}$ satisfies $\{\chi, \tilde{A}\} = 0$, at least on the new surface $\sum_1$. Thus $\tilde{A}$ is gauge invariant.

The operator $\mathbf{P}$ is central to the extraction of a gauge theory from the existing second class constraints, and has the following properties:

(1) $\mathbf{P}(\psi) = 0$.
(2) $\mathbf{P}(\text{constant B}) = B$.
(3) $\mathbf{P}(AB) = \mathbf{P}(A)\,\mathbf{P}(B)$.
(4) $\{\mathbf{P}(A), \mathbf{P}(B)\} = \mathbf{P}(\{A, B\}_{DB})$.
(5) $\mathbf{P}^2 = \mathbf{P}$.

In the fourth property above, the subscript DB implies the Dirac bracket. Note that the last property implies that $\mathbf{P}$ is a projection operator.

121.3 The Case of More Than Two Second Class Constraints

Consider a system with four second class constraints $\phi_i \approx 0$, $i = 1, 2, 3, 4$. The non-zero determinant of the matrix C with elements $\{\phi_i, \phi_j\}$ $i, j = 1, 2, 3, 4$ reveals the second class property of the ϕ_i. Going by the structure of this matrix, different cases can be considered.

121.3.1 Case 1 (the simplest)

$$
\begin{matrix} \{\phi_1, \phi_2\} = c_1, \\ \{\phi_3, \phi_4\} = c_2, \end{matrix} \qquad C = \begin{bmatrix} 0 & c_1 & 0 & 0 \\ -c_1 & 0 & 0 & 0 \\ 0 & 0 & 0 & c_2 \\ 0 & 0 & -c_2 & 0 \end{bmatrix} \qquad (121.4)
$$

with all other Poisson brackets zero. For simplicity, we take the c_1, c_2 to be constants.

It may be seen that there are two pairs of constraints, each pair being second class within itself: (ϕ_1, ϕ_2) and (ϕ_3, ϕ_4). This indicates that two P operators are possible:

$$
P_1 = : e^{-\psi_1 \hat{\chi}_1} :, \quad P_2 = : e^{-\psi_2 \hat{\chi}_2} :, \quad \chi_1 = \frac{1}{c_1}\phi_1, \quad \chi_2 = \frac{1}{c_2}\phi_3, \quad \psi_1 = \phi_2, \quad \psi_2 = \phi_4.
$$
$$(121.5)$$

Individually, both the P_1 and P_2 are projection operators. The other properties are also obeyed.

Moreover, the two operators commute with each other: $P_1 P_2 = P_2 P_1$. This can be seen by operating $P_1 P_2$ on some A, and the result can be rewritten as $P_2 P_1$ operating on the same A.

Finally, we see that the combined operator $P_1 P_2$ is also a projection operator: $(P_1 P_2)^2 = P_1 P_2$.

121.3.2 Case 2

Let us include one more non-zero Poisson bracket among the four constraints; say $\{\phi_2, \phi_3\} = \alpha$, with α a constant. Then, the C matrix appears as

$$
C = \begin{bmatrix} 0 & c_1 & 0 & 0 \\ -c_1 & 0 & \alpha & 0 \\ 0 & -\alpha & 0 & c_2 \\ 0 & 0 & -c_2 & 0 \end{bmatrix}, \qquad \text{with } \{\chi_2, \psi_1\} = -\frac{\alpha}{c_2}. \qquad (121.6)
$$

The operators P_1, P_2 defined earlier, **no longer commute**—$P_1 P_2 \neq P_2 P_1$. Consequently, the combined operator $P_1 P_2$ is not a projection operator.

However, we can define a different operator. First defining $\psi_1' = \left(\psi_1 + \frac{\alpha}{c_2} \psi_2 \right)$, we construct the following operator, $P_1' = : e^{-\psi_1' \hat{\chi}_1} :$. This is a projection operator—$(P_1')^2 = P_1'$. Further, this operator P_1' *commutes* with P_2—$P_1' P_2 = P_2 P_1'$. The combined operator $P_1' P_2$ is also a projection operator—$(P_1' P_2)^2 = P_1' P_2$. The new operator P_1' contains only quantities gauge invariant with respect to the χ_2.

Further, we also find the result

$$P_1 P_2(A) = \sum_{k=0}^{\infty} \frac{1}{k!} \left(\frac{\alpha}{c_2} \psi_2 \right)^k P_2 P_1' \left(\hat{\chi}_1^k A \right). \tag{121.7}$$

121.3.3   Case 3

Instead of $\{\phi_2, \phi_3\} = \alpha$, let us consider $\{\phi_1, \phi_4\} = \beta$, with β a constant. Then,

$$C = \begin{bmatrix} 0 & c_1 & 0 & \beta \\ -c_1 & 0 & 0 & 0 \\ 0 & 0 & 0 & c_2 \\ -\beta & 0 & -c_2 & 0 \end{bmatrix}, \qquad \{\chi_1, \psi_2\} = \frac{\beta}{c_1}. \tag{121.8}$$

Here also P_1, P_2 **no longer commute**—$P_1 P_2 \neq P_2 P_1$. Again, $P_1 P_2$ is not a projection operator. To overcome this, the P_2 is modified to $P_2' = : e^{-\psi_2' \hat{\chi}_2} :$, with $\psi_2' = \left(\psi_2 - \frac{\beta}{c_1} \psi_1 \right)$. Then we have $(P_2')^2 = P_2'$ and also $P_2' P_1 = P_1 P_2'$. Further, $(P_2' P_1)^2 = P_2' P_1$. The new operator P_2' contains only quantities gauge invariant with respect to the χ_1. We also have

$$P_2 P_1 (A) = \sum_{k=0}^{\infty} \frac{(-1)^k}{k!} \left(\frac{\beta}{c_1} \psi_1 \right)^k P_1 P_2' \left(\hat{\chi}_2^k A \right). \tag{121.9}$$

121.3.4   Case 4

Both the cases above can be combined to get,

$$C = \begin{bmatrix} 0 & c_1 & 0 & \beta \\ -c_1 & 0 & \alpha & 0 \\ 0 & -\alpha & 0 & c_2 \\ -\beta & 0 & -c_2 & 0 \end{bmatrix}, \qquad \{\chi_2, \psi_1\} = -\frac{\alpha}{c_2}, \qquad \{\chi_1, \psi_2\} = \frac{\beta}{c_1}. \tag{121.10}$$

Needless to say the $\mathbf{P}_1$ and $\mathbf{P}_2$ do not commute with each other. We then replace both ψ_1 and ψ_2 by $\psi_1' = \psi_1 + \frac{\alpha\psi_2}{c_2}$ and $\psi_2' = \psi_2 - \frac{\beta\psi_1}{c_1}$ respectively, and construct operators $\mathbf{P}_1'$ and $\mathbf{P}_2'$ which commute with each other. Then, the operator $\mathbf{P}_1'\mathbf{P}_2'$ will also be a projection operator.

We also have the useful relation $\quad \mathbf{P}_1\mathbf{P}_2(\mathsf{A}) = \sum_{k=0}^{\infty} \frac{1}{k!}\left(\frac{\alpha}{c_2}\psi_2\right)^k \mathbf{P}_2'\mathbf{P}_1'\, (\hat{\chi}_1^k \mathsf{A}).$

121.4 Conclusions and Discussion

When combining two or more projection operators, the combined operator may not be a projection operator. The original projection operators have to be modified. These modified (projection) operators, when combined, result in more projection operators. The requirement of gauge invariance is the guiding factor in such modifications. These results can be applied to both finite-dimensional systems and to field theories.

We also mention a completely different approach. In general, as in Case 4 above, we can construct a different combined operator as follows: we first demand that the ψ_i $(i = 1, 2)$ **always stay outside** the Poisson brackets; then

$$\mathbf{P}_{\mathsf{comb}} = \,:\, e^{-\psi_1\hat{\chi}_1 - \psi_2\hat{\chi}_2}\,: \tag{121.11}$$

can be used and shown to obtain gauge invariant observables and a gauge theory.

More generalisations of the above results can be considered. One such generalisation is the case of the c_1, c_2, α, β being variables. A second generalisation arises when the two first class constraints form a non-Abelian set. Work along these lines is in progress.

Acknowledgements The author thanks DAE-BRNS, India and IIT Madras, India for the opportunity to present this work at the 23rd DAE-BRNS High Energy Physics Symposium. Funding under DST-PURSE Scheme of Department of Science & Technology, Government of India, is acknowledged.
The author wishes to acknowledge Dr. R Anishetty for raising the issue of Projection operators. He also thanks the University of Madras and the Head of Department of Theoretical Physics for constant encouragement.

Reference

1. A.S. Vytheeswaran, Annals of Phys. **236**, 297–324 (1994)

Chapter 122
Same Sign WW Studies Using EFT at LHC

Geetanjali Chaudhary, Jan Kalinowski, Manjit Kaur, Paweł Kozów, Kaur Sandeep, Michał Szleper, and Sławomir Tkaczyk

Abstract Study of discovery potential of new physics in the Effective Field Theory (EFT) framework for the same-sign WW Vector Boson Scattering (VBS) process at the High-Luminosity (HL) and High-Energy Large Hadron Collider (HE-LHC) is done. We focus on purely leptonic decays of the W and as a final state in proton-proton collisions, look for $pp \rightarrow 2jets + W^+W^+ \rightarrow jjl^+\nu l'^+\nu'$ decay at the LHC and the WW invariant mass cannot be determined experimentally. We study the effect of a single dim-8 operator that alters $WWWW$ quartic gauge coupling and for hints of new physics, we look for the deviations from the Standard Model case. Work at 27 TeV at HE-LHC is reported for both positive and negative values of operators and is compared with previously done work at 14 TeV with $3ab^{-1}$ luminosity. In this paper, we present comparison results for two dim-8 operators. It is observed that discovery regions for the individual dim-8 operators at HE-LHC shift to lower values of the Wilson coefficients but the overall discovery potential does not get significantly enhanced.

122.1 Introduction

At the LHC, no physics Beyond Standard Model (BSM) has been discovered so far. Lack of direct observation of new physics necessitates the indirect BSM searches. Any deviation in data from the SM predictions would be a signal of new physics. The

G. Chaudhary (✉) · M. Kaur · K. Sandeep
Panjab University, Chandigarh, India
e-mail: geetanjali.chaudhary@cern.ch

J. Kalinowski · P. Kozów
Faculty of Physics, University of Warsaw, ul. Pasteura 5, 02-093 Warsaw, Poland

M. Szleper
National Center for Nuclear Research, High Energy Physics Department, ul. Pasteura 7, 02-093 Warsaw, Poland

S. Tkaczyk
Fermi National Accelerator Laboratory, Batavia, IL 60510, USA

© Springer Nature Singapore Pte Ltd. 2021   869
P. K. Behera et al. (eds.), *XXIII DAE High Energy Physics Symposium*,
Springer Proceedings in Physics 261,
https://doi.org/10.1007/978-981-33-4408-2_122

BSM contributions are effectively parameterized in terms of higher dimension operators in EFT and the information of the scale Λ of new physics and the strength C of the coupling is encoded in the Wilson coefficients of the higher dimension operators, $f_i = C^m/\Lambda^n$. The extended SM Lagrangian with higher dimension operators looks like $L = L_{SM} + \sum_i f_i^{(6)} O_i^{(6)} + \sum_i f_i^{(8)} O_i^{(8)} + \cdots$, where $f_i^{(6)}$ and $f_i^{(8)}$ are the Wilson Coefficients for dim-6 and dim-8 operators, respectively.

For any value of Wilson coefficient, the EFT formulation is valid for $M_{WW} < \Lambda \leq M^U$, as in WW gauge Boson scattering (Feynman diagrams shown in Fig. 122.1), Wilson coefficients of dim-8 operators affect the $WWWW$ quartic vertex in such a way that above certain M_{WW}, the scattering amplitude violates unitarity. This scale where unitarity gets violated is denoted by M^U. Also, EFT is an expansion of E/Λ where E is the energy scale of the process and Λ is scale of new physics (E is identified with M_{WW} in our case of WW scattering). Hence M_{WW} must fulfil $M_{WW} < \Lambda$ condition otherwise expansion in powers of M_{WW}/Λ is not convergent. To test the EFT "model" ($O_i^{(d)}$, $f_i^{(d)}$), the BSM signal S is defined as the deviation from SM predictions observed in the distribution of some observable D_i,

$$S = D_i^{model} - D_i^{SM} \tag{122.1}$$

A quantitative estimate of the signal can be written as

$$D_i^{model} = \int_{2M_W}^{\Lambda} \frac{d\sigma}{dM}|_{Model} dM + \int_{\Lambda}^{M_{max}} \frac{d\sigma}{dM}|_{SM} dM \tag{122.2}$$

where M_{max} is the kinematic limit of the WW invariant mass. Equation (122.2) defines the signal coming uniquely from the EFT "model" in its range of validity and assumes only the SM contribution in the region $M_{WW} > \Lambda$. BSM contribution from the region above Λ may enhance the signal, but it may also prevent the proper description of the data in the "EFT model". To effectively describe BSM physics within EFT framework, the additional contribution above Λ should be small enough compared to the contribution from the validity region of EFT "model". For a quantitative estimate of this contribution, a second estimate has been defined in which all the helicity amplitudes above Λ are assumed to remain constant at their respective values when they reach Λ. A quantitative estimate of the signal can be written as

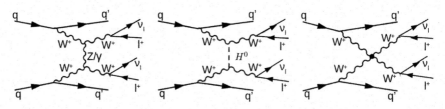

Fig. 122.1 Feynman diagrams of WW scattering

$$D_i^{model} = \int_{2M_W}^{\Lambda} \frac{d\sigma}{dM}|_{Model} dM + \int_{\Lambda}^{M_{max}} \frac{d\sigma}{dM}|_{\Lambda=const} dM \qquad (122.3)$$

This estimate regularizes the helicity amplitudes above Λ that violate unitarity at M^U. Following this criterion, the EFT "model" is tested for values of $(\Lambda \leq M^U, f_i)$ and we check whether the signals computed from (122.2) are statistically consistent within 2σ with the signals computed with (122.3). BSM observability imposes some minimum value of f_{min}, while the description within the EFT "model" imposes some maximum value of f_{max} such that signal estimates computed from (122.2) and (122.3) remain statistically consistent. For $\Lambda = M^U$ a finite interval of f_i values is possible, while for $\Lambda \leq M^U$ the respective limits on f_i depend on the actual value of Λ. As a result, a triangle in the $(\Lambda \leq M^U, f_i)$ plane is formed which is bounded from above by the unitarity bound $M^U(fi)$, from the left by the signal significance of 5σ and from the right by the consistency within 2σ from (122.3). Following this consistency criterion, discovery potential of BSM physics can be explored within EFT framework as long as the "EFT triangle" is not empty. The physics potential of the single-operator EFT approach has already been tested on a hypothetical new physics signal observed in the same-sign WW scattering process at the 14 TeV HL-LHC [1].

122.2   Simulations

We present generator level study aimed at finding the EFT triangles for individual dim-8 operators at the HE-LHC. Event samples of the process $pp \rightarrow jj\mu^+\mu^+\nu\nu$ at 27 TeV were generated at Leading Order (LO) for each dim-8 operator O_i that modifies the $WWWW$ quartic coupling (i = S0, S1, T0, T1, T2, M0, M1, M6, M7) using the MadGraph5 aMC@NLO v5.2.6.2 generator [2], with the appropriate UFO files containing additional vertices involving the desired dim-8 operators. A scan of f_i values for each operator was made using the MadGraph reweight command, including $f_i = 0$ to represent the SM case. The Pythia package v6.4.1.9 [3] was used for hadronization as well as initial and final state radiation processes. Unitarity limits, M^U, for different helicity amplitudes are different and were determined using the VBFNLO calculator v1.4.0 [4]. Cross sections at the output of MadGraph were multiplied by a factor 4 to account for all the lepton (electron and/or muon) combinations in the final state. Only signal samples were generated and the SM case was treated as irreducible background in the study of possible BSM effects. Standard VBS cuts are applied to require at least two reconstructed jets and exactly two leptons (muons or electrons) satisfying the following conditions: $M_{jj} > 500$ GeV, $\Delta\eta_{jj} > 2.5$, $p_T^j > 30$ GeV, $|\Delta\eta_j| < 5$, $p_T^l > 25$ GeV and $|\Delta\eta_l| < 2.5$. According to (122.3), the total BSM signal was estimated by suppressing the high-mass tail above the calculated value of Λ. This was achieved by applying an additional weight of the form $(\Lambda/M_{WW})^4$ to each generated event in this region. For each dim-8 operator, signal significance expressed in standard deviations(σ) is calculated as the square root of a χ^2 resulting

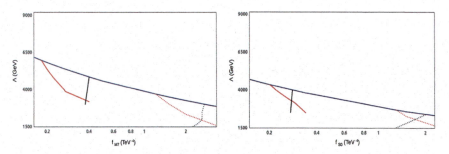

Fig. 122.2 Regions in the Λ versus f space for f_{M7} and f_{S0} operators

from a bin-by-bin comparison of the event yields in a given dim-8 scenario and in the SM. For each dim-8 operator, the kinematic distribution that produced highest χ^2 is considered. The most sensitive kinematic variables are R_{P_T} for f_{S0} and f_{S1} operators, while M_{01} for the remaining operators where

$$R_{P_T} \equiv \frac{p_T^{l1} p_T^{l2}}{p_T^{j1} p_T^{j2}} \tag{122.4}$$

$$M_{01} \equiv \sqrt{(|\vec{p_T^{l1}}| + |\vec{p_T^{l2}}| + |\vec{p_T^{miss}}|)^2 - (p_T^{l1} + p_T^{l2} + p_T^{miss})^2} \tag{122.5}$$

These kinematic distributions tell about the deviation from SM predictions. EFT triangles are made in the Λ versus f space for dim-8 operators and the comparison results are presented for f_{M7} and f_{S0} (Fig. 122.2) at the HE-LHC and HL-LHC case, assuming in each case an integrated luminosity of 3 ab^{-1}. In these triangles, 5σ BSM signal can be observed and the EFT is applicable. The unitarity limit is shown in blue, the lower limits for a 5σ signal significance from (122.3) (red) and the upper limit on 2σ EFT consistency (black). The solid (dotted) lines correspond to $\sqrt{s} = 27$ (14) TeV.

122.3 Results and Conclusions

We have analyzed the physics potential of EFT models defined by the choice of single dim-8 operators in the same-sign WW scattering process in the purely leptonic decay modes. It is observed that increasing the proton beam energy allows exploring much lower values of the Wilson coefficients, as lower limits for a 5σ BSM discovery are being shifted. On the other hand, the upper limit on consistent EFT description shifts likewise by a similar amount. This is due to the fact that by increasing the collision energy more and more events come from the region, where $M_{WW} > \Lambda$ and therefore shrinking the range of Wilson coefficients that satisfy our EFT consistency criterion. Thus, the area of the actual EFT triangle does not get significantly larger for 27 TeV compared to 14 TeV, even when viewed in a log scale and the triangles turned to be rather narrow. This result concludes that the study of BSM effects by means of varying single Wilson coefficients has little physics potential and future data analysis should be rather focussed on simultaneous fits of many operators to the combined data from all VBS processes. We find this conclusion to hold equally regardless of the actual beam energy.

Acknowledgements We would like to acknowledge the funding agency, Department of Science and Technology (DST) and Council of Scientific and Industrial Research (CSIR), New Delhi for providing generous financial support.

References

1. J. Kalinowski, P. Kozów, S. Pokorski, J. Rosiek, M. Szleper, S. Tkaczyk, Same-sign WW scattering at the LHC: can we discover BSM effects before discovering new states? Eur. Phys. J. C **78**(5), 403 (2018)
2. J. Alwall et al., JHEP **07**, 079 (2014). https://doi.org/10.1007/JHEP07(2014)079. arXiv:1405.0301 [hep-ph]
3. T. Sjostrand, S. Mrenna, P. Skands, JHEP **0605**, 026 (2016). https://doi.org/10.1088/1126-6708/2006/05/026. arXiv:hep-ph/0603175
4. K. Arnold et al., Comput. Phys. Commun. **180**, 1661 (2009). https://doi.org/10.1016/j.cpc.2009.03.006. arXiv:0811.4559 [hep-ph]. J. Baglio et al., arXiv:1107.4038 [hep-ph], arXiv:1404.3940 [hep-ph]

Chapter 123
Effect of Non-universal Z' Boson on $B \rightarrow K^* \tau^+ \tau^-$ Decay

P. Maji and S. Sahoo

Abstract Recently, the LHCb collaboration has detected some hints of lepton flavour universality violation while measuring the parameter R_{k^*}, i.e. the ratio of branching fraction of μ-channel to that of e-channel of $B \rightarrow K^* l^+ l^-$ decay mode. This lepton flavour non-universality parameter R_{k^*} has unit value in the standard model (SM) but the experiment shows a clear deviation of ~3σ from its SM value. It is expected that the heavy leptons could provide some clear explanations for various anomalies that are being observed in recent times. In this work, we are interested to study the $B \rightarrow K^* \tau^+ \tau^-$ mode. We predict the branching fraction for $B \rightarrow K^* \tau^+ \tau^-$ decay mode in the SM and in the family non-universal Z' model. We also look towards the ratios of branching fractions of τ-channel to μ- and e-channels, i.e. $R_{K^*}^{\tau\mu}$ and $R_{K^*}^{\tau e}$, respectively, to find out whether any lepton flavour non-universality is present in τ sector.

123.1 Introduction

The standard model (SM) of particle physics is the most promising theory for maximum of the experimental observations. But some of the information like neutrino mass, lepton flavour universality (LFU) violation, etc. could not be extracted from SM itself. So, we need to extend our theory to some new direction towards physics beyond the SM.

Rare B meson decays provide some stringent way to test different descriptions of flavour within SM. Recently, some hints of LFU have been observed at LHC through the decay channel $B \rightarrow K^{(*)} l \bar{l}$ [1–3]. Various decay modes containing other light leptons like μ and e are well-measured at different experiments. But the channels having heavy τ-leptons in its final state are still far from experimental reach. Taking hints of new physics (NP) into account, we might expect LFU violation in

P. Maji (✉) · S. Sahoo
Department of Physics, National Institute of Technology Durgapur, Durgapur 713209, West Bengal, India
e-mail: majipriya@gmail.com

© Springer Nature Singapore Pte Ltd. 2021
P. K. Behera et al. (eds.), *XXIII DAE High Energy Physics Symposium*,
Springer Proceedings in Physics 261,
https://doi.org/10.1007/978-981-33-4408-2_123

the neutral current transitions involving tau leptons, such as, in $b \rightarrow s\tau^+\tau^-$ [4, 5] or $b \rightarrow d\tau^+\tau^-$ [6] processes. But till now the decay modes that consist of the above transitions have not been observed experimentally. So, the possibility for NP remains hidden under poor constraints. In some [7–9], it has been shown that there might be an increment of three orders of magnitude compared to the SM predictions in $b \rightarrow s\tau^+\tau^-$ processes, whereas, only LHCb [10] has given the upper limit for $B_s \rightarrow \tau^+\tau^-$ as $Br(B_s \rightarrow \tau^+\tau^-)$[LHCb] $\leq 6.8 \times 10^{-3}$ which is far from SM prediction of $\mathcal{O}(10^{-7})$. BaBar [11] has given the upper limit for $B \rightarrow K\tau^+\tau^-$ channel as $Br(B \rightarrow K\tau^+\tau^-)$[BaBar] $\leq 2.25 \times 10^{-3}$. The SM prediction for branching ratio of the exclusive channels $B \rightarrow K^*\tau^+\tau^-$, $B \rightarrow \phi\tau^+\tau^-$ has been assessed in [12–14] and that of the inclusive mode $B \rightarrow X_s\tau^+\tau^-$ has been measured in [4, 15]. The authors in [5] have updated the SM values for the branching ratio of these decays as following: $Br(B \rightarrow K\tau^+\tau^-)_{SM} = (1.20\pm0.12) \times 10^{-7}$, $Br(B \rightarrow K^*\tau^+\tau^-)_{SM} = (0.98 \pm 0.10) \times 10^{-7}$, $Br(B \rightarrow \phi\tau^+\tau^-)_{SM} = (0.86 \pm 0.06) \times 10^{-7}$. In [4], the constraints on NP contributions are found to be much loose for these processes if one can correlate the effects in $b \rightarrow s\tau^+\tau^-$ and $b \rightarrow d\tau^+\tau^-$ transitions to avoid the stringent bounds coming from $\Delta\Gamma_s/\Delta\Gamma_d$.

123.2 Theoretical Framework

In the SM neglecting the doubly Cabibo-suppressed contributions, the effective Hamiltonian describing the $b \rightarrow s\tau^+\tau^-$ transitions can be written as [16]

$$\mathcal{H}_{eff} = \frac{G_F}{\sqrt{2}} V_{tb} V_{ts}^* \sum_i C_i(\mu) O_i(\mu). \tag{123.1}$$

The above Hamiltonian leads to the following free quark matrix element for our desired decay mode [17]

$$n_{eff}(b \rightarrow s\tau^+\tau^-) = \frac{G_F}{\sqrt{2}} \frac{\alpha_{em}}{2\pi} V_{tb} V_{ts}^* \left[-2\frac{C_{7\gamma}(m_b)}{q^2} \left((m_b + m_s)(\bar{s}i\sigma_{\mu\nu}q^\nu b) + (m_b - m_s)(\bar{s}i\sigma_{\mu\nu}q^\nu \gamma_5 b) \right)(\bar{\tau}\gamma_\mu \tau) \right.$$
$$\left. + C_{9V}^{eff}(m_b, q^2)(\bar{s}\gamma_\mu(1 - \gamma_5)b)(\bar{\tau}\gamma_\mu \tau) + C_{10A}(m_b)(\bar{s}\gamma_\mu(1 - \gamma_5)b)(\bar{\tau}\gamma_\mu \gamma_5 \tau) \right]. \tag{123.2}$$

Introducing the dimensionless kinematical variables $\hat{s} = q^2/M_B^2$, we have got the expressions for dilepton mass distribution (or differential decay rate) such as [18]

$$\frac{d\Gamma(B \rightarrow K^*\tau^+\tau^-)}{d\hat{s}} = \frac{G_F^2 M_B^5 |V_{ts}^* V_{tb}|^2 \alpha_{em}^2}{1536\pi^5} \sqrt{1 - \frac{4\hat{m}}{\hat{s}}} \lambda^{\frac{1}{2}}$$
$$(1, \hat{s}, \hat{r})[\beta_V \left(1 + \frac{2\hat{m}}{\hat{s}} \right) + 12\hat{m}\delta_V]. \tag{123.3}$$

We have introduced a new lepton flavour non-universality parameter in the form of $R_{K^*}^{\tau l}$ which is defined as

$$R_{K^*}^{\tau l} = \frac{BR(B \to K^* \tau^+ \tau^-)}{BR(B \to K^* l^+ l^-)} \text{(where } l = e, \mu). \tag{123.4}$$

As we know two operators O_9 and O_{10} are mainly responsible for semileptonic transitions, so it is much convenient for the NP model to change the two Wilson coefficients C_9^{eff} and C_{10} corresponding to those semileptonic operators. In the presence of non-universal Z', neglecting $Z - Z'$ mixing and considering the Z' couplings with right-handed quarks to be diagonal, new effective Hamiltonian for FCNC transitions mediated by Z' boson could be written as [19]

$$\mathcal{H}_{eff}^{Z'} = -\frac{4G_F}{\sqrt{2}} V_{tb} V_{ts}^* \left[\Lambda_{sb} C_9^{Z'} O_9 + \Lambda_{sb} C_{10}^{Z'} O_{10} \right], \tag{123.5}$$

where $\Lambda_{sb} = \frac{4\pi e^{-i\phi_{sb}}}{\alpha_{em} V_{tb} V_{ts}^*}$, $C_9^{Z'} = |B_{sb}| S_{ll}^{LR}$, $C_{10}^{Z'} = |B_{sb}| D_{ll}^{LR}$
with $S_{ll}^{LR} = B_{ll}^L + B_{ll}^R$, $D_{ll}^{LR} = B_{ll}^L - B_{ll}^R$.

Here, B_{sb} corresponds to off-diagonal left-handed coupling of Z' with quarks, B_{ll}^L and B_{ll}^R are left- and right-handed couplings for Z' with leptons. ϕ_{sb} is the new weak phase angle. It is very much convenient to use Z' model because in this model, the operator basis remains the same as in the SM; only the modifications are done for C_9 and C_{10} while C_7^{eff} remains unchanged. The new Wilson coefficients C_9 and C_{10} with the total contributions of SM and Z' model are written as

$$C_9^{Total} = C_9^{eff} + C_9^{NP}, C_{10}^{Total} = C_{10} + C_{10}^{NP}, \tag{123.6}$$

where $C_9^{NP} = \Lambda_{sb} C_9^{Z'}$ and $C_{10}^{NP} = \Lambda_{sb} C_{10}^{Z'}$.

The numerical values of the Z' couplings suffer from several constraints that arise due to different exclusive and inclusive B decays that have been found in the [20] and also from other low energy experiments. We have used three scenarios in our calculation. From the available data of $B_s - \bar{B}_s$ mixing, UTfit collaboration has found three different fitting values of new weak phase angle ϕ_{sb} which arise due to the measurement uncertainties. These three are referred to as S_1, S_2 and S_3. The values of input parameters are set by UTfit collaborations [21] and recollected in Table 123.1.

Table 123.1 The numerical values of the Z' parameters [22]

	$|B_{sb}| \times 10^{-3}$	ϕ_{sb} (Degree)	$S_{ll}^{LR} \times 10^{-2}$	$D_{ll}^{LR} \times 10^{-2}$
S_1	1.09 ± 0.22	-72 ± 7	-2.8 ± 3.9	-6.7 ± 2.6
S_2	2.20 ± 0.15	-82 ± 4	1.2 ± 1.4	-2.5 ± 0.9
S_3	4.0 ± 1.5	150 ± 10 (or -150 ± 10)	0.8	-2.6

123.3 Result and Discussion

We have predicted the branching ratio for $B \rightarrow (K, K^*)\tau^+\tau^-$ decay channels in the SM as well as for the two scenarios of Z' model as we discussed before. The distributions for differential branching ratio indicate some interesting phenomena of the corresponding channels. The resonant contributions coming from $c\bar{c}$ loop are highly peaked in narrow regions around their masses. So, outside of these regions, one can neglect their effects. For τ-channels, we observe only ψ' peak within the allowed kinematical region, i.e. $4m_\tau^2 < q^2 < (m_B - m_{K^{(*)}})^2$.

To get the branching ratio for the full kinematic region, we integrate the decay width above the resonance regime to the low recoil endpoint and neglect the region near the high recoil endpoint due to high uncertainty. For each of the modes except S_3, we have seen some enhancement due to the presence of Z' which we summarize in Table 123.2. For S_3, the branching ratio and LFU parameter have been decreased. This might happen due to the different couplings of leptons and quarks with Z' boson. We are expecting the mode $B \rightarrow K^*\tau^+\tau^-$ to be seen experimentally and thus branching ratio should not be decreased in any scenario. For this reason, we might neglect S_3 in this study. From the above table, it is clear that the values of all the parameters are deviated from their SM values. This fact could be a way of investigation of LFU violation between the tau and other lepton families.

Table 123.2 Branching ratio and predicted values for $R_{K^*}^{\tau l}$ for $B \rightarrow K^*\tau^+\tau^-$ decay channel

Parameter	Bin size	Z' model prediction		
		S_1	S_2	S_3
Branching ratio ($\times 10^{-7}$)	14–19 GeV^2	2.63 ± 0.22	2.11 ± 0.33	0.83 ± 0.63
$R_{K^*}^{\tau l}$	14–19 GeV^2	1.02 ± 0.07	0.82 ± 0.16	0.32 ± 0.29

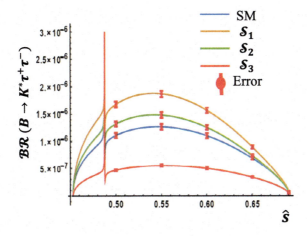

Acknowledgements P. Maji acknowledges the DST, Govt. of India for providing INSPIRE Fellowship through IF160115 for her research work. S. Sahoo would like to thank SERB, DST, Govt. of India for financial support through grant no. EMR/2015/000817.

References

1. R. Aaij et al., Phys. Rev. Lett. **111**, 191801 (2013) (LHCb Collaboration). arXiv:1308.1707
2. R. Aaij et al., Phys. Rev. Lett. **113**, 151601 (2014) (LHCb Collaboration). arXiv:1406.6482
3. R. Aaij et al., J. High Energy Phys. **8**, 055 (2017) (LHCb Collaboration). arXiv:1705.05802
4. C. Bobeth, U. Haisch, Acta Phys. Pol. B **44**, 127 (2013). arXiv:1109.1826
5. B. Capdevila et al., Phys. Rev. Lett. **120**, 181802 (2018)
6. C. Bobeth, U. Haisch, A. Lenz, B. Pecjak, G. Tetlalmatzi-Xolocotzi, J. High Energy Phys. **06**, 040 (2014). arXiv:1404.2531
7. R. Alonso, B. Grinstein, J.M. Camalich, J. High Energy Phys. **10**, 184 (2015). arXiv:1505.05164.
8. A. Crivellin, D. Mueller, T. Ota, J. High Energy Phys. **9**, 040 (2017). arXiv:1703.09226
9. L. Calibbi, A. Crivellin, T. Li, Phys. Rev. D **98**, 115002 (2018). arXiv:1709.00692
10. R. Aaij et al., Phys. Rev. Lett. **118**, 251802 (2017) (LHCb Collaboration). arXiv:1703.02508
11. J.P. Lees et al., Phys. Rev. Lett. **118**, 031802 (2017) (BaBar Collaboration). arXiv:1605.09637
12. J.F. Kamenik, S. Monteil, A. Semkiv, L.V. Silva, Eur. Phys. J. C **77**, 701 (2017). arXiv:1705.11106.
13. J.L. Hewett, Phys. Rev. D **53**, 4964 (1996). arXiv:hep-ph/9506289
14. C. Bouchard et al., Phys. Rev. Lett. **111**, 162002 (2013) [Erratum: Phys. Rev. Lett. **112**, 149902 (2014)], arXiv:1306.0434.
15. D. Guetta, E. Nardi, Phys. Rev. D **58**, 012001 (1998). arXiv:hep-ph/9707371
16. B. Grinstein, M.J. Savage, M.B. Wise, Nucl. Phys. B **319**, 271 (1989)
17. N. Shirkhanghah, Acta Phys. Pol. B **42**, 1219 (2011)
18. D. Melikhov, N. Nikitin, S. Simula, Phys. Rev. D **57**, 6814 (1998)
19. V. Barger et al., Phys. Rev. D **80**, 055008 (2009)
20. P. Langacker, Rev. Mod. Phys. **81**, 1199 (2009)
21. M. Bona et al., PMC Phys. A **3**, 6 (2009) (UTfit Collaborations)
22. I. Ahmed, A. Rehman, Chin. Phys. C **42**, 063103 (2018). arXiv:1703.09627

Chapter 124
Inferring the Covariant Θ-Exact Noncommutative Coupling at Linear Colliders

J. Selvaganapathy and Partha Konar

Abstract The covariant Θ-exact noncommutative standard model (NCSM) provides a novel non-minimal interaction of neutral right-handed fermion and abelian gauge field as well as it satisfies the Very Special Relativity (VSR) Lorentz subgroup symmetry, which opens an avenue to study the top quark pair production at linear colliders. In this work, we consider helicity basis technique further, the realistic electron and positron beam polarization are taken into account to measure the model parameters NC scale Λ and non-minimal coupling κ.

124.1 Introduction

At high energy nearly Planck scale, when gravity becomes strong, the spacetime coordinates become an operator $\hat{x}_\mu$. They no longer commute, satisfying the algebra

$$[\hat{x}_\mu, \hat{x}_\nu] = i\Theta_{\mu\nu} = \frac{ic_{\mu\nu}}{\Lambda^2} \tag{124.1}$$

Here $\Theta_{\mu\nu}$ (of mass dimension -2) is real and antisymmetric tensor. $c_{\mu\nu}$ is the antisymmetric constant and Λ, the noncommutative scale. In the noncommutative space, the ordinary product between fields is replaced by Moyal-Weyl (MW) [1] star ($\star$) product defined by

$$(f \star g)(x) = exp\left(\frac{i}{2}\Theta_{\mu\nu}\partial_{x^\mu}\partial_{y^\nu}\right)f(x)g(y)|_{y=x}. \tag{124.2}$$

J. Selvaganapathy (✉) · P. Konar
Theoretical Physics Group, Physical Research Laboratory, Ahmedabad 380009, India
e-mail: jselva@prl.res.in

P. Konar
e-mail: konar@prl.res.in

© Springer Nature Singapore Pte Ltd. 2021   881
P. K. Behera et al. (eds.), *XXIII DAE High Energy Physics Symposium*,
Springer Proceedings in Physics 261,
https://doi.org/10.1007/978-981-33-4408-2_124

The non-zero commutator of the $U(1)_Y$ abelian gauge field and SM fermion provides the non-minimal interaction under the background antisymmetric tensor field. Such type of background tensor interaction equally acts on the SM fermion field e.g. quarks, leptons or left handed, right handed particle, irrespective of their group representation. In this scenario, although the neutral current interactions get modified, the charged current interaction does not. Thus the NC fields takes hybrid Seiberg-Witten (SW) map in the Θ-exact approach which constructs the Θ-exact non-minimal covariant NCSM [6]. The Θ-exact hybrid SW map for NC fields are as follows,

$$\widehat{\Psi}_L = \Psi_L - \frac{\Theta^{\mu\nu}}{2}\left(g\,A_\mu^a T^a + Y_L\,g_Y B_\mu\right) \bullet \partial_\nu \Psi_L - \Theta^{\mu\nu}\,\kappa\,g_Y\,B_\mu \circledast \partial_\nu \Psi_L + \mathcal{O}(V^2)\Psi_L$$

$$\widehat{l}_R = l_R - \frac{\Theta^{\mu\nu}}{2}(Y_R\,g_Y B_\mu) \bullet \partial_\nu l_R - \Theta^{\mu\nu}\,\kappa\,g_Y\,B_\mu \circledast \partial_\nu l_R + \mathcal{O}(V^2)l_R$$

$$\widehat{v}_R = v_R - \Theta^{\mu\nu}\,\kappa\,g_Y\,B_\mu \circledast \partial_\nu v_R + \mathcal{O}(V^2)v_R \qquad (124.3)$$

The products (used above) are defined as $f \circledast g = f\left(\dfrac{\sin\left(\frac{1}{2}\overleftarrow{\partial}_\mu \Theta^{\mu\nu}\overrightarrow{\partial}_\nu\right)}{\frac{1}{2}\overleftarrow{\partial}_\mu \Theta^{\mu\nu}\overrightarrow{\partial}_\nu}\right) g$

$$f \star g = f\left(e^{\frac{i}{2}\overleftarrow{\partial}_\mu \Theta^{\mu\nu}\overrightarrow{\partial}_\nu}\right) g\,; \quad f \bullet g = f\left(\frac{e^{\frac{i}{2}\overleftarrow{\partial}_\mu \Theta^{\mu\nu}\overrightarrow{\partial}_\nu} - 1}{\frac{i}{2}\overleftarrow{\partial}_\mu \Theta^{\mu\nu}\overrightarrow{\partial}_\nu}\right) g\,; \qquad (124.4)$$

In the (124.4), $\Theta^{\mu\nu}$ is the constant antisymmetric tensor which can have arbitrary structure. But when we impose Cohen-Glashow Very Special Relativity (VSR) [2] on $\Theta^{\mu\nu}$ which provides the relation $\Theta^{0i} = -\Theta^{3i}$ Where $i = 1, 2$. We the structure of $\Theta^{\mu\nu}$ which is given in [7], admits translational symmetry and azimuthal anisotropy by virtue of broken rotational symmetry. In the present work we predominantly focusing the top quark pair production on certain center of mass energy of the electron-positron collision which is future plan of the CLIC [3].

124.2 Helicity Correlation

Considering the top pair production, the measured spin correlation depends on the choice of the spin basis. Thereby we work on helicity basis which admits the center of mass frame in which the top spin axis is defined. The helicity correlation factor defined as [4]

$$C_{t\bar{t}} = \frac{\sigma_{LL} + \sigma_{RR} - \sigma_{LR} - \sigma_{RL}}{\sigma_{LL} + \sigma_{RR} + \sigma_{LR} + \sigma_{RL}} \qquad (124.5)$$

Here σ_{ij} $(i, j = L, R)$ represents the total cross section of the final state top, anti-top helicity. The top quark SM helicity correlation as shown in Fig.124.1, which are $C_{t\bar{t}} = -0.9518(-0.9894)$ at $\sqrt{s} = 1.4(3.0)$ TeV respectively. Here the negative sign

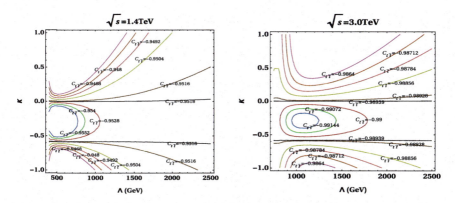

Fig. 124.1 Helicity correlation between the top and anti-top quark

emphasize that the opposite final state helicity cross section are usually dominant over the one with same helicity final state cross section in the $e^- e^+$ colliders. The helicity region $0 > \kappa > -0.596$, the value of $C_{t\bar{t}}$ reduces gradually when the NC scale increases those are greater than the SM helicity correlation. We have notice that the region $\Lambda < 350$ GeV at $\sqrt{s} = 1.4$ TeV and $\Lambda < 800$ GeV at $\sqrt{s} = 3.0$ TeV are excluded because which are arises due to unphysical oscillatory behaviour near lower values of NC scale. The non-minimal coupling κ can be arbitrary in the positive region which can be restricted by statistical significance of the experimental results but in the negative region it can be taken as $\kappa_{\max} = -0.296$.

124.3 Polarized Beam Analysis

In general, the electron and positron beams can have two types of polarizations which are transverse and longitudinal polarization. Since we consider the bunch of massless electrons as a beam at linear colliders, the transverse polarizations are negligible, thus one can define the total cross section with arbitrary longitudinal polarization (P_{e^-}, P_{e^+}) given by [5]

$$\sigma_{P_{e^-} P_{e^+}} = (1 - P_{e^-} P_{e^+})\sigma_0 (1 - P_{eff} A_{LR}) \qquad (124.6)$$

Here the unpolarized total cross section $\sigma_0 = (\sigma_{LR} + \sigma_{RL})/4$, left-right asymmetry $A_{LR} = (\sigma_{LR} - \sigma_{RL})/(\sigma_{LR} + \sigma_{RL})$ and effective polarization $P_{eff} = (P_{e^-} - P_{e^+})/(1 - P_{e^-} P_{e^+})$. In our analysis σ_0 and A_{LR} are function of Λ. We made χ^2 test for polarized beam analysis by keeping the azimuthal anisotropy as an observable shown in Fig. 124.2. The polarization enhances the lower limit on NC scale and non minimal coupling $\kappa_{\max}$ also which are given in the Table 124.1 for certain values of integrated luminosity.

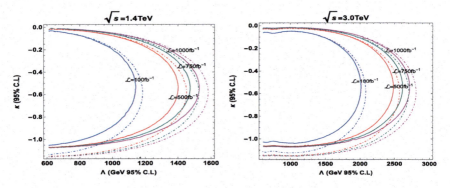

Fig. 124.2 The figure depicts χ^2 statistical test (95% C.L) of polarized the noncommutative signal event arises due to azimuthal anisotropy only considering the $-^{ve}$ region of the κ. The color line corresponds to $P_{e^-} = -80\%$, $P_{e^+} = 30\%$ and dot-dashed color line corresponds to $P_{e^-} = -80\%$, $P_{e^+} = 60\%$ for various integrated luminosity $\int \mathcal{L}\,dt =$ 100 (*blue*), 500 (*red*), 750 (*cyan*) fb^{-1} and 1000 (*megenta*) fb^{-1} at CLIC $\sqrt{s} = 1.4$ TeV (left) and $\sqrt{s} = 3.0$ TeV (right)

Table 124.1 The lower bound on noncommutative scale Λ (95% C.L) and $\kappa_{max} = \{-0.5445, -0.607\}$ obtained by χ^2 analysis when $P_{\{e^-,e^+\}} = \{-0.8, 0.3\}$ and $P_{\{e^-,e^+\}} = \{-0.8, 0.6\}$ for four different integrated luminosity

Integrated luminosity ($\mathcal{L}$)	100 fb^{-1}	500 fb^{-1}	750 fb^{-1}	1000 fb^{-1}
$\kappa_{max} = -0.5445$ and $P_{\{e^-,e^+\}} = \{-0.8, 0.3\}$				
Lower limit on Λ: $\sqrt{s} = 1.4$ TeV	1.131 TeV	1.390 TeV	1.462 TeV	1.517 TeV
Lower limit on Λ: $\sqrt{s} = 3.0$ TeV	2.000 TeV	2.474 TeV	2.606 TeV	2.703 TeV
$\kappa_{max} = -0.607$ and $P_{\{e^-,e^+\}} = \{-0.8, 0.6\}$				
Lower limit on Λ: $\sqrt{s} = 1.4$ TeV	1.172 TeV	1.440 TeV	1.514 TeV	1.570 TeV
Lower limit on Λ: $\sqrt{s} = 3.0$ TeV	2.075 TeV	2.560 TeV	2.70 TeV	2.80 TeV

124.4 Summary and Conclusion

We study the top quark pair production at the TeV energy linear collider in the non-minimal NCSM within the framework of covariant Θ-exact Seiberg-Witten approach [7]. We present the helicity correlation $C_{t\bar{t}}$ of the final state top quark and anti-top quark produced with a certain helicity and found that such a correlation which is constant at $\sqrt{s} = 1.4(3.0)$ TeV i.e. $C_{t\bar{t}}^{SM} = -0.9518$, (-0.9894) respectively, varies with the NC scale Λ for different coupling constant κ in the NCSM. Further, we perform a detailed χ^2 analysis for the polarized electron-polarized beam with

$P_{e^-} = -80\%$ and $P_{e^+} = 60\%$ corresponding to the machine energy $\sqrt{s} = 1.4$ (3.0) TeV for different machine luminosities. The results are given in the following table as well as in [7].

Acknowledgements   The work of SJ and PK is supported by Physical Research Laboratory (PRL), Department of Space, Government of India and also acknowledge the computational support from Vikram-100 HPC at PRL.

References

1. M.R. Douglas, N. Nekrasov, Rev. Mod. Phys. **73**, 977 (2001). N. Seiberg, E. Witten, JHEP **09**, 032 (1999)
2. A.G. Cohen, S.L. Glashow, Phys. Rev. Lett. **97**, 021601 (2006). M.M. Sheikh-Jabbari, A. Tureanu, Phys. Rev. Lett. **101**, 261601 (2008)
3. CLICdp Collaboration, Eur. Phys. J. C **77**, 475 (2017). arXiv:1807.02441 [hep-ex]
4. K. Jiro, N. Takashi, S. Parke, Phys. Rev. D **59**, 014023 (1998), G. Mahlon, S. Parke, Phys. Rev. D **81**, 074024 (2010)
5. H.E. Haber, arXiv:hep-ph/9405376v1. G. Moortgat, et al., Phys. Rep. **460**, 131 (2008)
6. R. Horvat, A. Ilakovac, P. Schupp, J. Trampetic, J. You, Phys. Lett. B **715**, 340 (2012)
7. J. Selvaganapathy, P. Konar, P.K. Das ArXiv:1903.03478v3 [hep-ph]. Accepted in JHEP

Chapter 125
$\mathcal{N} = 2$ Supersymmetric Theory with Lifshitz Scaling

Akhila Mohan, Kallingalthodi Madhu, and V. Sunilkumar

Abstract We construct an $\mathcal{N} = 2$ supersymmetric gauge theory in Lifshitz spacetimes by starting from a $\mathcal{N} = 1$ supersymmetric lagrangian and imposing an additional $\mathbb{Z}_4$ subgroup of R-symmetry.

125.1 Introduction

It has been pointed out that supersymmetry can be realized in Lifshitz scaling spacetime [1–5]. It is important to know how supersymmetry is broken and Lorentz symmetry is restored at low energy, potentially of interest in beyond standard model scenarios. With the aim of studying these aspects we set up an $\mathcal{N} = 2$ supersymmetric theory in a spacetime with Lifshitz type anisotropic scaling and demonstrate its invariance under the two sets of supersymmetric variations and R-symmetry.

125.2 $\mathcal{N} = 2$ Supersymmetry in Lifshitz Case

The $\mathcal{N} = 1$ supersymmetry with Lifshitz scaling has chiral superfield and vector superfield as the irreducible representations. The chiral superfield is of the form

$$\tilde{\Phi} = \phi + i\theta\sigma^\mu\bar{\theta}\tilde{\partial}_\mu\phi + \frac{1}{4}\theta\theta\bar{\theta}\bar{\theta}\Box\phi + \sqrt{2}\theta\psi - \frac{i}{\sqrt{2}}\theta\theta\tilde{\partial}_\mu\psi\sigma^\mu\bar{\theta} + \theta\theta F \quad (125.1)$$

A. Mohan (✉) · K. Madhu
Department of Physics, BITS-Pilani, KK Birla Goa Campus, Zuarinagar, India
e-mail: mohan.akhila90@gmail.com

V. Sunilkumar
Department of Physics, UC College, Aluva, Kerala, India

© Springer Nature Singapore Pte Ltd. 2021
P. K. Behera et al. (eds.), *XXIII DAE High Energy Physics Symposium*,
Springer Proceedings in Physics 261,
https://doi.org/10.1007/978-981-33-4408-2_125

while the vector superfield

$$V = -\theta\sigma^\mu\bar\theta\tilde A_\mu + i\theta\theta\bar\theta\bar\lambda - i\bar\theta\bar\theta\theta\lambda + \frac{1}{2}\theta\theta\bar\theta\bar\theta D \tag{125.2}$$

Here $\tilde A_\mu = [A_0, (c + i\eta\Delta^{\frac{z-1}{2}})A_i]$ $\quad \tilde\Box = \tilde\partial_0^2 - \tilde\partial_i\tilde\partial_i = \partial_0^2 + c^2\Delta + 2c\eta\Delta^{\frac{z+1}{2}} + \eta^2\Delta^z$.
Superspace derivatives take the form

$$D_\alpha = \frac{\partial}{\partial\theta^\alpha} + i\sigma^\mu_{\alpha\dot\alpha}\bar\theta^{\dot\alpha}\tilde\partial_\mu$$

$$\bar D_{\dot\alpha} = -\frac{\partial}{\partial\bar\theta^{\dot\alpha}} - i\theta^\alpha\sigma^\mu_{\alpha\dot\alpha}\tilde\partial_\mu \tag{125.3}$$

Lifshitz scaling is realized as $x^i \to bx^i, t \to b^z t$ for the spatial and time coordinates respectively.

The $\mathcal{N} = 2$ supersymmetry consists of vector multiplet and hypermultiplet. These multiplets are constructed out of $\mathcal{N} = 1$ superfields. Vector multiplet consists of $\mathcal{N} = 1$ chiral superfield $\Phi(\phi, \zeta, F)$ and real vector field $V(\tilde A_\mu, \lambda, D)$ which are in the adjoint representation. Hypermultiplet consists of $\mathcal{N} = 1$ chiral superfield $P(P, \psi, K)$ and anti-chiral field $Q(Q, \xi, L)$ which are expressed in the fundamental representation.

The vector multiplet lagrangian in terms of superfields is given by,

$$\mathcal{L}_V = \int d^4\theta \, tr[\Phi^\dagger e^{-gV}\Phi e^{gV}] + \frac{Im}{64\pi}[\tau\int d^2\theta \, tr W_\alpha W^\alpha + cc.] \tag{125.4}$$

where $\tau = \frac{4\pi i}{g^2} + \frac{\theta}{2\pi}$.

For $z = 4n + 1$, the vector multiplet Lagrangian in terms of component fields

$$\mathcal{L}_V = tr[\phi^\dagger\tilde D_\mu\tilde D^\mu\phi + \sqrt{2}ig[\phi^\dagger, \lambda]\zeta + \sqrt{2}ig\bar\lambda[\bar\zeta, \phi] + gD[\phi, \phi^\dagger]$$

$$+ i\tilde D_\mu\zeta\sigma^\mu\bar\zeta + F^\dagger F + i\tilde D_\mu\lambda\sigma^\mu\bar\lambda + \frac{1}{2}D^2 - \frac{1}{4}\tilde A_{\mu\nu}\tilde A^{\mu\nu}] \tag{125.5}$$

The hypermultiplet lagrangian in terms of the $\mathcal{N} = 1$ superfields is,

$$\mathcal{L}_{hyper} = \int d^4\theta(P^{\dagger i}e^V P_i + Q^i e^{-V} Q_i^\dagger) + (-\frac{i}{\sqrt{2}}\int d^2\theta Q^i \Phi P_i + cc.)$$

$$+ (\int d^2\theta\mu^i_j Q^j P_i + cc.) \tag{125.6}$$

In terms of component fields (for $z = 4n + 1$ case),

$$\mathcal{L}_{hyper} = P^{\dagger i} \tilde{D}_\mu \tilde{D}^\mu P_i + \frac{i}{\sqrt{2}} P^{\dagger i} \lambda \psi_i - \frac{i}{\sqrt{2}} \bar{\psi}^i \bar{\lambda} P_i + P^{\dagger i} D P_i + K^{\dagger i} K_i - i \psi^i \sigma^\mu \tilde{D}_\mu \bar{\psi}_i$$

$$+ Q^i \tilde{D}_\mu \tilde{D}^\mu Q_i^\dagger + \frac{i}{\sqrt{2}} Q^i \bar{\lambda} \bar{\xi}_i - \frac{i}{\sqrt{2}} \xi^i \lambda Q_i^\dagger + L L^\dagger - \frac{1}{2} Q^i D Q_i^\dagger - i \xi \sigma^\mu \tilde{D}_\mu \bar{\xi}_i$$

$$- \frac{i}{\sqrt{2}} [Q^i \phi K_i - Q^i \zeta \psi + Q^i F P_i - \xi^i \phi \psi_i - \xi^i \zeta P_i + L^i \phi P_i]$$

$$+ \frac{i}{\sqrt{2}} [P^{\dagger i} \phi^\dagger L_i^\dagger - P^{\dagger i} \bar{\zeta} \bar{\xi} + P^{\dagger i} F^\dagger Q_i^\dagger - \bar{\psi}^i \phi^\dagger \bar{\xi}_i - \bar{\psi}^i \bar{\zeta} Q_i^\dagger + K^{\dagger i} \phi^\dagger Q_i^\dagger]$$

$$+ \mu_j^i [Q^j K_i - \xi^j \psi_i + L^j P_i] + \mu_j^{\dagger i} [P^{\dagger j} L_i^\dagger - \bar{\psi}^j \bar{\xi}_i + K^{\dagger j} Q_i^\dagger] \tag{125.7}$$

The on-shell conditions for auxiliary fields are

$$F^a = \frac{i}{\sqrt{2}} Q^{\dagger i} t^a P_i \; ; \quad F^{\dagger a} = -\frac{i}{\sqrt{2}} P^{\dagger i} t^a Q_i$$

$$D^a = -g[\phi, \phi^\dagger]^a + \frac{1}{2} Q^i t^a Q^\dagger - \frac{1}{2} P^{\dagger i} t^a P_i$$

$$K_i = -\frac{i}{\sqrt{2}} \phi^{\dagger a} t^a Q_i^\dagger - \mu_i^{\dagger j} Q_j^\dagger \; ; \quad K^{\dagger i} = \frac{i}{\sqrt{2}} Q^i \phi^a t^a - \mu_j^i Q^j$$

$$L^i = \frac{-i}{\sqrt{2}} P^{\dagger i} \phi^{\dagger a} t^a - \mu_j^{\dagger i} P^{\dagger j} \; ; \quad L_i^\dagger = \frac{i}{\sqrt{2}} \phi^a t^a P_i - \mu_i^j P_i \tag{125.8}$$

The on-shell Lagrangian of $\mathcal{N} = 2$ gauge theory with matter fields can be written as $\mathcal{L} = \mathcal{L}_V + \mathcal{L}_{hyper}$ after integrating out the auxiliary fields. The couplings in the Lagrangian are so arranged that under the following $\mathbb{Z}_4$ transformations of certain component fields, Lagrangian in is invariant.

$$\lambda \to \zeta; \quad \bar{\lambda} \to \bar{\zeta}; \quad \zeta \to -\lambda, \quad \bar{\zeta} \to -\bar{\lambda}$$

$$P \to Q^\dagger; \quad P^\dagger \to Q; \quad Q^\dagger \to -P; \quad Q \to -P^\dagger \tag{125.9}$$

This $\mathbb{Z}_4$ is a subgroup of the R-symmetry.

125.3 Conclusion

We have constructed the $\mathcal{N} = 2$ supersymmetric gauge theory with a Lifshitz scaling. We observe that this could be achieved by imposing additional symmetries to a $\mathcal{N} = 1$ theory similar to the Lorentz invariant case by introducing a $\mathbb{Z}_4$ symmetry involving certain components of the chiral and vector superfields. The present Lagrangian now provides us with the ability to study phenomenologically interesting questions like that of partial breaking of supersymmetry and restoration of Lorentz invariance.

Acknowledgements Akhila Mohan is grateful to the Department of Physics, UC College Aluva for the hospitality during her visit. We would like to acknowledge the generous support of the people of India for research in the basic sciences.

References

1. S. Chapman, Y. Oz, A. Raviv-Moshe, On supersymmetric Lifshitz field theories. JHEP **10**, 162 (2015)
2. M. Gomes, J.R. Nascimento, A.Y. Petrov, A.J. da Silva, Horava-Lifshitz-like extensions of supersymmetric theories. Phys. Rev. D **90**(12), 125022 (2014)
3. P.R.S. Gomes, M. Gomes, On higher spatial derivative field theories. Phys. Rev. D **85**, 085018 (2012)
4. M. Pospelov, C. Tamarit, Lifshitz-sector mediated SUSY breaking. JHEP **01**, 048 (2014)
5. M. Gomes, J. Queiruga, A.J. da Silva, Lorentz breaking supersymmetry and Horava-Lifshitz-like models. Phys. Rev. D **92**(2), 025050 (2015)

Chapter 126
Majorana Dark Matter in a Minimal S_3 Symmetric Model

Subhasmita Mishra and Anjan K. Giri

Abstract The Standard Model is extended with S_3 and Z_2 to explain the neutrino phenomenology under the framework of type I seesaw mechanism. Along with the Standard Model particles we consider three right-handed Majorana neutrinos and two extra Higgs doublets with one of them being inert. We explain the Majorana Dark matter of TeV scale mass in this model with a correct relic density which demands the Yukawa coupling to be large enough satisfying both neutrino mass and Dark Matter relic density.

126.1 Introduction

Despite all its success, Standard Model (SM) is considered not to be a complete theory to explain all experimental evidences. Of them, we focus on neutrino mass and dark matter in this model. Experiments based on solar and atmospheric neutrino oscillation have already confirmed the non-zero masses of neutrinos and the improvement in the sensitivity of oscillation data have achieved many more information regarding the neutrinos, like the non-zero mixing angles and furthermore, the possible existence of fourth generation sterile neutrino. But still the mass hierarchy and the absolute mass scales of neutrinos remain open questions to be solved. Like the visible sector, there are a lot of experimental evidences, hints towards the existence of Dark Matter, occupying one-fourth of the universe energy budget. Through many decades theorist are trying to connect the visible sector with the Dark sector to find the existence of new physics that will open an window towards the new era of HEP. This model is an extension of SM with simplest permutation symmetry S_3 and Z_2 along-with three right-handed neutrinos and two scalar doublets to explain the neutrino phenomenology and DM relic density simultaneously [1].

S. Mishra (✉) · A. K. Giri
IIT Hyderabad, Hyderabad, India
e-mail: subhasmita.mishra92@gmail.com

A. K. Giri
e-mail: giria@iith.ac.in

© Springer Nature Singapore Pte Ltd. 2021
P. K. Behera et al. (eds.), *XXIII DAE High Energy Physics Symposium*,
Springer Proceedings in Physics 261,
https://doi.org/10.1007/978-981-33-4408-2_126

126.2   The Model

In order to discuss the Neutrino phenomenology, we start with the particle contents and their corresponding quantum numbers with respect to the SM and S_3 symmetry in Table 126.1. The $SM \times S_3 \times Z_2$ invariant interaction Lagrangian in charged and neutral lepton sector is given by [1–3]

$$
\begin{aligned}
\mathscr{L}_{D+M} = &-y_1 \left[\overline{L_e} \tilde{H}_2 N_{1R} + \overline{L_\mu} \tilde{H}_1 N_{1R} + \overline{L_e} \tilde{H}_1 N_{2R} - \overline{L_\mu} \tilde{H}_2 N_{2R} \right] \\
&-y_2 \left[\overline{L_\tau} \ \tilde{H}_1 N_{1R} + \overline{L_\tau} \ \tilde{H}_2 N_{2R} \right] - y5[\overline{L_\tau} \tilde{H}_3 N_{3R}] \\
&-y_{l2} \left[\overline{L_e} H_2 E_{1R} + \overline{L_\mu} H_1 E_{1R} + \overline{L_e} H_1 E_{2R} - \overline{L_\mu} H_2 E_{2R} \right] \\
&-y_{l4} \left[\overline{L_\tau} \ H_1 E_{1R} + \overline{L_\tau} \ H_2 E_{2R} \right] - y_{l5}[\overline{L_e} H_1 E_{3R} + \overline{L_\mu} H_2 E_{3R}] \\
&-\frac{1}{2} \overline{N}^c_{iR} M_1 N_{iR} - \frac{1}{2} \overline{N}^c_{3R} M_{DM} N_{3R} + \text{h.c.} \tag{126.1}
\end{aligned}
$$

This model includes 3 scalar doublets H_1, H_2 and H_3, Of them first two retain non-zero VEVs, $\langle H_1 \rangle = v_1$ and $\langle H_2 \rangle = v_2$, after electroweak symmetry breaking and the third one doesn't gain any VEV due to the remnant Z_2 symmetry. As H_1 and H_2 couple to both quark and lepton sectors, giving rise to tree level FCNCs, which can be suppressed by making the other Higgs, except the SM like Higgs, to be heavy. We explicitly add a soft breaking term in the potential $V_{SB} = \mu_{SB}{}^2 \left(H_1{}^\dagger H_2 + H_2{}^\dagger H_1 \right)$. Hence from the mixing between these two Higgs fields, i.e. $H_L = H_1 \cos \beta + H_2 \sin \beta$ and $H_H = -H_1 \sin \beta + H_2 \cos \beta$ with $\tan \beta = \frac{v_2}{v_1}$, and $v^2 = v_1{}^2 + v_2{}^2 = 246\,\text{GeV}$ is the VEV of SM like Higgs, we can have the SM like Higgs (H_L) with mass 125 GeV and other Higgs (charged and neutral) with mass of order TeV in mass eigenstate. The TeV scale mass is achieved by finetuning the symmetry breaking parameter. In order to discuss the Neutrino masses and mixing we start with the mass matrix, where the full mass matrix for neutral leptons is given by in the basis $\tilde{N} = (v_L^c, N_R)^T$ with the expression for type I seesaw mass is as the following

$$
\mathscr{M} = \begin{pmatrix} 0 & m_D \\ m_D^T & M_R \end{pmatrix}, \quad m_\nu = M_D M_R^{-1} (M_D)^T \tag{126.2}
$$

Looking at the Lagrangian, as described above in (126.1), we can write the flavor structure of Dirac mass matrix for neutral and charged leptons as following.

Table 126.1 Particle contents and quantum numbers under SM and S_3

Particles	(L_e, L_μ)	L_τ	(E_{1R}, E_{2R})	E_{3R}	(N_{1R}, N_{2R})	N_{3R}	(H_1, H_2)	H_3
G_{SM}	$(1, 2, -1)$	$(1, 2, -1)$	$(1, 1, -2)$	$(1, 1, -2)$	$(1, 1, 0)$	$(1, 1, 0)$	$(0, 2, 1)$	$(0, 2, 1)$
S_3	2	1	2	1	2	1	2	1
Z_2	+	+	+	+	+	−	+	−

$$M_D = \begin{pmatrix} y_1 v \sin\beta & y_1 v \cos\beta & 0 \\ y_1 v \cos\beta & -y_1 v \sin\beta & 0 \\ y_2 v \cos\beta & y_2 v \sin\beta & 0 \end{pmatrix} , \quad M_l = \begin{pmatrix} y_{l2} v \sin\beta & y_{l2} v \cos\beta & y_{l5} v \cos\beta \\ y_{l2} v \cos\beta & -y_{l2} v \sin\beta & y_{l5} v \sin\beta \\ y_{l4} v \cos\beta & y_{l4} v \sin\beta & 0 \end{pmatrix}$$

$$(126.3)$$

Where we get $m_{\nu_1} = \frac{2 y_1^2 v^2}{M_l}$, $m_{\nu_2} = \frac{2(y_1^2 + y_2^2) v^2}{M_l}$ and $m_{\nu_3} = 0$ after diagonalization. Hence the third generation neutrino can get mass through radiative correction as described in the Ma model [5].

$$m_{\nu_{33}} = \frac{y_5^2 M_{DM}}{16\pi^2} \left[\frac{m_R^2}{m_R^2 - M_{DM}^2} ln \left(\frac{m_R^2}{M_{DM}^2} \right) - \frac{m_I^2}{m_I^2 - M_3^2} ln \left(\frac{m_I^2}{M_{DM}^2} \right) \right]$$

$$(126.4)$$

where m_R and m_I are the masses of CP even (h_3) and CP odd (a_3) components of H_3, M_{DM} is the mass of lightest Majorana neutrino N_3. The simple block diagonal form of mass matrix is easy to diagonalize, where the mixing matrices for both neutrino and charged leptons are given as following [1, 4]

$$U_\nu = \begin{pmatrix} \cos\theta & \sin\theta e^{-i\phi_\nu} & 0 \\ -\sin\theta e^{i\phi_\nu} & \cos\theta & 0 \\ 0 & 0 & 1 \end{pmatrix} , \quad U_{el} = \begin{pmatrix} \frac{x}{\sqrt{2(1-x^2)}} & \frac{1}{\sqrt{2(1+x^2)}} & \frac{1}{\sqrt{2(1+\sqrt{z})}} \\ \frac{-x}{\sqrt{2(1-x^2)}} & \frac{-1}{\sqrt{2(1+x^2)}} & \frac{1}{\sqrt{2(1+\sqrt{z})}} \\ \frac{\sqrt{1-2x^2}}{\sqrt{1-x^2}} & \frac{x}{\sqrt{1+x^2}} & \frac{\sqrt{z}}{\sqrt{(1+\sqrt{z})}} \end{pmatrix} .$$

Where $x = \frac{m_e}{m_\mu}$ and $z = \frac{m_e^2 m_\mu^2}{m_\tau^4}$, $U_\nu^T M_\nu U_\nu = Diag(m_{\nu_1} e^{i\phi_1}, m_{\nu_2} e^{i\phi_2}, m_{\nu_3})$ and $U_{el} M_l M_l^\dagger U_{el}^\dagger = Diag(m_e^2, m_\mu^2, m_\tau^2)$. From the above parameterization, we find $\sin\theta_{13} \approx 0.004$, $\tan\theta_{12} \approx 0.56$ for $\theta = \frac{\pi}{6}$ and $\tan\theta_{23} \approx 0.707$ and the sum of neutrino mass can be obtained within the cosmological bound (< 0.12 eV) by fixing y_1 and y_2 of order 10^{-2} and $M_1 \approx O(10^{12})$ GeV.

126.3 Dark Matter

This model allows the lightest right handed neutrino to be a dark matter candidate stabilized by the Z_2 symmetry. Here the dark matter is allowed to have only scalar and lepton mediated t-channel annihilation process and hence doesn't have any s-channel resonance, which contribute to the relic density, is shown in Fig. 126.1.

The relic abundance expression for the dark matter is given by

$$\Omega h^2 = \frac{2.14 \times 10^9 \text{ GeV}^{-1}}{M_{pl} \sqrt{g_\star} J_f}, \quad J_f = \int_{x_f}^\infty \frac{<\sigma v>}{x^2} dx, \quad \text{and} \quad M_{pl} = 1.22 \times 10^{19} GeV.$$

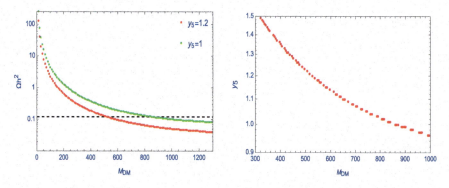

Fig. 126.1 Feynman diagrams of the channel contribute to DM relic

Fig. 126.2 First figure shows the variation of dark matter mass with the relic density for different Yukawa couplings and 2nd figure shows the parameter space for Yukawa coupling and dark matter mass as per the observed relic density ($\Omega h^2 = 0.12$)

And the freezeout parameter is given by [6] $x_f = \ln \left(\frac{0.038 g M_{pl} M_{DM} \langle \sigma v \rangle}{(g_* x_f)^{0.5}} \right)$.

We can find from Fig. 126.2, the DM mass of order TeV with a large Yukawa coupling of order 1, satisfy the correct DM relic density and the Yukawa coupling decreases with the DM mass within the allowed experimental value of Ωh^2.

126.4 Conclusion

In this model we discussed the neutrino masses and mixing by generating the mass for third generation neutrino by radiative correction, satisfying the 3σ experimental values. Dark matter phenomenology is explained in this framework with correct relic density with Majorana mass of the order of about 1.2 Tev, which demands the corresponding Yukawa coupling to be large for which we assumed the quartic coupling, responsible for the mass splitting of the real and imaginary part of H_3 to be very small to satisfy the neutrino mass. The model is discussed in detail with additional information in [7].

Acknowledgements Subhasmita would like to acknowledge DST Inspire for its financial support.

References

1. T. Araki, J. Kubo, E.A. Paschos, Eur. Phys. J. C **45**, 465 (2006). https://doi.org/10.1140/epjc/s2005-02434-3 [hep-ph/0502164]
2. C. Patrignani et al., Particle data group. Chin. Phys. C **40**, 100001 (2016)
3. S.F. King, J. Phys. Conf. Ser. **631**(1), 012005 (2015). https://doi.org/10.1088/1742-6596/631/1/012005
4. A.A. Cruz, M. Mondragn, arXiv:1701.07929 [hep-ph]
5. E. Ma, Phys. Rev. D **73**, 077301 (2006). https://doi.org/10.1103/PhysRevD.73.077301 [hep-ph/0601225]
6. D. Borah, D. Nanda, N. Narendra, N. Sahu, arXiv:1810.12920 [hep-ph]
7. S. Mishra, A. Giri (to be submitted)

Chapter 127
Magnetic Affect in the Rotation of Boson Star

Bharti Jarwal and S. Somorendro Singh

Abstract Boson star made of bosonic particles are hypothetical objects of early universe. We calculated the possible presence of these objects by calculating the rotation of it in our earlier paper [1]. Now we look forward for strong confirmation in their rotation considering magnetic field. This may indicate the possible parameter for its existence in the universe as the content of the universe have large number of charge particles. So introducing the magnetic field in it, the rotation of boson star is slightly increased from the rotation of boson star without magnetic field. The slight increment in rotation predicts the presence of this hypothetical objects ….

127.1 Introduction

Magnetic fields play an important role in the life history of astrophysical objects especially of compact relativistic stars which posses surface magnetic fields of 10^{12} G. Magnetic fields of magnetars [2, 3] can reach up to $10^{15} \sim 10^{16}$ G and in the deep interior of compact stars, the magnetic field strength may be estimated up to 10^{18} G. The strength of compact star's magnetic field is one of the main quantities determining their observability, for example as pulsars through magneto-dipolar radiation. Therefore it is extremely important to study the effect of the different phenomena on evolution and behaviour of stellar interior and exterior magnetic fields.

The Astrophysical compact stars are providing us the opportunity to study the strongly interacting dense nuclear matter under the extreme condition in their interior, which has not yet reproduced in the laboratory environment. Theoretically, it is discussed that the composition of compact stars is ranging from the mixture of hadrons, leptons to various phases of superconducting quark matter under beta equilibrium.

B. Jarwal (✉) · S. S. Singh
Department of Physics and Astrophysics, University of Delhi, Delhi 110007, India
e-mail: jarwalbharti22@gmail.com

© Springer Nature Singapore Pte Ltd. 2021
P. K. Behera et al. (eds.), *XXIII DAE High Energy Physics Symposium*,
Springer Proceedings in Physics 261,
https://doi.org/10.1007/978-981-33-4408-2_127

In this context, Boson stars are hypothetical and astronomical objects formed of self gravitating field of bosonic particles. These objects were first studied by considering second quantized scalar field satisfying Klein-Gordon equation [4, 5] in non interacting systems. In this equation the semi-classical Einstein equation had been solved and gravitationally bound state of field was obtained. Quarks, antiquarks, gluons, fermions, pions etc. are also supposed to be present in the boson stars. In this work, the magnetic field due to the presence of kaon in the boson star is considered and the affect of this internal field on rotation of star is investigated.

127.1.1 Theoretical Treatment

BS is considered to be a complex scalar field which is influenced by couple to gravity [4]. The action principle of such system is discussed in detail under weak field approximation. The Lagrangian density of the system is modified by addition of magnetic term in it.

The modified Lagrangian density is given by

$$£ = \frac{R}{16\pi G} + g^{\mu\nu}\partial_\mu \Phi^* \partial_\nu \Phi - V(|\Phi|^2) \tag{127.1}$$

where, $V(|\Phi|^2) = M^2 \Phi^* \Phi + \frac{q\vec{L}.\vec{B}}{2v\mu}$, q is charge of the particle, μ is mass of particle and v is volume density.

The metric $g^{\mu\nu}$ is expanded as follows $g^{\mu\nu} = \eta^{\mu\nu} + h^{\mu\nu}$ with $|h^{\mu\nu} \ll 1|$ and $\eta^{\mu\nu} = diag(1, -1, -r^2, -r^2 sin\theta^2)$. The equation of motion for Φ is given by

$$\Box\Phi + M^2\Phi = 0. \tag{127.2}$$

The solution with stationary rotation for Φ will depend on t and ϕ only. Thus the solution is given as $\Phi(\mathbf{r}, t) = \phi(r, \theta)e^{i\omega t}e^{im\varphi}$. The second equation of motion is given as

$$\Box h_{\mu\nu} = -16\pi G S_{\mu\nu} \tag{127.3}$$

where $S_{\mu\nu} = T_{\mu\nu} - \frac{1}{2}\eta_{\mu\nu}T$

$$T_{\mu\nu} = \partial_\mu\Phi^*\partial_\nu\Phi + \partial_\nu\Phi^*\partial_\mu\Phi - \eta_{\mu\nu}[\eta^{\alpha\beta}\partial_\alpha\Phi^*\partial_\beta\Phi - M^2\Phi^*\Phi + \frac{q\vec{L}.\vec{B}}{2v\mu}]$$

.

By using weak field approximation of general relativity the further equations are simplified. Further rescaling of parameter solutions for stationary state and first excited state are obtained.

Fig. 127.1 The variation of ground state radial wavefunction, R with distance r with and without magnetic field

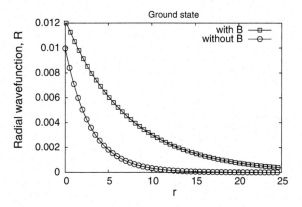

Fig. 127.2 The variation of first excited state (l = 1 and m = 0, 1) radial wavefunction, R with distance r with and without magnetic field

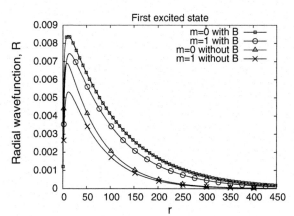

127.1.2 Results and Conclusion

The radial wave function for ground and first excited states are obtained to observe the effect of magnetic field on the rotation of boson stars. The results are shown in Figs. 127.1 and 127.2. There is change in energy which is observed for stationary and first excited state. For the calculation we use the mass of the BS used is around 30 GeV. Along with this the enhancement in the radial wave function R is observed for both states after including the magnetic term in Lagrangian.

It is clear from the graph that wave function for ground state attains maximum value around 0.012 after considering the internal magnetic effects. The radial wave function of excited state shows the similar behavior as seen in stationary state. In the case of first excited state the wave function R has larger amplitude when the value of m = 0 in comparison to m = 1. Figure 127.2 also shows that the radial wave function is attaining peak value of 0.0084 for m equals to 0 and for m = 1 case, its that maximum value at 0.0072 after considering the internal magnetic affects. It also indicates the minimization of the oscillating amplitude with the higher excited state. The amplitude of ground state wave function is higher than that of first excited state. Along with this the oscillation of boson star is more favored at the lower value of angular contribution for same excited state. This shifting of the peaks shows the rotating behavior of the BS after considering the effect of magnetic field.

References

1. B. Jarwal, S.S. Singh, Oscillation of boson star in Newtonian approximation. Modn. Phys. Lett. A **32**, 1750037 (2017). https://doi.org/10.1142/S0217732317500377. B. Jarwal, S.S. Singh, Rotating boson star under weak gravity potential, in *XXII DAE High Energy Physics Symposium, Springer Proceedings in Physics*, vol. 203, p. 789 (2018). https://doi.org/10.1007/978-3-319-73171-1_190. B. Jarwal, S.S. Singh, Boson star under cornell potential. Proc. DAE Symp. Nucl. Phys. **62** (2017). www.sympnp.org/proceedings
2. R.C. Duncan, C. Thompson, Formation of very strongly magnetized neutron stars—implications for gamma-ray bursts. Astrophys. J. **392**, L9 (1992). https://doi.org/10.1086/186413
3. C. Thompson, R.C. Duncan, The soft gamma repeaters as very strongly magnetized neutron stars—I. Radiative mechanism for outbursts. Mon. Not. R. Astron. Soc. **275**, 255 (1995). Bibliographic code: 1995MNRAS.275..255T
4. R. Ruffini, S. Bonazolla, Systems of self-gravitating particles in general relativity and the concept of an equation of state. Phys. Rev. **187**, 1767 (1969). https://doi.org/10.1103/PhysRev.187.1767
5. Kaup, Klein-Gordon Geon. Phys. Rev. **172**, 1333 (1968). https://doi.org/10.1103/PhysRev.172.1331

Chapter 128
Higher Derivative Theory for Curvature Term Coupling with Scalar Field

Pawan Joshi and Sukanta Panda

Abstract Higher order derivative theories, generally suffer from instabilities, known as Ostrogradsky instabilities. This issue can be resolved by removing any existing degeneracy present in such theories. We consider a model involving at most second-order derivatives of scalar field non-minimally coupled to curvature terms. Here we perform (3+1) decomposition of Lagrangian to separate second-order time derivative from the rest. This is useful to check the degeneracy hidden in the Lagrangian will help us to find conditions under which Ostrogradsky instability does not appear. In our case, we find no such non-trivial conditions which can stop the appearance of the Ostrogradsky ghost.

128.1 Introduction

Observations suggest that our current universe is in an accelerating phase [1] and explanation of this acceleration can be provided by dark energy. The dark energy is generally modelled by modifying gravity part of Einstein Hilbert action [2]. One of the brands of this modified gravity model is scalar-tensor theories [3, 4]. When we consider scalar-tensor theories with higher order derivative terms. If higher derivative Lagrangian is non-degenerate there exist a ghost-like instability known as Ostrogradsky instability [5] in which Hamiltonian contains such terms which are linear in momentum. In degenerate theory, the higher derivative is present in the Lagrangian but they cancel such a way that does not appear in the equation of motion [6]. In 1974 Horndeski proposed a general action for scalar field that contain the higher derivative term in the Lagrangian but the equation of motion is of second order [7]. Horndeski Lagrangian takes a form

P. Joshi (✉) · S. Panda
Department of Physics, Indian Institute of Science Education and Research Bhopal,
Bhopal, India
e-mail: pawanjoshi697@iiserb.ac.in

S. Panda
e-mail: sukanta@iiserb.ac.in

© Springer Nature Singapore Pte Ltd. 2021
P. K. Behera et al. (eds.), *XXIII DAE High Energy Physics Symposium*,
Springer Proceedings in Physics 261,
https://doi.org/10.1007/978-981-33-4408-2_128

$$L = L_2^H + L_3^H + L_4^H + L_5^H, \tag{128.1}$$

where

$$L_2^H = G_2(\phi, X), \quad L_3^H = G_3(\phi, X)\Box\phi, \quad L_4^H = G_4(\phi, X)R - 2G_{4,X}(\phi, X)(\Box\phi^2 - \phi_{\mu\nu}\phi^{\mu\nu}), \tag{128.2}$$

$$L_5^H = G_5(\phi, X)G_{\mu\nu}\phi^{\mu\nu} + \frac{1}{3}G_5(\phi, X)R - (\Box\phi^3 - 3\Box\phi\phi_{\mu\nu}\phi^{\mu\nu} + 2\phi_{\mu\nu}\phi_\rho^\nu\phi^{\mu\rho}), \tag{128.3}$$

and $\phi_\mu = \nabla_\mu\phi$, $\phi_{\mu\nu} = \nabla_\mu\nabla_\nu\phi$, $X = \nabla_\mu\phi\nabla^\mu\phi$, R, $R_{\mu\nu}$, $G_{\mu\nu}$ is Ricci scalar, Ricci Tensor and Einstein tensor, respectively. Here we notice that in L_2^H is some combination scalar field and its first derivative and L_3^H additionally contain the second derivative of the scalar field, L_4^H, L_5^H contain curvature term and first and second derivative of ϕ. In the case, we have constructed a Lagrangian containing higher derivatives of a scalar field with non-minimal coupling to curvature. Our motivation is to find a degeneracy condition to get rid of Ostrogradsky instability in our higher derivative Lagrangian.

128.2 Possible Terms for $\nabla_\mu\nabla_\nu\phi\nabla_\rho\nabla_\sigma\phi$

Consider an action of the form,

$$S = \int d^4x\sqrt{-g}\tilde{C}^{\mu\nu,\rho\sigma}\nabla_\mu\nabla_\nu\phi\nabla_\rho\nabla_\sigma\phi, \tag{128.4}$$

where $\tilde{C}^{\mu\nu,\rho\sigma}$ contain metric tensor and curvature term and it is only possible term for non-minimal coupling of second derivative of scalar field with the curvature term. Its simple form is

$$\tilde{C}^{\mu\nu,\rho\sigma} = (D_1 g^{\mu\rho}g^{\nu\sigma} + D_2 g^{\mu\sigma}g^{\nu\rho})R + (D_3 g^{\eta\rho}g^{\mu\nu}g^{\beta\sigma} + D_4 g^{\mu\eta}g^{\beta\rho}g^{\nu\sigma})R_{\eta\beta}$$
$$+ (D_5 g^{\mu\eta}g^{\nu\beta}g^{\gamma\rho}g^{\delta\sigma} + D_6 g^{\mu\eta}g^{\sigma\beta}g^{\gamma\rho}g^{\delta\nu})R_{\eta\beta\gamma\delta}.$$

Now we rewrite the action replace first derivative of scalar field by a new field A_μ, i.e. $\nabla_\mu\phi = A_\mu$. So the new form of action is given as

$$S = \int d^4x\sqrt{-g}[\tilde{C}^{\mu\nu,\rho\sigma}\nabla_\mu A_\nu\nabla_\rho A_\sigma + \lambda^\mu(\nabla_\mu\phi - A_\mu)], \tag{128.5}$$

this action can be written as

$$S = \int d^4x\sqrt{-g}[\mathcal{L}_1 + \mathcal{L}_2 + \mathcal{L}_3 + \mathcal{L}_4 + \mathcal{L}_5 + \mathcal{L}_6] + \lambda^\mu(\nabla_\mu\phi - A_\mu)], \tag{128.6}$$

where

$$\mathcal{L}_1 = D_1 R g^{\mu\rho} g^{\nu\sigma} \nabla_\mu A_\nu \nabla_\rho A_\sigma, \tag{128.7}$$

$$\mathcal{L}_2 = D_2 R g^{\mu\nu} g^{\rho\sigma} \nabla_\mu A_\nu \nabla_\rho A_\sigma, \tag{128.8}$$

$$\mathcal{L}_3 = D_3 g^{\eta\mu} g^{\beta\rho} g^{\nu\sigma} R_{\eta\beta} \nabla_\mu A_\nu \nabla_\rho A_\sigma, \tag{128.9}$$

$$\mathcal{L}_4 = D_4 g^{\eta\rho} g^{\mu\nu} g^{\beta\sigma} R_{\eta\beta} \nabla_\mu A_\nu \nabla_\rho A_\sigma, \tag{128.10}$$

$$\mathcal{L}_5 = D_5 g^{\mu\eta} g^{\nu\beta} g^{\gamma\rho} g^{\delta\sigma} R_{\eta\beta\gamma\delta} \nabla_\mu A_\nu \nabla_\rho A_\sigma, \tag{128.11}$$

$$\mathcal{L}_6 = D_6 g^{\mu\eta} g^{\nu\beta} g^{\gamma\rho} g^{\delta\sigma} R_{\eta\beta\gamma\delta} \nabla_\mu A_\nu \nabla_\rho A_\sigma, \tag{128.12}$$

and λ^μ are Lagrange multiplier. Here it is noted that due to the symmetry property of Riemann tensor L_5 term vanishes.

128.3   Representation of Action in (3+1) Decomposition

In this section, we want to find degeneracy condition on Lagrangian, using a (3+1) decomposition. Our (3+1) convention and notation similar to [8, 9] in (3+1) decomposition $\nabla_a A_b$ is

$$\nabla_a A_b = \mathcal{D}_a \mathcal{A}_b - A_* K_{ab} + n_a (K_{bc} \mathcal{A}^c - \mathcal{D}_b A_*)$$
$$+ n_b (K_{ac} \mathcal{A}^c - \mathcal{D}_a A_*) + \frac{1}{N} n_a n_b (\dot{A}_* - N^c \mathcal{D}_c A_* - N \mathcal{A}_c a^c). \tag{128.13}$$

Now we introduced $X_{ab} = (\mathcal{D}_a \mathcal{A}_b - A_* K_{ab})$, $Y_b = (K_{bc} \mathcal{A}^c - \mathcal{D}_b A_*)$ and $Z = (N^c \mathcal{D}_c A_* + N \mathcal{A}_c a^c)$ for simplifying the calculation so (128.13) becomes

$$\nabla_a A_b = \frac{1}{N} n_a n_b (\dot{A}_* - Z) - X_{ab} - n_a Y_b - n_b Y_a, \tag{128.14}$$

where $\mathcal{D}_a$ spatial derivative associated with spatial metric h_{ab}. $\mathcal{A}_a$ and A_* spatial and normal projection of 4-vector A_a given as $\mathcal{A}_a = h_a^b A_b$ and $A_* = n^a A_a$, N^a is shift vector, N is lapse function, $a_b = n^c \nabla_c n_b$ is the acceleration vector and K_{ab} is Extrinsic curvature tensor related to first derivative of metric. In (3+1) formalism R, $R_{\mu\nu}$, $R_{\mu\nu\rho\sigma}$ are

$$R = \mathcal{R} + K^2 - 3K_{ab}K^{ab} + 2h^{ab}L_n K_{ab} - 2\mathcal{D}_b a^b - 2a_b a^b, \tag{128.15}$$

$$\perp R_{ab} = \mathcal{R}_{ab} + K_{ab}K - K_{as}K_b^s + L_n K_{ab} - \mathcal{D}_a a_b - a_a a_b, \tag{128.16}$$

$$\perp R_{bn} = \mathcal{D}_s K_b^s - \mathcal{D}_b K, \tag{128.17}$$

$$\perp R_{nn} = K_{st}K^{st} - h^{st}L_n K_{st} + \mathcal{D}_s a^s + a_s a^s, \tag{128.18}$$

$$\perp R_{abcd} = \mathcal{R}_{abcd} + K_{ac}K_{ad} - K_{ad}K_{bc}, \tag{128.19}$$

$$\perp R_{abcn} = \mathcal{D}_a K_{bc} - \mathcal{D}_b K_{ac}, \tag{128.20}$$

$$\perp R_{abnn} = K_{au} K_b^u - L_n K_{ab} + \mathcal{D}_a a_b - a_a a_b, \tag{128.21}$$

where $L_n K_{ab}$ is Lie derivative of Extrinsic curvature tensor and related to second-order derivative of metric. $\perp R_{ab}$, $\perp R_{bn}$, $\perp R_{nn}$ is spatial, one normal and two normal projection of Ricci tensor and $\perp R_{abcd}$, $\perp R_{abcn}$, $\perp R_{abnn}$ is spatial, one normal and two normal projection of Riemann tensor known as Gauss, Codazzi, and Ricci relations, respectively.

By using these relation we decompose the action (128.6) in (3+1) formalism and separate out second-order derivative of metric. Next we derive conditions that no second-order derivative of metric appear in the action. Here we are not analysing Ostrogradsky instability arising from the higher derivative of scalar field.

128.4 Condition for No Second-Order Derivative of Metric

In this section, after substituting (128.13–128.20) in (128.6) and keeping the terms that are second derivative of metric.

$$\mathcal{L} = h^{ab} L_n K_{ab} \left(\frac{\dot{A}_*^2}{N^2} - \frac{2\dot{A}_* Z}{N^2} + \frac{Z^2}{N^2} \right) (2D_1 + 2D_2 + D_3 + D_4) + h^{ab} L_n K_{ab} Y_c Y^c (-4D_1 - D_3) + h^{ab} L_n K_{ab}$$

$$\frac{(\dot{A}_* - Z) X}{N} (4D_2 + D_4) + 2 h^{ab} L_n K_{ab} (D_1 X^{cd} X_{cd} + D_2 X^2) + L_n K_{ab} Y^a Y^b (-D_3 - 2D_6)$$

$$+ L_n K_{ab} \frac{(\dot{A}_* - Z) X^{ab}}{N} (D_4 - 2D_6) + L_n K_{ab} (D_3 X_d^a X^{bd} + D_4 X X^{ab}) + \text{other terms.} \tag{128.22}$$

To be free from terms containing second derivative of metric in the Lagrangian, we require the coefficient of $h^{ab} L_n K_{ab}$ and $L_n K_{ab}$ to vanish. This amounts to a trivial condition $D_1 = D_2 = D_3 = D_4 = D_5 = D_6 = 0$. There is no non-trivial condition exists in this case.

128.5 Unitary Gauge

Here we check the possibility to get rid of second derivative of metric in unitary gauge, this gauge gives the condition $\phi(x, t) = \phi_0(t)$. In this case constant time hypersurfaces coincide with uniform scalar field hypersurfaces. When we apply unitary gauge ($\mathcal{A}^a = 0$), demand that the values of X_{ab}, Y_b and Z become, $X_{ab} = -A_* K_{ab}$, $Y_b = -D_b A_*$ and $Z = 0$, respectively, after substituting this result into (128.22), we get

$$\mathcal{L} = h^{ab} L_n K_{ab} \frac{\dot{A}_*^2}{N^2} (2D_1 + 2D_2 + D_3 + D_4) + \frac{A_*}{N} \dot{A}_* K_c^c (4D_2 + D_4) + 2 h^{ab} L_n K_{ab} A_*^2 (D_1 K_c^d K_d^c$$

$$+ D_2 A_*^2 K_c^c K_d^d) + D_4 L_n K_{ab} \frac{A_*}{N} \dot{A}_* K^{ab} (D_4 + 2D_6) + L_n K_{ab} (D_3 A_*^2 K_c^a K^{bc} + D_4 A_*^2 K_c^c K^{ab}) + \text{other terms.} \tag{128.23}$$

In this case also, we have find condition on the condition $D_1 = D_2 = D_3 = D_4 = D_5 = D_6 = 0$. There is no non-trivial condition exists in this case.

128.6   Conclusion

In this paper, we work with higher derivative model where both second derivative of metric and scalar field arise in the Lagrangian. Then using (3+1) decomposition, we have shown that no non-trivial conditions can be found under which all the terms containing second-order derivative of metric disappear from the Lagrangian.

Acknowledgements This work was partially funded by DST (Govt. of India), Grant No. SERB/ PHY/2017041. Calculations were performed using xAct packages of Mathematica.

References

1. A.G. Riess et al., (Supernova Search Team), Astron. J. **116**, 1009 (1998). arXiv:astro-ph/9805201 [astro-ph]
2. S. Nojiri, S.D. Odintsov, Theoretical physics: current mathematical topics in gravitation and cosmology, in *Proceedings, 42nd Karpacz Winter School*, Ladek, Poland, 6–11 February 2006, eConf C0602061, 06 (2006) [Int. J. Geom. Meth. Mod. Phys. **4**, 115 (2007)], arXiv:hep-th/0601213 [hep-th]
3. J. Gleyzes, D. Langlois, F. Vernizzi, Int. J. Mod. Phys. D **23**, 1443010 (2015). arXiv:1411.3712 [hep-th]
4. B. Ratra, P.J.E. Peebles, Phys. Rev. D **37**, 3406 (1988)
5. M. Ostrogradsky, Mem. Acad. St. Petersbourg **6**, 385 (1850)
6. A. Nicolis, R. Rattazzi, E. Trincherini, Phys. Rev. D **79**, 064036 (2009). arXiv:0811.2197 [hep-th]
7. G.W. Horndeski, Int. J. Theo. Phys. **10**, 363 (1974)
8. T.W. Baumgarte, S.L. Shapiro, *Numerical Relativity: Solving Einstein's Equations on the Computer* (Cambridge University Press, 2010)
9. D. Langlois, K. Noui, J. Cosmo. Astropart. Phys. 034 (2016)

Chapter 129
Status of σ_8 Tension in Different Cosmological Models

Priyank Parashari, Sampurn Anand, Prakrut Chaubal, Gaetano Lambiase, Subhendra Mohanty, Arindam Mazumdar, and Ashish Narang

Abstract ΛCDM model provides the most plausible theoretical framework for our Universe. However, there are some drawbacks within this model, one of them is the σ_8 tension between CMB and LSS observations. We study two models, namely, Hu-Sawicki Model (HS) and Chavallier-Polarski-Linder (CPL) parametrization of dynamical dark energy (DDE). We found that σ_8 tension increases in HS model whereas it is somewhat alleviated in DDE model. Recently, it has also been shown that viscosity in dark matter can resolve this tension. Modified cosmological models change the matter power spectrum which also depends upon the neutrino mass. As massive neutrinos suppress the matter power spectrum on the small length scales, bounds on neutrino mass also get modified in these models.

129.1 Introduction

The ΛCDM cosmology is the most plausible theoretical framework which is invoked to explain Cosmic Microwave Background (CMB) temperature anisotropy and Large-Scale Structure (LSS) observations. However, there are some tension between these two observations, within ΛCDM paradigm. One of the problem is the value of σ_8, the r.m.s fluctuation of perturbation $8h^{-1}$ Mpc scale, inferred from the CMB and LSS experiments are not in agreement with each other [1, 2]. Many generalisations of the ΛCDM model were proposed to resolve this tension. Recently, it has been shown that viscous cold dark matter can resolve this tension [3]. In this article, we analyse different cosmological models in order to resolve this tension.

P. Parashari (✉) · S. Anand · P. Chaubal · S. Mohanty · A. Mazumdar · A. Narang
Physical Research Laboratory, Ahmedabad 380009, India
e-mail: priyank.du94@gmail.com

P. Parashari · A. Narang
Indian Institute of Technology Gandhinagar, Gandhinagar 382355, India

G. Lambiase
Dipartimento di Fisica "E.R. Caianiello", Universitá di Salerno, 84084 Fisciano (SA), Italy

© Springer Nature Singapore Pte Ltd. 2021
P. K. Behera et al. (eds.), *XXIII DAE High Energy Physics Symposium*,
Springer Proceedings in Physics 261,
https://doi.org/10.1007/978-981-33-4408-2_129

We have analysed the Hu-Sawicki (HS) model [4] of $f(R)$ cosmology, Chavallier-Polarski-Linder (CPL) [5, 6] parametrization of Dynamical Dark Energy (DDE) and viscous ΛCDM model. We find that tension in σ_8 between Planck CMB and LSS observations worsens in the HS model compared to the ΛCDM model, whereas it is somewhat alleviated in DDE model as compared to ΛCDM model. Neutrino oscillation experiments have established the fact that neutrinos have mass. Since massive neutrinos play a crucial role in the background evolution as well as formation of structures in the universe, cosmological observations can also constraint neutrino mass. We, therefore, do the analysis with massive neutrinos and obtain the bound on the neutrino mass allowed in these models. The details of these work can be found in [7, 8]. This article is structured as follows: we first started by briefly explaining the different cosmological models. We then present our results in the next section.

129.2 Different Cosmological Models

In this section, we will discuss three different cosmological models, namely, HS model, DDE model, and viscous ΛCDM model.

129.2.1 Dynamical Dark Energy Model

The current measurements of cosmic expansion indicate that the present Universe is dominated by dark energy (DE). The most common dark energy candidate is cosmological constant Λ representing a constant energy density occupying the space homogeneously. However, a constant Λ makes the near coincidence of Ω_Λ and Ω_m in the present epoch hard to explain naturally. One of the approach to explain this is the DDE. We use the Chavallier-Polarski-Linder (CPL) [5, 6] parametrization of DDE, in which

$$w_{DE}(z) = w_0 + w_a \frac{z}{z+1}, \qquad (129.1)$$

where w_0 and w_a are the CPL parameters. This parametrization describes a non phantom field when $w_a + w_0 \geq -1$ and $w_0 \geq -1$.

129.2.2 F(R) Theory: Hu-Sawicki Model

We consider the Hu-Sawicki model, which explains DE while evading the stringent tests from solar system observations. In HS model

$$f(R) = R - 2\Lambda - \frac{f_{R_0}}{n} \frac{R_0^{n+1}}{R^n}, \qquad (129.2)$$

where $R \geq R_0$ and $f_R = \partial f / \partial R$. We use the following parametrization which was introduced in [9] to include the effect of modified gravity.

$$k^2 \Psi = -4\pi G a^2 \mu(k, a)\rho\delta \quad \text{and} \quad \frac{\Phi}{\Psi} = \gamma\,(k, a), \quad\quad (129.3)$$

where $\mu(k, a)$ and $\gamma(k, a)$ are two scale and time dependent functions, which are given as [10]

$$\mu(k, a) = A^2(\phi)(1 + \epsilon(k, a)), \quad \text{and} \quad \gamma\,(k, a) = \frac{1 - \epsilon(k, a)}{1 + \epsilon(k, a)}, \quad (129.4)$$

where

$$\epsilon(k, a) = \frac{2\beta^2(a)}{1 + m^2(a)\frac{a^2}{k^2}}. \quad\quad (129.5)$$

Coupling function $\beta(a)$ is constant for all the $f(R)$ models and is equal to $\frac{1}{\sqrt{6}}$, whereas mass function $m(a)$ is a model dependent quantity, which is given as

$$m(a) = H_0 \sqrt{\frac{4\Omega_\Lambda + \Omega_m}{(n + 1)|f_{R_0}|}} \left(\frac{4\Omega_\Lambda + \Omega_m a^{-3}}{4\Omega_\Lambda + \Omega_m} \right)^{(n+2)/2}. \quad (129.6)$$

Modification in the evolution of Ψ and Φ in turn modifies the evolution of matter perturbation to as

$$\delta'' + \mathcal{H}\delta' - \frac{3}{2}\Omega_m \mathcal{H}^2 \mu(k, a)\delta = 0. \quad\quad (129.7)$$

129.2.3   Viscous Cold Dark Matter

We write the stress-energy tensor for non-ideal CDM fluid as

$$T^{\mu\nu} = \rho u^\mu u^\nu + (p + p_b)\Delta^{\mu\nu} - 2\eta \left[\frac{1}{2}\left(\Delta^{\mu\alpha}\nabla_\alpha u^\nu + \Delta^{\nu\alpha}\nabla_\alpha u^\mu \right) - \frac{1}{3}\Delta^{\mu\nu}\left(\nabla_\alpha u^\alpha \right) \right]$$

$$(129.8)$$

where η is the coefficient of shear viscosity. We introduce perturbation and derive the evolution equations for background and perturbed quantities. Using $T^{\mu\nu}_{;\mu} = 0$ for the background quantities, we get the continuity equation of each species in the following form

$$\dot{\rho}_i + 3\mathcal{H}\,(\rho_i + p_i) = 0. \quad\quad (129.9)$$

The perturbed part of continuity equation for the cold dark matter provides the density and velocity perturbation equations in viscous ΛCDM model [3, 8].

$$\dot{\delta} = -\left(1 - \frac{\tilde{\zeta}\, a}{\Omega_{\rm cdm}\, \tilde{\mathcal{H}}}\right)(\theta - 3\dot{\phi}) + \left(\frac{\tilde{\zeta}\, a}{\Omega_{\rm cdm}\, \tilde{\mathcal{H}}}\right)\theta - \left(\frac{3\,\mathcal{H}\,\tilde{\zeta}\, a}{\Omega_{\rm cdm}}\right)\delta \quad (129.10)$$

and

$$\dot{\theta} = -\mathcal{H}\theta + k^2\psi - \frac{k^2\, a\, \theta}{3\,\mathcal{H}\,(\Omega_{\rm cdm}\, \tilde{\mathcal{H}} - \tilde{\zeta}\, a)}\left(\tilde{\zeta} + \frac{4\tilde{\eta}}{3}\right) - 6\mathcal{H}\theta\left(1 - \frac{\Omega_{\rm cdm}}{4}\right)\left(\frac{\tilde{\zeta}\, a}{\Omega_{\rm cdm}\, \tilde{\mathcal{H}}}\right),$$
$$(129.11)$$

where $\tilde{\eta} = \frac{8\pi G \eta}{\mathcal{H}_0}$ and $\tilde{\zeta} = \frac{8\pi G \zeta}{\mathcal{H}_0}$ are the dimensionless parameters constructed from the viscosity coefficients.

129.3   Result and Discussion

In this paper we analyse ΛCDM, HS, and DDE model using Planck CMB observations and LSS observations. We have studied the effect of viscosity, massive neutrino, HS and DDE model parameters on the matter power spectrum and found that the effect of viscous CDM and massive neutrinos have some similarities as both suppress $P(k)$ on small scales. On the other hand, for HS model the power increases slightly on small length scales, however, for DDE, $P(k)$ gets suppress at all length scales, though its effect is dominant on small length scales. We then perform MCMC analysis for all these models with the Planck and LSS datasets and found that there is tension between the values of σ_8 inferred from both observations within ΛCDM cosmology. This tension seems to be worsen in HS model, however, is somewhat alleviated in DDE model(see Fig. 129.1). As it was shown previously, we also show that this tension can be resolved if we include viscosity in the dark matter fluid (see Fig. 129.2). We then performed MCMC analysis within viscous cosmological frame-

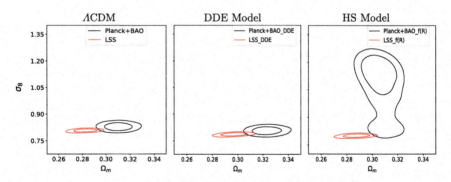

Fig. 129.1 It is shown that the σ_8 discrepancy worsens in the HS model whereas in DDE model the discrepancy is somewhat relieved

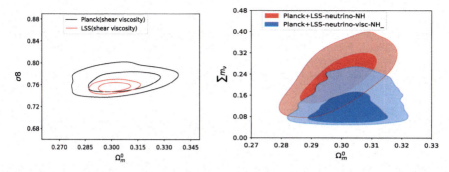

Fig. 129.2 There is no mismatch between σ_8 values obtained from CMB and LSS observations in viscous ΛCDM model(left side). Bounds on neutrino mass are more stringent in the viscous cosmological framework (right side)

work with neutrino mass as free parameter and we found that we can indeed constrain neutrino mass in the more stringent manner (see Fig. 129.2). This is expected result as both viscosity and neutrino affect the power spectrum in the similar fashion.

The tensions between the results of LSS and Planck CMB observation are well studied in the literature. These tensions were believed to be the signature of some unknown, interesting, and exotic physics. In conclusion we see that σ_8 measurement from CMB and LSS experiments can be used as a probe of different cosmological models. Future observations of CMB and LSS may shrink the parameter space for $\sigma_8 - \Omega_m$ and then help in selecting the correct $f(R)$ and DDE theory. We acknowledge the computation facility, 100TFLOP HPC Cluster, Vikram-100, at Physical Research Laboratory, Ahmedabad, India.

References

1. R.A. Battye, T. Charnock, A. Moss, Tension between the power spectrum of density perturbations measured on large and small scales. Phys. Rev. D **91**(10), 103508 (2015)
2. N. MacCrann, J. Zuntz, S. Bridle, B. Jain, M.R. Becker, Cosmic discordance: are Planck CMB and CFHTLenS weak lensing measurements out of tune? Mon. Not. Roy. Astron. Soc. **451**(3), 2877–2888 (2015)
3. S. Anand, P. Chaubal, A. Mazumdar, S. Mohanty, Cosmic viscosity as a remedy for tension between PLANCK and LSS data. JCAP **1711**(11), 005 (2017)
4. H. Wayne, I. Sawicki, Models of f(R) Cosmic acceleration that Evade solar-system tests. Phys. Rev. D **76**, 064004 (2007)
5. M. Chevallier, D. Polarski, Accelerating universes with scaling dark matter. Int. J. Mod. Phys. D **10**, 213–224 (2001)
6. E.V. Linder, Exploring the expansion history of the universe. Phys. Rev. Lett. **90**, 091301 (2003)
7. G. Lambiase, S. Mohanty, A. Narang, P. Parashari, Testing dark energy models in the light of σ_8 tension. Eur. Phys. J. C **79**(2), 141 (2019)

8. S. Anand, P. Chaubal, A. Mazumdar, S. Mohanty, P. Parashari, Bounds on neutrino mass in viscous cosmology. JCAP **1805**(05), 031 (2018)
9. G.-B. Zhao, L. Pogosian, A. Silvestri, J. Zylberberg, Searching for modified growth patterns with tomographic surveys. Phys. Rev. D **79**, 083513 (2009)
10. P. Brax, P. Valageas, Impact on the power spectrum of Screening in modified gravity scenarios. Phys. Rev. D **88**(2), 023527 (2013)

Chapter 130
Comparative Study of Bulk and Surface Pressure of Charged AdS Black Hole

K. V. Rajani and Deepak Vaid

Abstract One of the most interesting areas of research is the study of thermo-dynamic properties of black holes. In the extended phase, cosmological constant has been identified with the thermodynamic pressure of the black hole and phase transitions similar to Van der waals fluid is observed. On other side, fluid gravity correspondence which gives the connection between fluid on a d dimensional boundary and gravity theory in d + 1-dimensional bulk relates Einstein's equations in the bulk with Navier-Stokes equations on the boundary. Projecting Einsteins equation onto a black hole horizon gives Damours Navier-Stokes equation and fluid pressure is found to be $\frac{\kappa}{8\pi}$. In our present work, we are comparing the thermodynamic quantities in the bulk and boundary of charged AdS black holes.

130.1 Introduction

The existence of black hole was first predicted by Einsteins theory of general relativity. Classically black holes never emit anything. There exists a singularity at the horizon. The horizon behaves like a boundary, which obstruct all the information to flow out from the black hole. They have zero temperature. By considering the quantum effect, Bekenstein claim [1] the existence of non zero entropy of the black hole in 1973, and hawking [2, 3] showed that black hole has temperature proportional to its surface gravity and entropy proportional to area of its event horizon.

K. V. Rajani (✉) · D. Vaid
Department of Physics, National Institute of Technology Karnataka (NITK), Surathkal, Mangalore 575025, India
e-mail: rajanikv10@gmail.com

D. Vaid
e-mail: dvaid79@gmail.com

© Springer Nature Singapore Pte Ltd. 2021
P. K. Behera et al. (eds.), *XXIII DAE High Energy Physics Symposium*,
Springer Proceedings in Physics 261,
https://doi.org/10.1007/978-981-33-4408-2_130

$$T = \frac{\kappa}{2\pi} \tag{130.1}$$

$$S = \frac{A}{4}, \tag{130.2}$$

where κ is the surface gravity and A is the area of the event horizon.

130.1.1   Thermodynamics of Black Holes

Classical mechanics of black holes analogous to thermodynamics. Bardeen et al. [3] formulated the black hole thermodynamic laws. For a rotating charged black hole with mass M, angular momentum J and charge Q, the first law of black hole thermodynamics states that

$$dM = TdS + \Omega dJ + \Phi dQ. \tag{130.3}$$

Corresponding Smarr-Gibbs-Duhem relation

$$M = 2(TS + \Omega J) + \phi Q. \tag{130.4}$$

There is no pressure and volume term in this relation. Teitelboim and Brown [4, 5] proposed that Λ itself be a dynamical variable. Then the first law of black hole thermodynamics become

$$dM = TdS + VdP + \Omega dJ + \Phi dQ. \tag{130.5}$$

When we included the cosmological constant term in the Einsteins equation, we will get the notion of pressure. It is associated with the negative cosmological constant Λ

Pressure

$$P = -\frac{\Lambda}{8\pi} \tag{130.6}$$

When $\Lambda < 0$ we will get a positive pressure in the space time, because, for a black hole with AdS radius ℓ

$$\Lambda = -\frac{(d-1)(d-2)}{2\ell^2}, \tag{130.7}$$

where d is the dimension of the spacetime.

130.1.2 Damour-Navier-Stoke Equation on the Black Hole Boundary

Fluid gravity correspondence will give the connection between fluid on a d dimensional boundary and gravity theory in $d + 1$-dimensional bulk. Mathematically, it relates Einstein's equations in $d + 1$ dimension with Navier-Stokes equation in d dimension [6–9]. An example of such a boundary is the horizon of a black hole. Projecting Einsteins equation onto a black hole horizon will give Damours Navier-Stokes equation

$$\left(\partial_0 + v^B \partial_B\right)\left(\frac{-\omega_A}{8\pi}\right) = 2\frac{1}{16\pi}\partial_B \sigma_A^B - \partial_A\left(\frac{\kappa}{8\pi}\right) - \frac{1}{16\pi}\partial_A\theta - T_{mA}l^m. \quad (130.8)$$

Comparing this equation with the Navier-Stokes equation of fluid dynamics, one can obtain the expression for pressure on the boundary fluid of the black hole. We can see that which is proportional to surface gravity κ.

$$P_{bou} = \frac{\kappa}{8\pi}. \quad (130.9)$$

130.2 Thermodynamics of Charged AdS Black Hole in the Bulk and the Boundary

Anti-de Sitter space is the maximally symmetric solution of Einsteins equation with constant negative curvature. The AdS black hole metric is

$$dS^2 = -f(r)dt^2 + \frac{1}{f(r)}dr^2 + r^2 d\Omega^2 \quad (130.10)$$

$f(r) = 1 - \frac{2M}{r} + \frac{Q^2}{r^2} + \frac{r^2}{\ell^2}$, Q is the black hole charge and $d\Omega^2$ is the line element of the two sphere. Black hole event horizon at $f(r_+) = 0$. Mass of the black hole is

$$M = \frac{r_+}{2}\left(1 + \frac{Q^2}{r_+^2} + \frac{r_+^2}{\ell^2}\right). \quad (130.11)$$

Temperature $T = \frac{\kappa}{2\pi} = \frac{f'(r_+)}{4\pi}$

$$T = \frac{M}{2\pi r_+^2} - \frac{Q^2}{2\pi r_+^3} + \frac{r_+}{2\pi\ell^2}. \quad (130.12)$$

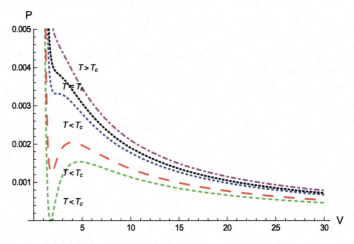

Fig. 130.1 Charged AdS black hole P-v diagram in the bulk

Substituting the value of M

$$T = \frac{1}{4\pi r_+} + \frac{3r_+}{4\pi \ell^2} - \frac{Q^2}{4\pi r_+^3}. \tag{130.13}$$

For charged SAdS black hole the cosmological constant is $-\frac{3}{\ell^2}$. The thermodynamic pressure is

$$P_{bulk} = -\frac{\Lambda}{8\pi} = \frac{3}{8\pi \ell^2} = \frac{6M}{8\pi r_+^3} - \frac{3Q^2}{8\pi r_+^4} - \frac{3}{8\pi r_+^2}. \tag{130.14}$$

In terms of temperature

$$P_{bulk} = \frac{T}{2r_+} - \frac{1}{8\pi r_+^2} + \frac{Q^2}{8\pi r_+^4} = \frac{T}{v} - \frac{1}{2\pi v^2} + \frac{2Q^2}{\pi v^4}, \tag{130.15}$$

where $v = 2r_+$ is the specific volume of the black hole. From Fig. 130.1, we can calculate the critical values.

$$v_c = 2\sqrt{6}Q \quad T_c = \frac{\sqrt{6}}{18\pi Q} \quad P_c = \frac{1}{96\pi Q^2} \tag{130.16}$$

and the ratio

$$\frac{P_c v_c}{T_c} = \frac{3}{8} \tag{130.17}$$

one can see that the ratio is exactly matches with vander Waals system.

According to NS equation

$$P_{bou} = \frac{\kappa}{8\pi} \tag{130.18}$$

$$P_{bou} = \frac{M}{8\pi r_+^2} - \frac{Q^2}{8\pi r_+^3} + \frac{r_+}{8\pi \ell^2} = \frac{1}{16\pi r_+} - \frac{Q^2}{16\pi r_+^3} + \frac{3r_+}{16\pi \ell^2}. \tag{130.19}$$

From this we can see that

$$P_{bou} = \frac{T}{4} \tag{130.20}$$

130.3   Conclusion

In this work we studied the thermodynamics of charged AdS black holes. We calculated the pressure for both the extended bulk phase space, where the cosmological constant plays the role of pressure, and also for the surface phase space. We found that even though the temperature is constant, pressure is different in both bulk and boundary of the black holes.

References

1. J.D. Bekenstein, Phys. Rev. D **7**, 2333 (1973). Phys. Rev. D **9**, 3293 (1974)
2. S.W. Hawking, Commun. Math. Phys. **43**, 199 (1975)
3. J.M. Bardeen, B. Carter, S.W. Hawking, Commun. Math. Phys. **31**, 161 (1973)
4. C. Teitelboim, The cosmological constant as a thermodynamic black hole parameter. Phys. Lett. B **158**(4), 293 (1985)
5. J. Brown, C. Teitelboim, Neutralization of the cosmological constant by membrane creation. Nuclear Phys. B **297**(4), 787 (1988)
6. T. Damour, Surface effects in Black-Hole physics, in *Marcel Grossmann Meeting: General Relativity*, ed. R. Ruffini (1982)
7. T. Padmanabhan, Entropy density of spacetime and the Navier-Stokes fluid dynamics of null surfaces. Phys. Rev. D—Part. Fields Gravit. Cosmol. **83**(4) (2011)
8. V.E. Hubeny, The fluid/gravity correspondence: a new perspective on the membrane paradigm. Class. Quant. Grav. **28**(11), 114007+ (2011)
9. J. Skakala, S. Shankaranarayanan, Black hole thermodynamics as seen through a microscopic model of a relativistic Bose gas, p. 8. arXiv:1406.2477 (2014)

Chapter 131
Dark Energy Perturbation and Power Spectrum in Slotheon Model

Upala Mukhopadhyay, Debasish Majumdar, and Debabrata Adak

Abstract We explore the perturbationsn in Dark Energy by taking the Slotheon field model. The Slotheon field model follows from the extra-dimensional DGP (Dvali, Gabadadze, and Porrati) model and based on the Galileon transformation in curved space time. In this model a scalar field π in DGP theory is subjected to a shift symmetry in such a way that the final theory is invariant under this shift symmetry in curved space time. We consider that the accelerated expansion and Dark Energy are driven by the scalar field π with a potential $V(\pi)$. Then we explore the Dark Energy perturbations within the framework of this Slotheon model. Using these perturbations we compute the matter power spectrum.

131.1 Introduction

In this work we consider a scalar field model inspired by a model in theories of extra dimensions as an alternative way to explain the late time acceleration of the universe. At the limit when $M_{pl} \to \infty$ and $r_c \to \infty$ where M_{pl} denotes reduced Planck mass and r_c is the cross over scale for transition from 4d to 5d, the Dvali Gabadadze Poratti (DGP) [1] model (an extra-dimensional model) in Minkowski spacetime can be described by a scalar field [2] which obey the Galileon shift symmetry $\pi \to \pi + a + b_\mu x^\mu$. A suitable scalar field π which respects this symmetry when extended

U. Mukhopadhyay (✉) · D. Majumdar
Astroparticle Physics and Cosmology Division, Saha Institute of Nuclear Physics, HBNI,
1/AF Bidhannagar, Kolkata 700064, India
e-mail: upala.mukhopadhyay@saha.ac.in

D. Majumdar
e-mail: debasish.majumdar@saha.ac.in

D. Adak
Department of Physics, Government General Degree College, Singur Hooghly,
West Bengal 712409, India
e-mail: debabrata.adak.sinp@gmail.com

© Springer Nature Singapore Pte Ltd. 2021   919
P. K. Behera et al. (eds.), *XXIII DAE High Energy Physics Symposium*,
Springer Proceedings in Physics 261,
https://doi.org/10.1007/978-981-33-4408-2_131

to curved spacetime is termed as Slotheon field (π) [3]. The field π moves slower in this theory than the quintessence theory and for this reason π is called Slotheon.

In the present work we obtain density parameters ($\Omega = \frac{\rho}{\rho_c}$) and equation of state (ω) of Dark Energy considering Slotheon field Dark Energy model. We also calculate general relativistic perturbations for this model and observe evolutions of density fluctuations and gravitational potential. Finally the matter power spectra for Slotheon field are computed.

131.2 Background Evolutions

The action of the Slotheon field is given as [4]

$$S = \int d^4x \sqrt{-g} \left[\frac{1}{2} \left(M_{\rm pl}^2 R - \left(g^{\mu\nu} - \frac{G^{\mu\nu}}{M^2} \right) \pi_{;\mu} \pi_{;\nu} \right) - V(\pi) \right] + S_m, \quad (131.1)$$

where R, $g_{\mu\nu}$ and $G^{\mu\nu}$ are respectively Ricci scalar, metric and Einstein tensor. In the above M is an energy scale and S_m is the action corresponding to the matter field. Here $\pi_{;\mu}$ denotes the covariant derivative of π and $V(\pi)$ denotes the scalar field potential. It can be noted that without the term $\frac{G^{\mu\nu}}{M^2}\pi_{;\mu}\pi_{;\nu}$, the action of (131.1) is same as the action of standard quintessence field [5]. From the action (131.1) energy momentum tensor for Slotheon field is obtained as

$$
\begin{aligned}
T_{\mu\nu}^{(\pi)} = {} & \pi_{;\mu}\pi_{;\nu} - \frac{1}{2}g_{\mu\nu}(\nabla\pi)^2 - g_{\mu\nu}V(\pi) \\
& + \frac{1}{M^2}\left(\frac{1}{2}\pi_{;\mu}\pi_{;\nu}R - 2\pi_{;\alpha}\pi_{(;\mu}R_\nu^\alpha) + \frac{1}{2}\pi_{;\alpha}\pi^{;\alpha}G_{\mu\nu} \right. \\
& - \pi^{;\alpha}\pi^{;\beta}R_{\mu\alpha\nu\beta} - \pi_{;\alpha\mu}\pi_{;\nu}^\alpha \\
& \left. + \pi_{;\mu\nu}\pi_{;\alpha}^\alpha + \frac{1}{2}g_{\mu\nu}\left(\pi_{;\alpha\beta}\pi^{;\alpha\beta} - (\pi_{;\alpha}^\alpha)^2 + 2\pi_{;\alpha}\pi_{;\beta}R^{\alpha\beta} \right) \right).
\end{aligned}
\quad (131.2)
$$

In the above, $R_{\mu\nu\alpha\beta}$ denotes Riemann curvature tensor. For flat Friedmann Robertson Walker (FRW) universe the Einstein equations and equation of motion for π are obtained as follows

$$3M_{\rm pl}^2 H^2 = \rho_m + \frac{\dot\pi^2}{2} + \frac{9H^2\dot\pi^2}{2M^2} + V(\pi) \quad (131.3)$$

$$M_{\rm pl}^2(2\dot H + 3H^2) = -\frac{\dot\pi^2}{2} + V(\pi) + (2\dot H + 3H^2)\frac{\dot\pi^2}{2M^2} + \frac{2H\dot\pi\ddot\pi}{M^2}, \quad (131.4)$$

$$0 = \ddot\pi + 3H\dot\pi + \frac{3H^2}{M^2}\left(\ddot\pi + 3H\dot\pi + \frac{2\dot H\dot\pi}{H} \right) + V_\pi. \quad (131.5)$$

In the above, derivative of $V(\pi)$ w.r.t. π is denoted by V_π. For present analysis we consider exponential form of $V(\pi)$, given by $V(\pi) = V_0 \exp\left(-\dfrac{\lambda \pi}{M_{\text{pl}}}\right)$.

131.3 General Relativistic Perturbations

In the present work the scalar perturbed metric under longitudinal gauge or Newtonian gauge is taken which is obtained as

$$ds^2 = -(1 + 2\Phi)dt^2 + a^2(t)(1 + 2\Psi)\delta_{ij}dx^i dx^j, \qquad (131.6)$$

where Φ and Ψ denote the gravitational potential and perturbation in the spatial curvature, respectively, and $a(t)$ is the scale factor of the universe. The perturbed Einstein's equations can be written in the Fourier space in the following forms

$$3H^2\Phi + 3H\dot{\Phi} + \frac{k^2\Phi}{a^2} = -4\pi G \sum_i \delta\rho_i, \qquad (131.7)$$

$$k^2(\dot{\Phi} + H\Phi) = 4\pi G a \sum_i (\bar{\rho}_i + \bar{p}_i)\theta_i, \qquad (131.8)$$

$$\ddot{\Phi} + 4H\dot{\Phi} + 2\dot{H}\Phi + 3H^2\Phi = 4\pi G \sum_i \delta p_i, \qquad (131.9)$$

where $\theta = i\,\vec{k} \cdot \vec{v}$ and $k = \frac{2\pi}{\lambda_p}$ is defined as the wave number with λ_p being the length scale of the perturbations. The perturbed energy density ($\delta\rho$), peculiar velocity (v_i), perturbed pressure density (δp) for Slotheon field are calculated from $T^{(\pi)}_{\mu\nu}$ (131.2).

Now solving the equations of Sects. 131.2 and 131.3 with proper initial conditions we obtain the evolutions of the cosmological density parameter (Ω), the Dark Energy equation of state (ω), Dark Energy density fluctuations $\delta_\pi(= \frac{\delta\rho_\pi}{\bar{\rho}_\pi})$ and matter density fluctuations $\delta_m(= \frac{\delta\rho_m}{\bar{\rho}_m})$.

In order to solve the above equations it is convenient to introduce the following dimensionless variables:

$$x = \frac{\dot{\pi}}{\sqrt{6}H M_{\text{pl}}}, \quad y = \frac{\sqrt{V(\pi)}}{\sqrt{3}H M_{\text{pl}}}, \quad \lambda = -M_{\text{pl}}\frac{V_\pi}{V(\pi)}, \quad \epsilon = \frac{H^2}{2M^2}, \quad q = \delta\pi / \frac{d\pi}{dN}.$$

With this framework we explore the effect of the Slotheon field Dark Energy perturbations on the matter power spectrum of the universe. We calculate the matter power spectrum with the Slotheon field and compare it with the same obtained from ΛCDM model [5]. To this end we define a percentage suppression X as

$$\frac{Pm_{\Lambda\text{CDM}} - Pm_{\text{slotheon}}}{Pm_{\Lambda\text{CDM}}} \times 100 = \frac{\Delta Pm}{Pm_{\Lambda\text{CDM}}} \times 100 = X, \qquad (131.10)$$

where matter power spectrum Pm is defined as

$$Pm = \langle |\delta_m(k,a)|^2 \rangle. \tag{131.11}$$

131.4   Results and Discussions

We compute quantities given above using the formalism described in Sect. 131.3 and the results are plotted in Fig. 131.1a–d. From Fig. 131.1a it can be observed that at early matter dominated epoch $\Omega_m = 1$ and $\Omega_\pi = 0$ but Ω_m starts to decrease and Ω_π starts to increase with time and in the present epoch ($a = 1$ or $N = 0$) the values evolve to $\Omega_m \sim 0.3$ and $\Omega_\pi \sim 0.7$. It may be noted from Fig. 131.1a that Dark Energy domination is a recent phenomenon. Figure 131.1b shows the thawing Dark energy behaviour of Slotheon field Dark Energy and quintessence Dark Energy. In Fig. 131.1c we make a comparison between ΛCDM model and Slotheon model. It can be noted from Fig. 131.1c that gravitational potentials of Slotheon field Dark energy and ΛCDM model are identical in early universe but they are not same in later

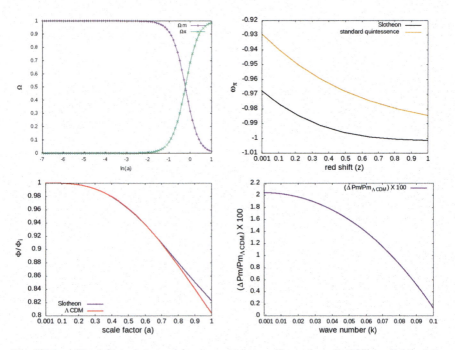

Fig. 131.1 a (Top left) variations of density parameters with number of e-foldings N (=$\ln a$). **b** (Top right) Variations of equation of state parameters with redshift z. **c** (Bottom left) Evolution of gravitational potential with scale factor a. **d** (Bottom right) Power spectrum ratio

time. It may also be noted that in matter dominated epoch Φ is almost constant and start to decrease when Dark Energy component becomes significant. In Fig. 131.1d the matter power spectrum as given in (131.10) are shown. One can conclude from Fig. 131.1d that matter power spectrums considering Slotheon Dark Energy model are close to the same considering ΛCDM model. It can be observed that for larger size of perturbations (smaller values of k) difference between these two models are larger than the smaller scale perturbations.

References

1. G. Dvali, G. Gabadadze, M. Porrati, Phys. Lett. B **485**, 208 (2000)
2. M.A. Luty, M. Porrati, R. Rattazzi, JHEP **0309**, 029 (2003), B. Jain, J. Khoury, Ann. Phys. **325**, 1479 (2010)
3. C. Germani, L. Martucci, P. Moyassari, Phys. Rev. D **85**, 103501 (2012)
4. D. Adak, A. Ali, D. Majumdar, Phys. Rev. D **88**, 024007 (2013)
5. E.J. Copeland, M. Sami, S. Tsujikawa, Int. J. Mod. Phys. D **15**, 1753 (2006)

Chapter 132
Geodetic Motion Around Rotating Black Hole in Nonlocal Gravity

Utkarsh Kumar, Sukanta Panda, and Avani Patel

Abstract Recently the nonlocal gravity theory has come out to be a good candidate for an effective field theory of quantum gravity and also it can provide rich phenomenology to understand late time accelerating expansion of the universe. For any valid theory of gravity, it has to surmount solar system tests as well as strong field tests. Having motivations to prepare the framework for the strong field test of the modified gravity using Extreme Mass Ratio Inspirals (EMRIs), here we try to obtain the metric for Kerr-like black hole for a nonlocal gravity model known as RR model and calculate the shift in orbital frequencies of a test particle moving around the black hole. We also derive the metric for a rotating object in the weak gravity regime for the same model.

132.1 Introduction

Nonlocal gravity theories have recently gained attention because of its ability to explain the late time cosmology unified with inflationary era. Especially so-called *RR* model proposed by [1] has been shown to explain CMB+BAO+SNIa+RSD data as well as ΛCDM [2]. Let us first write the action for the *RR* model :

$$S = \frac{1}{2\kappa^2} \int d^4x \sqrt{-g} \left[R + \frac{\mu^2}{3} R \frac{1}{\Box^2} R \right] + \mathcal{L}_m. \qquad (132.1)$$

U. Kumar · S. Panda
Department of Physics, Ariel University, Ariel 40700, Israel
e-mail: kumaru@ariel.ac.il

S. Panda
e-mail: sukanta@iiserb.ac.in

S. Panda · A. Patel (✉)
Indian Institute of Science Education and Research Bhopal, Bhopal 462066, India
e-mail: avani@iiserb.ac.in

© Springer Nature Singapore Pte Ltd. 2021
P. K. Behera et al. (eds.), *XXIII DAE High Energy Physics Symposium*,
Springer Proceedings in Physics 261,
https://doi.org/10.1007/978-981-33-4408-2_132

For this model we require a additional mass term, μ, that gives a constraint $\mu = 0.283H_0$, where H_0 is Hubble parameter today, to give the viable cosmology. Equation of motion corresponding to action (132.1) is

$$\kappa^2 T_{\alpha\beta} = G_{\alpha\beta} - \frac{\mu^2}{3}\left\{2\left(G_{\alpha\beta} - \nabla_\alpha\nabla_\beta + g_{\alpha\beta}\Box\right)S\right.$$

$$\left. + g_{\alpha\beta}\nabla^\gamma U\nabla_\gamma S - \nabla_{(\alpha}U\nabla_{\beta)}S - \frac{1}{2}g_{\alpha\beta}U^2\right\}, \qquad (132.2)$$

with $U = -\frac{1}{\Box}R$ and $S = -\frac{1}{\Box}U$, where $G_{\alpha\beta}$ is the Einstein tensor and $T_{\alpha\beta}$ is the energy-momentum tensor of the matter. The gravitational waves propagation in *RR* model has been studied in [3].

The orbits traversed by the small compact object(SCO) of mass m orbiting around a Super Massive Black Hole (SMBH) of mass M is generally known as Extreme Mass Ratio Inspirals (EMRIs). Since the mass ratio M/m of two objects is $\sim 10^4 - 10^8 M_\odot$, the motion of the SCO around the SMBH can be approximated as trajectory of a point particle along the geodesics of the SMBH spacetime. The structure of the spacetime can be reflected in orbital frequencies of the point particle whose imprints finally can be seen in gravitational waves emitted by SCO. Here, we aim to calculate the shift in orbital frequencies of geodesic motion of the test particle around rotating black hole due to nonlocal correction in RR model following the treatment prescribed in [4].

132.2 Rotating Object in the Weak Gravity Regime in RR Model

We consider the linearized gravity limit of the field equation written in (132.2). In linearized gravity limit we take $g_{\alpha\beta}$ as $g_{\alpha\beta} = \eta_{\alpha\beta} + h_{\alpha\beta}$, $|h| \ll 1$, where $h(\alpha\beta)$ is a small perturbation around Minkowski background $\eta_{\alpha\beta}$. Under this approximation, one can find the expressions for Riemann Tensor, Ricci Tensor, and Ricci scalar as follows

$$R_{\gamma\alpha\delta\beta} = \frac{1}{2}\left(\partial_\delta\partial_\alpha h_{\gamma\beta} + \partial_\beta\partial_\gamma h_{\alpha\delta} - \partial_\beta\partial_\alpha h_{\gamma\delta} - \partial_\delta\partial_\gamma h_{\beta\alpha}\right), \qquad (132.3)$$

$$R_{\alpha\beta} = \frac{1}{2}\left(\partial^\gamma\partial_\alpha h_{\gamma\beta} + \partial_\beta\partial_\gamma h_\alpha^\gamma - \partial_\beta\partial_\alpha h - \Box h_{\alpha\beta}\right), \qquad (132.4)$$

$$R = \partial_\alpha\partial_\beta h^{\alpha\beta} - \Box h. \qquad (132.5)$$

Then the field (132.2) becomes

$$
2\kappa^2 T_{\alpha\beta} = -\left[\Box h_{\alpha\beta} - \partial_\gamma \partial_{(\alpha} h^\gamma_{\beta)} + \left(1 - \frac{2M^2}{3}\Box^{-1}\right)(\partial_\alpha \partial_\beta h + \eta_{\alpha\beta}\partial_\gamma \partial_\delta h^{\gamma\delta})\right.
$$
$$
\left. - \left(-1 + \frac{2M^2}{3}\Box^{-1}\right)\eta_{\alpha\beta}\Box h + \frac{2M^2}{3}\Box^{-2}\nabla_\alpha \nabla_\beta \partial_\gamma \partial_\delta h^{\gamma\delta}\right].
$$

$$(132.6)$$

132.2.1   Spacetime Solution Around Rotating Object in Linearized Gravity Limit

Starting with a generic spherically symmetric and static metric

$$
ds^2 = -(1 + 2\Phi)dt^2 + 2\overrightarrow{h}.dxdt + (1 - 2\Psi)dx^2 \tag{132.7}
$$

and the stress-energy tensor for the rotating object having energy density $\rho = M\delta^3(\overrightarrow{r})$ with mass M and angular velocity v_i given by $T_{00} = \rho$, $T_{0i} = -\rho v_i$, we can solve the field equation in (132.2) and convert the solution into Boyer-Lindquist coordinates (t, r, θ, ϕ) to obtain the rotating metric as

$$
ds^2 = -(1 + 2\Phi)dt^2 + 4\frac{J\sin^2\theta}{M}(\Phi + \Psi)d\phi dt
$$
$$
+ (1 - 2\Psi)(dr^2 + r^2 d\theta^2 + r^2 \sin^2\theta d\phi^2), \tag{132.8}
$$

where J is angular momentum, defined as $v = \frac{r \times J}{Mr^2}$ and Φ and Ψ are given by [5]

$$
\Phi(r) = \frac{GM}{r}\left(\frac{e^{-\mu r} - 4}{3}\right), \quad \Psi(r) = \frac{GM}{r}\left(\frac{-e^{-\mu r} - 2}{3}\right). \tag{132.9}
$$

132.3   Geodetic Motion Around Rotating Black hole in Nonlocal Gravity

The metric for the spacetime around rotating black hole in RR model was obtained by applying Demiański-Janis-Newman algorithm [6, 7] on the spherically symmetric static solution [1] of the RR model in [8] as (in the form of $g_{\alpha\beta} = g_{\alpha\beta}^{Kerr} + b_{\alpha\beta}$)

$$ds^2 = \left[-1 + \frac{2GMr}{\Sigma} \left(1 + \frac{1}{6}\mu^2\Sigma \right) \right] dt^2$$
$$-2a\sin^2\theta \left[\frac{2GMr}{\Sigma} \left(1 + \frac{1}{6}\mu^2\Sigma \right) \right] dt\, d\phi$$
$$+\frac{\Sigma}{\Delta} dr^2 + \Sigma\, d\theta^2 + \sin^2\theta$$
$$\left[\Sigma + \left(1 + \frac{2GMr}{\Sigma} \left(1 + \frac{1}{6}\mu^2\Sigma \right) \right) a^2\sin^2\theta \right] d\phi^2, \quad (132.10)$$

where $\Sigma = r^2 + a^2\cos^2\theta$ and $\Delta = \Sigma - 2GMr\left(1 + \frac{1}{6}\mu^2\Sigma\right) + a^2\sin^2\theta$.

Since the Kerr metric is independent of t and ϕ it has two apparent symmetry and possesses two constants of motion, i.e., energy E as measured by observer at spatial infinity and axial component of angular momentum L_z. The equation of motion of a point mass m moving along the geodesics of the Kerr metric is given by geodesic equation $\dot{u}^\mu + \Gamma^\mu_{\alpha\beta}u^\alpha u^\beta = 0$, where u^μ is the four-velocity of the point mass and overdot denotes the derivative w.r.t. proper time τ. Since the nonlocal correction term added to Einstein-Hilbert (EH) action is very small compared to EH term we can use canonical perturbation theory to solve the geodesic equations of nearly Kerr-like black hole of nonlocal gravity model. A relativistic version of Hamilton-Jacobi method for the motion of a test mass in Kerr spacetime as proposed by Carter [9] shows that the motion is separable and it is attributed by three constants of motion: the Carter constant Q in addition with E and L_z. A further scheme of relativistic action-angle formalism to calculate the fundamental frequencies of the orbital motion in the Kerr geometry was provided by Schmidt [10]. According to canonical perturbation theory, if $\hat{\omega}^r$, $\hat{\omega}^\theta$ and $\hat{\omega}^\phi$ are orbital frequencies for Kerr black hole then the orbital frequencies for the Kerr-like BH in nonlocal gravity are given by $\omega^i = \hat{\omega}^i + \delta\omega^i$, where $m\delta\omega^i = \partial\langle H_1\rangle/\partial\hat{J}_i$ [4] and $\hat{J}_i$ is the action variable for the Kerr spacetime. The Hamiltonian of the system (test mass and SMBH) is given by $H = H_{Kerr} + H_1$, where $H_1 = -(m^2/2)b_{\alpha\beta}(dx^\alpha/d\tau)(dx^\beta/d\tau)$. $\langle H_1\rangle$ shows the averaged Hamiltonian H_1 over a period of the orbit in background spacetime. We numerically calculate the orbital frequencies for Kerr spacetime (shown in left plot of Fig. 132.1) and shift in frequencies from the Kerr frequencies due to nonlocal correction of $\mu^2 R\frac{1}{\Box^2}R$ (shown in right plot of Fig. 132.1). Equivalent to set of (E, L_z, Q), the orbital motion can be parametrized by set of parameters (p, e, θ_{min}), where p, e and θ_{min} are respectively semilatus rectum, eccentricity and turning point of θ-motion of the orbit. The detailed discussion on calculation of shift in orbital frequencies and solution of geodesic equations of the spacetime which is slightly different from the Kerr spacetime will be done in our future work.

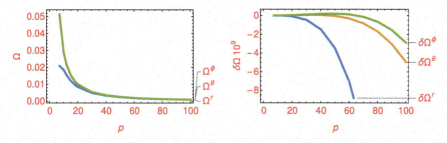

Fig. 132.1 $\Omega^i(=\omega^i/\omega^t)$ versus p. Left: Observable orbital frequencies of the orbits in Kerr space-time. Right: shift in observable orbital frequencies due to nonlocal correction

132.4   Conclusion

We calculate the axially symmetric stationary metric around the rotating object in RR model of nonlocal gravity by solving field equations in linearized gravity limit for axial symmetry. Using canonical perturbation theory we calculate the orbital frequencies of a test particle moving along the geodesic of the metric around rotating black hole in *RR* model of nonlocal gravity.

Acknowledgements This work was partially funded by DST (Govt. of India), Grant No. SERB/PHY/2017041.

References

1. M. Maggiore, M. Mancarella, J. Phys. Rev. D **90**, 023005 (2014), arXiv:1402.0448[hep-th]
2. L. Amendola, Y. Dirian, H. Nersisyan, S. Park, JCAP **1903** 045 (2019), arXiv:1901.07832[astro-ph.CO]
3. E. Belgacem, Y. Dirian, S. Foffa, M. Maggiore, Phys. Rev. **D98**, 023510 (2018), arXiv:1805.08731[gr-qc]
4. S.J. Vigeland, S.A. Hughes, J. Phys. Rev. **D81**, 024030 (2010), arXiv:0911.1756[gr-qc]
5. U. Kumar, S. Panda, A. Pate, Phys. Rev. **D98**, 124040 (2018), arXiv:1808.04569[gr-qc]
6. E.T. Newman, A.I. Janis, J. Math. Phys. **6**, 915 (1965)
7. M. Demiański, Phys. Lett. A **42**, 157 (1972)
8. U. Kumar, S. Panda, A. Patel, (2019), arXiv:1906.11714[gr-qc]
9. B. Carter, J. Phys. Rev. **174**, 1559 (1968)
10. W. Schmidt, Class. Quant. Grav. **19**, 2743 (2002), arXiv:gr-qc/0202090[gr-qc]

Chapter 133
On Modification of Phase-Space Measure in Theory with Invariant Planck Scale as an Ultraviolet Cut-Off

Dheeraj Kumar Mishra

Abstract Planck scale acts as a threshold where a new description of spacetime is expected to appear. Such a scale should be invariant which is achieved by modifying the algebra. For the exotic spacetimes appearing at Planck scale, the phase-space is also expected to modify. Considering such a modification in a relativistic theory with an invariant Planck scale as an ultraviolet cut-off, we study the thermodynamics of ideal gases. In case of ideal photon gas this leads to the modification of Planck energy density distribution and Wien's displacement law. We also study various equilibrium thermodynamic properties of blackbody radiation.

133.1 Deformed Algebra and Modified Dispersion Relation

In all the candidate Quantum Gravity theories Planck scale appears as a natural scale which has to be invariant for all the observers to observe the same scale where new description appears. We achieve this by modifying the standard algebra depending on the basis that we choose such as Classical, Bicross-product, and Magueijo-Smolin (MS). In MS basis [1, 2] the rotation generators remain intact but we modify the boost generators by adding a dilatation term in such a way to keep Lorentz sub-algebra intact but Poincare algebra modifies to κ−Poincare algebra to get along with c another invariant scale, κ acting as ultraviolet cut-off and a Modified Dispersion Relation (MDR). Correspondence principle gives the known usual relativistic theory in a limit $\kappa \to \infty$ as expected.

The standard Lorentz generators are $L_{\alpha\beta} = P_\alpha \frac{\partial}{\partial P^\beta} - P_\beta \frac{\partial}{\partial P^\alpha}$ with the deformed boost generator as $K^i := L_0^i + \frac{P^i}{\kappa} D$, where D is dilatation term given as $D = P_\alpha \frac{\partial}{\partial P^\alpha}$ and the rotation generator as $J^i := \epsilon^{ijk} L_{ij}$. This leads to the modified Poincare algebra,

D. K. Mishra (✉)
The Institute of Mathematical Sciences, Chennai 600113, India
e-mail: dkmishra@imsc.res.in

Homi Bhabha National Institute, Mumbai 400094, India

© Springer Nature Singapore Pte Ltd. 2021
P. K. Behera et al. (eds.), *XXIII DAE High Energy Physics Symposium*,
Springer Proceedings in Physics 261,
https://doi.org/10.1007/978-981-33-4408-2_133

$$[J^i, K^j] = i\epsilon^{ijk} K_k; \quad [K^i, K^j] = -i\epsilon^{ijk} J_k; \quad [J^i, J^j] = i\epsilon^{ijk} J_k; \quad (133.1)$$

$$[K^i, P^j] = i\left(\delta^{ij} P^0 - \frac{P^i P^j}{\kappa}\right); \quad [K^i, P^0] = i\left(1 - \frac{P^0}{\kappa}\right) P^i. \quad (133.2)$$

This in turn leads to the quadratic Casimir or MDR of the above-deformed algebra as $E^2 - P^2 = m^2 \left(1 - \frac{E}{\kappa}\right)^2$. For our purpose the energy of particle is always less than κ.

133.2 Modified Phase-Space for Exotic Spacetime

For exotic spacetimes while taking the large volume limit we expect the phase space to modify. In large volume limit the expected change is $\sum_\epsilon \rightarrow \frac{1}{(2\pi)^3} \int\int d^3x d^3p f(\mathbf{x}, \mathbf{p})$. Assuming spacetime as isotropic and f as Taylor expandable in powers of $\left(\frac{1}{r\kappa}\right)$ and $\left(\frac{\epsilon}{\kappa}\right)$ with κ as highest energy cut-off and $\frac{1}{\kappa}$ as the lowest length cut-off with the accessible volume given as $V_{ac} = V - \frac{4\pi}{3\kappa^3}$, for the most general possible modification independent of the underlying QG theory the integral of the form $\frac{1}{(2\pi)^3} \int\int d^3x d^3p F(\varepsilon)$ changes to $\frac{1}{(2\pi)^3} \int\int d^3x d^3p f(r, p) F(\varepsilon)$ as

$$\frac{1}{(2\pi)^3} \sum_{\substack{n=0, n'=0 \\ n' \neq 3}}^{\infty} \frac{a_{n,n'}}{n! n'! \kappa^{n+3}} \frac{4\pi}{(3-n')} \left[\left(\frac{3V\kappa^3}{4\pi}\right)^{\frac{3-n'}{3}} - 1\right] \int_{p=0}^{\kappa} d^3p \, (\varepsilon)^n \, F(\varepsilon)$$

$$+ \frac{1}{(2\pi)^3} \sum_{n=0}^{\infty} \frac{a_{n,3}}{n! 3! \kappa^{n+3}} \left(\frac{4\pi}{3}\right) \ln\left(\frac{3V\kappa^3}{4\pi}\right) \int_{p=0}^{\kappa} d^3p \, (\varepsilon)^n \, F(\varepsilon). \quad (133.3)$$

133.3 Modified Thermodynamics of Ideal Gases and Its Possible Implications

133.3.1 Classical Ideal Gas

The partition function for classical ideal gas in canonical ensemble obeying Maxwell-Boltzmann statistics is $Z_N (V_{ac}, T) = \sum_E \exp[-\beta E] = \frac{1}{N!}[Z_1 (V_{ac}, T)]^N$, where $Z_1(V_{ac}, T) = \sum_\varepsilon \exp[-\beta(\varepsilon - m_0)]$ is the single particle partition function. In the large volume limit we get

$$Z_1(V_{ac}, T) = \sum_{\substack{n=0, n'=0 \\ n' \neq 3}}^{\infty} \frac{a_{n,n'}}{n! n'! \kappa^n} \left(\frac{3}{(3-n')(\kappa^3 V_{ac})} \right) \left[\left(\frac{3V\kappa^3}{4\pi} \right)^{\frac{3-n'}{3}} - 1 \right] \left(m_0 - \frac{\partial}{\partial \beta} \right)^n Z_1^0(V_{ac}, T)$$

$$+ \sum_{n=0}^{\infty} \frac{a_{n,3}}{n! \kappa^n} \left(\frac{4\pi}{18\kappa^3 V_{ac}} \right) \ln \left(\frac{3V\kappa^3}{4\pi} \right) \left(m_0 - \frac{\partial}{\partial \beta} \right)^n Z_1^0(V_{ac}, T), \tag{133.4}$$

where $Z_1^0(V_{ac}, T)$ is the single particle partition function with the unmodified measure $Z_1^0(V_{ac}, T) = \frac{V_{ac}}{(2\pi)^3} \int_{p=0}^{\kappa} d^3 p \exp(-\beta(\varepsilon - m_0))$. The expression for $Z_1(V_{ac}, T)$ has now non-trivial dependence on V. Various thermodynamic quantities can be calculated using the results obtained in [3] and the standard relations for other thermodynamic quantities (Fig. 133.1).

133.3.2 Ideal Photon Gas

Considering the grand canonical ensemble of ideal photons obeying Bose-Einstein statistics. The energy density distribution modifies to

$$u(\omega)d\omega = \frac{1}{(\pi)^2} \sum_{\substack{n=0, n'=0 \\ n' \neq 3}}^{\infty} \frac{a_{n,n'}}{n! n'! \kappa^n} \frac{4\pi}{(3-n')} \left(\frac{1}{V_{ac}\kappa^3} \right) \left[\left(\frac{3V\kappa^3}{4\pi} \right)^{\frac{3-n'}{3}} - 1 \right] \frac{\omega^{n+3} d\omega}{e^{\frac{\omega}{T}} - 1}$$

$$+ \frac{1}{(\pi)^2} \sum_{n=0}^{\infty} \frac{a_{n,3}}{n! 3! \kappa^n} \left(\frac{4\pi}{3\kappa^3 V_{ac}} \right) \ln \left(\frac{3\kappa^3 V}{4\pi} \right) \frac{\omega^{n+3} d\omega}{e^{\frac{\omega}{T}} - 1} = \sum_{n=0}^{\infty} A_n \frac{\omega^{n+3}}{e^{\frac{\omega}{T}} - 1} d\omega. \tag{133.5}$$

Here A_n is constant and independent of both wavelength λ and T and is given as

$$A_n = \frac{1}{\pi^2} \sum_{n'=0, n' \neq 3}^{\infty} \frac{a_{n,n'}}{n! n'! \kappa^n} \left(\frac{4\pi}{3-n'} \right) \frac{1}{\kappa^3 V_{ac}} \left[\left(\frac{3V\kappa^3}{4\pi} \right)^{\frac{3-n'}{3}} - 1 \right]$$

$$+ \frac{1}{\pi^2} \frac{a_{n,3}}{n! 3! \kappa^n} \left(\frac{4\pi}{3\kappa^3 V_{ac}} \right) \ln \left(\frac{3V\kappa^3}{4\pi} \right). \tag{133.6}$$

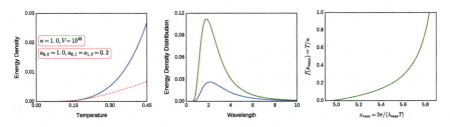

Fig. 133.1 The figure on the left represents the variation of the energy density with temperature, middle figure represents the modified Planck energy distribution and the figure on the right represents the modified Wien's law. Similar behaviour is seen in other thermodynamic quantities as well [4]. Blue: normal SR result; red dotted and green: the modified result

Writing $u(\omega)$ from (133.5) in terms of wavelength λ and obtaining maximum $u(\lambda)$ at $\lambda = \lambda_{max}$ using extremum condition $\left.\frac{du(\lambda)}{d\lambda}\right|_{\lambda_{max}} = 0$, the usual $5 = \frac{x_{max}}{1-e^{-x_{max}}}$ modifies to $\sum_{n=0}^{\infty} T^n x_{max}^n A_n \left[n + 5 - \frac{x_{max}}{1-e^{-x_{max}}}\right] = 0$, where $x_{max} = \frac{2\pi}{\lambda_{max}T}$. In this case x_{max} is a monotonic increasing function of T. Keeping the leading order terms in $\frac{T}{\kappa}$ and $\frac{1}{V^{1/3}\kappa}$ and neglecting all the higher order terms we get $\frac{T}{\kappa} = -\frac{1}{x_{max}a_{1,0}}\left(\frac{5-\frac{x_{max}}{1-e^{-x_{max}}}}{6-\frac{x_{max}}{1-e^{-x_{max}}}}\right) = f(x_{max})$. We note that the significant change in value of x_{max} occurs if order of change in temperature is non-negligible with respect to κ. Since $f^{-1}\left(\frac{T}{\kappa}\right) > 4.965$ so $\frac{(\lambda_{max})_{DSR}}{(\lambda_{max})_{SR}} = \frac{4.965}{f^{-1}\left(\frac{T}{\kappa}\right)} \leq 1$. The frequency at which the energy density distribution at a given temperature peaks gets a positive correction. Now if we demand at least 1% correction, i.e. $\frac{(\lambda_{max})_{DSR}}{(\lambda_{max})_{SR}} = \frac{4.965}{f^{-1}\left(\frac{T}{\kappa}\right)} = \frac{99}{100}$ then we get $f^{-1}\left(\frac{T}{\kappa}\right) = 5.015$.

The energy density $u \equiv \frac{U}{V_{ac}} = \int_0^{\kappa} u(\omega)d\omega$ is given as

$$u = \frac{1}{(\pi)^2} \sum_{\substack{n=0,n'=0 \\ n'\neq 3}}^{\infty} \frac{a_{n,n'}}{n!n'!\kappa^n} \frac{4\pi}{(3-n')}\left(\frac{T^{n+4}}{V_{ac}\kappa^3}\right)\left[\left(\frac{3V\kappa^3}{4\pi}\right)^{\frac{3-n'}{3}} - 1\right]$$

$$\Gamma(n+4)\left[Z_{n+4}(0) - Z_{n+4}\left(\frac{\kappa}{T}\right)\right]$$

$$+ \frac{1}{(\pi)^2} \sum_{n=0}^{\infty} \frac{a_{n,3}}{n!3!\kappa^n}\left(\frac{4\pi T^{n+4}}{3V_{ac}\kappa^3}\right)\ln\left(\frac{3\kappa^3 V}{4\pi}\right)$$

$$\Gamma(n+4)\left[Z_{n+4}(0) - Z_{n+4}\left(\frac{\kappa}{T}\right)\right] = \sum_{n=0,n'=0}^{\infty} u_{n,n'}. \qquad (133.7)$$

The radiation pressure modifies to

$$P = \sum_{\substack{n=0,n'=0 \\ n'\neq 3}}^{\infty} \left[\frac{1}{(\pi)^2}\frac{a_{n,n'}}{n!n'!\kappa^n}\frac{4\pi}{(3-n')}\left(\frac{T}{V_{ac}\kappa^3}\right)\left[\left(\frac{3V\kappa^3}{4\pi}\right)^{\frac{3-n'}{3}} - 1\right]\right.$$

$$\left\{-\ln(1-e^{-\frac{\kappa}{T}})\frac{\kappa^{n+3}}{n+3}\right\} + \left.\frac{u_{n,n'}}{(n+3)}\right]$$

$$+ \sum_{n=0}^{\infty}\left[\frac{1}{(\pi)^2}\frac{a_{n,3}}{n!3!\kappa^n}\left(\frac{4\pi T}{3V_{ac}\kappa^3}\right)\ln\left(\frac{3\kappa^3 V}{4\pi}\right)\right.$$

$$\left.\left\{-\ln(1-e^{-\frac{\kappa}{T}})\frac{\kappa^{n+3}}{n+3}\right\} + \frac{u_{n,3}}{(n+3)}\right]. \qquad (133.8)$$

The equilibrium number of photons is given as

$$\bar{N} = \frac{1}{(\pi)^2} \sum_{\substack{n=0,n'=0 \\ n' \neq 3}}^{\infty} \frac{a_{n,n'}}{n!n'!k^{n+3}} \frac{4\pi T^{n+3}}{(3-n')} \left[\left(\frac{3V\kappa^3}{4\pi} \right)^{\frac{3-n'}{3}} - 1 \right]$$

$$\Gamma(n+3) \left[Z_{n+3}(0) - Z_{n+3}\left(\frac{\kappa}{T} \right) \right]$$

$$+ \frac{1}{(\pi)^2} \sum_{n=0}^{\infty} \frac{a_{n,3}}{n!3!\kappa^{n+3}} \left(\frac{4\pi T^{n+3}}{3} \right) \ln \left(\frac{3\kappa^3 V}{4\pi} \right)$$

$$\Gamma(n+3) \left[Z_{n+3}(0) - Z_{n+3}\left(\frac{\kappa}{T} \right) \right]. \tag{133.9}$$

Other thermodynamic quantities can be similarly calculated using above results. For massive quantum gases, i.e. bosons and fermions the q-potential modifies to

$$q = \frac{g}{(2\pi)^3} \sum_{\substack{n=0,n'=0 \\ n' \neq 3}}^{\infty} \frac{a_{n,n'}}{n!n'!\kappa^{n+3}} \frac{4\pi}{(3-n')} \left[\left(\frac{3V\kappa^3}{4\pi} \right)^{\frac{3-n'}{3}} - 1 \right] \left(\frac{4\pi}{a} \right)$$

$$\int_{\varepsilon=0}^{\kappa} \left[\varepsilon + \frac{m^2}{\kappa} \left(1 - \frac{\varepsilon}{\kappa} \right) \right] \left[\varepsilon^2 - m^2 \left(1 - \frac{\varepsilon}{\kappa} \right)^2 \right]^{\frac{1}{2}} (\varepsilon)^n \ln \left[1 + aze^{-\beta\varepsilon(p)} \right] d\varepsilon$$

$$+ \frac{g}{(2\pi)^3} \sum_{n=0}^{\infty} \frac{a_{n,3}}{n!3!\kappa^{n+3}} \left(\frac{4\pi}{3} \right) \ln \left(\frac{3V\kappa^3}{4\pi} \right) \left(\frac{4\pi}{a} \right)$$

$$\int_{\varepsilon=0}^{\kappa} \left[\varepsilon + \frac{m^2}{\kappa} \left(1 - \frac{\varepsilon}{\kappa} \right) \right] \left[\varepsilon^2 - m^2 \left(1 - \frac{\varepsilon}{\kappa} \right)^2 \right]^{\frac{1}{2}} (\varepsilon)^n \ln \left[1 + aze^{-\beta\varepsilon(p)} \right] d\varepsilon \tag{133.10}$$

References

1. J. Magueijo, L. Smolin, Phys. Rev. Lett. **88**, 190403 (2002)
2. J. Magueijo, L. Smolin, Phys. Rev. D **67**, 044017 (2003)
3. N. Chandra, S. Chatterjee, Phys. Rev. D **85**, 045012 (2012)
4. D.K. Mishra, N. Chandra, V. Vaibhav, Ann. Phys. **385**, 605 (2017)

Chapter 134
Thermodynamic Geometry of Regular Black Hole Surrounded by Quintessence

C. L. Ahmed Rizwan, A. Naveena Kumara, and Deepak Vaid

Abstract We investigate thermodynamics of regular Bardeen AdS black hole surrounded by quintessence. Pressure-Volume (P-V) and temperature-entropy (T-S) plots are obtained from the first law of black hole thermodynamics, shows a critical behaviour. This is reflected in the divergence of specific heat against entropy plots. Using the thermodynamic geometry, we have tried to affirm this critical property. From the Ruppeiner and Weinhold geometries, we have calculated the thermodynamic curvature scalar R_R and R_W in the quintessence dark energy regime. It is found that in our case these thermodynamic scalars can only identify the critical behaviour and fail to show divergence at the phase transition points observed in specific heat study.

134.1 Introduction

Black hole thermodynamics has remained as a hot topic in high energy physics for past five decades. The motivations behind this are that the black hole physics forms a bridge between quantum gravity and general relativity. From the work of Hawking and Bekenstein, black hole temperature was found proportional to the surface gravity and entropy related to the area of event horizon. Using these facts, four laws of black hole thermodynamics were formulated parallel to laws in classical thermodynamics. In 1980s, Hawking and Page studied black hole thermodynamics in the asymptotically anti-de Sitter (AdS) geometry. The AdS black holes exhibited rich phase structure with thermal radiation, small and large black hole phases. Again through the paper of Maldecena, AdS black hole caught attention by an introduction of a correspondence between classical field in the bulk and a conformal quantum field theory (CFT) in the AdS boundary. This AdS/CFT correspondence triggered research in AdS black holes. Further progress in black hole thermodynamics happened when

C. L. Ahmed Rizwan (✉) · A. Naveena Kumara · D. Vaid
Department of Physics, National Institute of Technology Karnataka (NITK), Surathkal, Mangaluru 575025, India
e-mail: ahmedrizwancl@gmail.com

© Springer Nature Singapore Pte Ltd. 2021
P. K. Behera et al. (eds.), *XXIII DAE High Energy Physics Symposium*,
Springer Proceedings in Physics 261,
https://doi.org/10.1007/978-981-33-4408-2_134

phase space was extended by identification of cosmological constant with thermo-dynamic pressure. Thermodynamic study in the extended phase space showed close resemblance with van der Waals gas. These studies give the information about the microscopic constituencies of the black hole which is not completely understood yet. In another context a geometric approach to thermodynamics and phase transitions were introduced by Weinhold and Ruppeiner. Using the thermodynamic geometry a metric is constructed in the equilibrium thermodynamic state space. From that metric one can write the curvature scalar that encodes the information about the microscopic interactions. The critical behaviour of the black hole can be seen in the divergence behaviour of this curvature scalar. A regular black hole surrounded by quintessence is studied in the context of black hole thermodynamics and thermodynamic geometry. Quintessence is a potential candidate for explaining the dark energy responsible for accelerating universe. The cosmic source for inflation has equation of state $p_q = \omega \rho_q$ ($-1 < \omega < -1/3$) and $\omega = -2/3$ corresponds to quintessence dark energy regime.

The metric for a regular AdS black hole surrounded by quintessence is given by

$$ds^2 = -f(r)dt^2 + \frac{dr^2}{f(r)} + r^2 d\theta^2 + r^2 \sin^2 \theta d\phi^2 \tag{134.1}$$

with $f(r) = 1 - \frac{2M(r)}{r} + \frac{a}{r^{2\omega+1}} - \frac{\Lambda r^2}{3}$ and $M(r) = \frac{mr^3}{(r^2+\beta^2)^{3/2}}$, where β is the monopole charge of a self-gravitating magnetic field described by a non-linear electromagnetic source, m is the mass of the black hole, Λ is the cosmological constant and a is the normalisation constant related to quintessence density. The mass corresponds to the above metric is

$$M = \frac{1}{6} r_h^{-3(1+\omega)} (\beta^2 + r_h^2)^{3/2} (-3a + r_h^{1+3\omega}(3 + 8P\pi r_h^2)). \tag{134.2}$$

In the extended phase space cosmological constant is considered as thermodynamic pressure. Entropy of black hole is given by area of event horizon. Both reads as follows:

$$P = -\frac{\Lambda}{8\pi}, \quad S = \pi r_h^2. \tag{134.3}$$

First law of thermodynamics for this black hole can be written as

$$dM = TdS + \Phi dQ + VdP. \tag{134.4}$$

The Hawking temperature of the black hole can be derived from the first law as

$$T = \left(\frac{\partial M}{\partial S}\right)_{\Psi, P, a} = \frac{(\beta^2 + \frac{S}{\pi})^{\frac{1}{2}}}{4} S^{-\frac{3\omega}{2} - \frac{5}{2}} \Big[3a\pi^{\frac{3\omega}{2} + \frac{1}{2}} (\pi \beta^2(\omega + 1)$$
$$+ S\omega) + S^{\frac{3\omega}{2} + \frac{1}{2}} (-2\pi\beta^2 + 8PS^2 + S) \Big]. \tag{134.5}$$

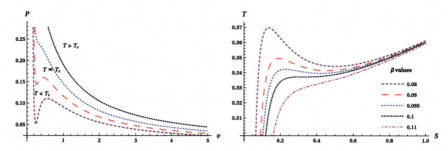

Fig. 134.1 To the left we have P-V diagram for regular AdS black hole surrounded by quintessence ($a = 0.07$, $\beta = 0.1$, $\omega = -2/3$, $T_c = 0.36$). In the right side T-S plot for different values of β is shown

The equation of state is obtained as

$$P = \frac{1}{8\pi}\left[\frac{8\pi T}{\sqrt{4\beta^2+v^2}} + 32\beta^2 v^{-3\omega-5}\left(v^{3\omega+1} - 3a8^\omega(\omega+1)\right) - 3a8^{\omega+1}\omega v^{-3(\omega+1)} - \frac{4}{v^2}\right],$$

where $v = 2r_h$ is specific volume. Using the equation of state, the $P - v$ and $T - S$ curves are plotted in Figs. 134.1 and 134.1. The critical points are found using

$$\frac{\partial P}{\partial v} = 0 \quad, \quad \frac{\partial^2 P}{\partial v^2} = 0. \tag{134.6}$$

Below the critical point T_C, $P - v$ isotherm splits into three branches corresponding to small, intermediate, and large black holes. This behaviour is quite similar to liquid/gas transition in van der Waals fluids. For obtaining more details about phase transitions, the heat capacity is calculated at fixed value of parameters β, P and a given by

$$C_P = T\left(\frac{\partial S}{\partial T}\right)_P \tag{134.7}$$

$$= \frac{2S\left(\pi\beta^2 + S\right)\left(S\left(\sqrt{\pi}(8PS+1) - 2a\sqrt{S}\right) + \beta^2\left(\pi a\sqrt{S} - 2\pi^{3/2}\right)\right)}{\sqrt{\pi}\left(\beta^4\left(8\pi^2 - 3\pi^{3/2}a\sqrt{S}\right) + S^2(8PS-1) + 4\pi\beta^2 S\right)}. \tag{134.8}$$

Sign of heat capacity tells about thermodynamic stability, positive for stable and negative for unstable systems. From Fig. 134.1, we observe two stable region separated by an unstable one leads to *Reentrant phase transitions* Fig. 134.2.

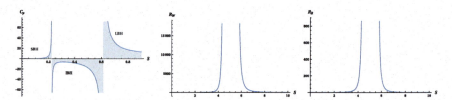

Fig. 134.2 Specific heat versus entropy diagram plotted below the critical point shown in Fig. 134.1. In the next two figures, Curvature divergence plots for Weinhold and Ruppeiner metric is shown. In all plots quintessence parameter and monopole charge is fixed, $a = 0.5$ and $\beta = 1$

134.1.1 Thermodynamic Geometry

The Weinhold and Ruppeiner geometries are used to confirm these critical nature of black holes. Weinhold metric is defined ad hoc in the thermodynamic equilibrium space as the Hessian of the internal energy M and Ruppeiner metric as Hessian of the entropy function S

$$g_{ij}^{W} = \partial_i \partial_j M(S, N^a) dx^i dx^j, \ g_{ij}^{R} = -\partial_i \partial_j S(M, N^\alpha), \ (i, j = 1, 2), \quad (134.9)$$

where N^a represents other thermodynamic extensive variables. From metric tensor g_{ij}^{W} and g_{ij}^{R}, one can calculate curvature scalar which is found to be a complicated expression, $R_W(S, P, b, \omega, a)$ and $R_R(S, P, b, \omega, a)$. Plotting the curvature scalars R_W and R_R versus entropy S, we have studied its divergence behaviour, which occurs at multiple points. Around the critical point ($P_c = 0.207$ for $a = 0.07$ and $\beta = 0.1$), scalars show multiple divergences which is different from that in specific heat plots.

134.2 Conclusion

In this paper, we studied thermodynamics of Regular Bardeen AdS black holes with quintessence. The P-v isotherms, T-S diagrams, and specific heat plots are obtained in quintessence dark energy regime $\omega = -2/3$. The critical behaviour exhibited is further confirmed in thermodynamic geometries. But from randomly located diverging points in thermodynamic scalars, we could infer only the critical behaviour of the system, but not the exact phase transition points.

References

1. V.V. Kiselev, Quintessence and black holes. Class. Quantum Gravity **20**(6), 1187 (2003)
2. George Ruppeiner, Riemannian geometry in thermodynamic fluctuation theory. Rev. Mod. Phys. **67**(3), 605 (1995)

3. M. Saleh, B.B. Thomas, T.C. Kofane, Thermodynamics and phase transition from regular bardeen black hole surrounded by quintessence. Int. J. Theor. Phys. **57**(9), 2640–2647 (2018)
4. Shao-Wen Wei, Qing-Tao Man, Yu. Hao, Thermodynamic geometry of charged ads black hole surrounded by quintessence. Commun. Theor. Phys. **69**(2), 173 (2018)
5. C.L. Ahmed Rizwan, A. Naveena Kumara, K.V. Rajani, D. Vaid, K.M. Ajith, Effect of dark energy in geometro thermodynamics and phase transitions of regular bardeen AdS black hole. arXiv e-prints, arXiv:1811.10838 (2018)

Chapter 135
Cosmological Constant and α-Quantization

Sovan Ghosh

Abstract Cosmological constant is related to the fine structure constant. Variation of the fine structure constant is an experimentally verified fact at present. It prompts us to think about the variation of the cosmological constant as well. Again fine structure constant is a key factor to different particle properties and alpha quantization helps to connect particles in a single chain. So a variation of the fine structure constant can contribute to the alpha-quantized behavior of the elementary particles. Here in this article, the cosmological constant and alpha quantization are brought together to provide a better realization of the nature of the tiny particles.

135.1 Introduction

The cosmological constant, the brainchild of Einstein [1] was almost abandoned by himself [2]. Again the cosmological constant was assumed to be zero for almost the rest of the time of the previous century [3]. With the evolution of the dark energy theory, the cosmological constant is regarded as one of the possible dark energy candidates [4, 5]. Beck has provided the expression of the same with the help of Khinchin axioms as

$$\Lambda = \frac{G^2}{\hbar^4}\left(\frac{m_e}{\alpha}\right)^6 \tag{135.1}$$

and his expression is in good agreement with the recent data [6]. Here G is the gravitational constant, m_e is the mass of the electron, $\hbar$ is the reduced Planck constant, and α is the fine structure constant. Vacuum energy density which is related to cosmological constant is expressed as

S. Ghosh (✉)
Post Graduate, Department of Physics, Vijaya College, R. V. Road, Bangalore 560004, India
e-mail: gsovan@gmail.com

© Springer Nature Singapore Pte Ltd. 2021
P. K. Behera et al. (eds.), *XXIII DAE High Energy Physics Symposium*,
Springer Proceedings in Physics 261,
https://doi.org/10.1007/978-981-33-4408-2_135

$$\rho_\Lambda = \frac{c^4}{8\pi G}\Lambda = \frac{G}{8\pi}\frac{c^4}{\hbar^4}\left(\frac{m_e}{\alpha}\right)^6 \qquad (135.2)$$

The expressions (135.1) and (135.2) hint clearly that the vacuum energy density and cosmological constant are inversely proportional to the sixth power of the fine structure constant. Hence any variation in α effects Λ and ρ_Λ.

135.2 Variation of Fundamental Constants

The variation of the fundamental constants was hinted long back by Milne and Dirac independently, where both of them indicated a time variation of G [6, 7]. Teller was the first to propose a temporal variation of the fine structure constant α [8]. Variation of α was the second time hinted by Gamow [9]. The variation was probed by renormalization in QED by Dyson [10]. The current theory of fine structure constant is standing on the foundation provided by Bekenstein [11] and modified as the BBSM model [12].

Here we must mention that the variation of the fine structure constant basically can take place by the variation of the electrical charge or speed of light or both. The varying speed of light of course is only in cosmic time [13]. Bekenstein has considered the varying electrical charge [11] as

$$e = e_0\varepsilon(x^\mu) \qquad (135.3)$$

In a previous article by the current author, the calculations are shown considering both of them [14]. With the help of (135.3) calculations are done following both Bekenstein and the varying speed of light (VSL) and the bridge between them is also framed here.

The variation of the fine structure according to the Bekenstein prescription and the VSL theory are shown in Table 135.1.

Similarly, the temporal variation of magnetic moment are also calculated and tabulated in Table 135.2.

As we have $\rho_\Lambda \propto \Lambda \propto \alpha^{-6}$, one can say

$$\frac{\dot{\rho}_\Lambda}{\rho_\Lambda} = \frac{\dot{\Lambda}}{\Lambda} = -6\frac{\dot{\alpha}}{\alpha} \qquad (135.4)$$

Using (135.2) we find that

Table 135.1 Variation of fine structure in time	Using Bekenstein Prescription	Using VSL	Connection between them
	$\frac{\dot{\alpha}}{\alpha} = 2\frac{\dot{\varepsilon}}{\varepsilon}$	$\frac{\dot{\alpha}}{\alpha} = 2\frac{\dot{\varepsilon}}{\varepsilon} - \frac{\dot{c}}{c}$	$\left(\frac{\dot{\alpha}}{\alpha}\right)_C - \frac{\dot{c}}{c}$

Table 135.2 Variation of the magnetic moment in time	Using Bekenstein Prescription	Using VSL	Connection between them
	$\dfrac{\dot{\mu}}{\mu} = \dfrac{\dot{\varepsilon}}{\varepsilon} + \dfrac{\dot{\alpha}}{2\pi\left(1+\frac{\alpha}{2\pi}\right)}$	$\dfrac{\dot{\mu}}{\mu} = \dfrac{\dot{\varepsilon}}{\varepsilon} + \dfrac{\dot{\alpha}}{2\pi\left(1+\frac{\alpha}{2\pi}\right)} - \dfrac{\dot{c}}{c}$	$\left(\dfrac{\dot{\mu}}{\mu}\right)_C - \dfrac{\dot{c}}{c}$

$$\frac{\dot{\Lambda}}{\Lambda} = -12\frac{\dot{\varepsilon}}{\varepsilon} \tag{135.5}$$

and with varying speed of light theory we get that

$$\frac{\dot{\Lambda}}{\Lambda} = -12\frac{\dot{\varepsilon}}{\varepsilon} + 6\frac{\dot{c}}{c} \tag{135.6}$$

Hence for a decreasing speed of light, the cosmological constant and the vacuum energy density decrease.

From (135.4), we have seen that $\frac{\dot{\rho}_\Lambda}{\rho_\Lambda} = \frac{\dot{\Lambda}}{\Lambda}$.

But again following the varying speed of light theory, we have a difference in that result as

$$\frac{\dot{\rho}_\Lambda}{\rho_\Lambda} = \frac{\dot{\Lambda}}{\Lambda} + 4\frac{\dot{c}}{c} \tag{135.7}$$

135.3 α-Quantization

The α-quantized energy relations of the leptons and quarks are provided by MacGregor [15–17] and the current author [18]. Boson channel is expressed as

$$E_b = \frac{E_e}{\alpha} \tag{135.8}$$

Using (135.8), we get

$$\Lambda = \frac{G^2}{\hbar^4} m_b^6 \tag{135.9}$$

where m_b is the mass of the corresponding boson. Similarly, the fermion channel is expressed by MacGregor as

$$E_f = \frac{3E_e}{2\alpha} \tag{135.10}$$

Alike (135.9) the expressions of Λ can be derived from (135.10) as

$$\Lambda = \frac{G^2}{\hbar^4}\left(\frac{2}{3}m_f\right) \tag{135.11}$$

For muon channel, we have

$$E_\mu = \left(\frac{3}{2\alpha}+1\right)E_e \tag{135.12}$$

that leads to the expression of the cosmological constant as

$$\Lambda = \frac{G^2}{\hbar^4}\left(\frac{2}{3}\Delta m_{\mu e}\right) \tag{135.13}$$

where $\Delta m_{\mu e} = m_\mu - m_e$. Similarly for proton channel,

$$E_p = \left(\frac{27}{2\alpha}+1\right)E_e \tag{135.14}$$

we get the expression of the cosmological constant as

$$\Lambda = \frac{G^2}{\hbar^4}\left(\frac{2}{27}\Delta m_{pe}\right) \tag{135.15}$$

where $\Delta m_{pe} = m_p - m_e$. Comparing the expressions of (135.13) and (135.15), we find that

$$\Delta m_{pe} = 9\Delta m_{\mu e} \tag{135.16}$$

which is in good approximation with experimental data and this proves the α-quantization of the mass of the particles, proposed by MacGregor is true.

135.4   Conclusion

Cosmological constant is discussed here on the basis of its relation with fine structure and the temporal variation of the fine structure constant. As the cosmological constant is one of the dark energy candidates, this variation can lead us in the study of dark energy also. Further using the relation of the cosmological constant and fine structure, the α-quantization of the mass of the particles is proved to be true.

Acknowledgements The author is grateful to his wife Moumita Ghosh and daughter Oishani for an untiring inspiration and support without which this work was not possible.

References

1. A. Einstein, Kosmologische Betrachtungen zur Allegemeinen Relativitastheorie, Sitzung-berchte der Koniglich Preubischen Akademie der Wissenschaften (berlin), part 1, 142–152 (1917)
2. G. Gamow, *My World Line* (Viking Press, New York, 1970), p. 44
3. H. Wei et al., Eur. Phys. J. C **77**, 14 (2017)
4. Y.B. Zel'dovich, Pis'ma Zh. Eksp. Teor. Fiz. **6** 883 (1967) (English translation in JETP Lett. **6** 316 (1967))
5. Y.B. Zel'dovich, Usp. Fiz. Nauk **95** 209 (1968) (English translation in Sov. Phys. Usp. **11** 381 (1968))
6. C. Beck, Phys. A **388** 3384(2009), arxiv:0810.0752
7. E.A. Milne, Proc. R. Soc. A **158**, 324 (1937)
8. P.A.M. Dirac, Nat. Lett. **139**, 323 (1937)
9. E. Teller, Phys. Rev. **73**, 801 (1948)
10. G. Gamow, Phys. Rev. Lett. **19**, 759 (1967)
11. F.J. Dyson, Phys. Rev. Lett. **19**, 1291 (1967)
12. J.D. Bekenstein, Phys. Rev. D. **25**, 1527 (1982)
13. H.B. Sandvik, J.D. Barrow, J. Magueijo, Phys. Rev. Lett. **88**, 031302 (2002)
14. A. Albrecht, J. Magueijo, Phys. Rev. D **59**, 043516 (1999)
15. S. Ghosh, XXII DAE High Energy Physics Symposium, Springer Proceedings in Physics 203, https://doi.org/https://doi.org/10.1007/978-3-319-73171-1_107 (2018)
16. M.H. MacGregor, The Poewr of α, World Scientific (2007)
17. M.H. MacGregor, Indian J. Phys. **91**, 1437 (2017)
18. M.H. MacGregor, Indian. J Phys. **92**, 1473 (2018)
19. S. Ghosh, A. Choudhury, J.K. Sarma, Indian J. Phys **86**, 481 (2012)

Chapter 136
Q-PET: PET with 3rd Eye Quantum Entanglement Based Positron Emission Tomography

Sunil Kumar, Sushil Singh Chauhan, and Vipin Bhatnagar

Abstract In the present ongoing study, we proposed a prototype model by the introduction of a new discriminatory window parameter, which can be a new-generation PET detection technique. Positron Emission Tomography (PET) detection technique involves a coincidence detection technique to correlate the two annihilation photons. We introduced polarization measurement of the annihilation photons as an additional parameter in our proposed prototype to correlate annihilation photons of particular annihilation event. The motivation behind this introduction is quantum entanglement relation between the two photons. Simulation studies for this research work are undergoing and some preliminary results were presented.

136.1 Introduction

Medical imaging is the field in which radiation is used for imaging the body of the diseased patient. For this purpose, many such systems have been developed to serve the mankind. X-ray radiography, Computed Tomography (CT), Ultrasonography, Magnetic Resonance Imaging, Nuclear Medicine Imaging and Positron Emission Tomography (PET) are some techniques generally used in medical imaging. Out of these, PET is widely used for the staging, restaging, drug and therapy response of patients diagnosed with cancer. PET enables us to get morphological imaging of bio-distribution of positron-emitting radionuclide or radiopharmaceutical injected in the body of the patient/animal. Current PET detection technique involves coincidence detection technique to correlate the two annihilation photons emitted in the almost

S. Kumar (✉) · S. S. Chauhan · V. Bhatnagar
Department of Physics, Panjab University, Chandigarh 160014, India
e-mail: s.kumar@cern.ch

S. S. Chauhan
e-mail: schauhan@cern.ch

V. Bhatnagar
e-mail: Vipin.Bhatnagar@cern.ch

© Springer Nature Singapore Pte Ltd. 2021   949
P. K. Behera et al. (eds.), *XXIII DAE High Energy Physics Symposium*,
Springer Proceedings in Physics 261,
https://doi.org/10.1007/978-981-33-4408-2_136

exactly opposite direction detected by a ring-type scintillation-based detection system [1].

To reconstruct the raw data in an informative PET image, one needs to investigate exactly the true events with paired photons from the random, scattered and multiple events [2].

False coincidences introduce noise and contrast lost in the reconstructed image and also enhance the chances of misinterpretation. PET systems involve the two windows for the selection of true events which are energy window and timing window. At present, time and energy windowing, which are applied over the primary coincidence data, discard all multiple events, as well as a considerable fraction of unscattered coincidence events due to the relatively poor energy resolution of current detectors [2].

Besides these parameters for the selection of true coincidence, there is another parameter which, possibly can accurately measure the true coincidence.

136.2 Motivation

The motivation towards this work comes from the fact that using Compton scattering, we can find the polarization of the photons and that can be used to identify the correlated true annihilation photons. Two photons emitted are linearly polarized such that the polarization vectors are orthogonal to each other, i.e. having a quantum entangled state in which both the photons have their planes of polarization perpendicular to each other [3]. This quantum entangled state can work as a discriminatory window for the identification of true annihilation events after clearing the first two windows. So if one can measure the polarization of each photon and can identify the orthogonal relation between them, this technique can work as a powerful tool to identify accurately the two annihilation photons.

136.3 Method

Compton cross section depends on the photon polarization and this dependence make us able to calculate polarization of the photons involved in the annihilation process [4]. It is also implied by Quantum Electrodynamics that the two photons emitted in an electron–positron annihilation process are polarized orthogonal(perpendicular) to each other [4]. In this study we have planned for using the angular correlation between Compton scattered photons. For this purpose, Klein–Nishina Model for Compton Scattering will be used as a theoretical basis. A GEANT4 Simulation will provide the requisite variables for the calculations of the angular correlation.

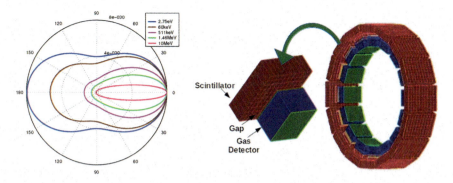

Fig. 136.1 Klein–Nishina Differential Cross Section for different energies (left) and a GEANT4 constructed geometry for the proposed prototype (right)

The differential cross section of Compton scattering [5] is expressed as

$$\frac{d\sigma}{d\Omega} = \frac{r_0^2}{2} \left(\frac{E'}{E}\right)^2 \left(\frac{E}{E'} + \frac{E'}{E} - 2\sin^2\theta\cos^2\eta\right), \tag{136.1}$$

where

r_0 = classical electron radius, E = energy of incident γ ray, E' = energy of scattered γ ray, θ = angle of scattering, η = angle between plane of scattering & plane of polarization.

136.4   Prototype

136.4.1   Geometry

Q-PET consists of 16 units of detector arranged in a ring of inner radius 30.78 cm and outer radius 47.04 cm. Each unit consists of one gaseous detector (Scatterer/Tracker) and one scintillation block.

136.4.1.1   Tracker/Scatterer

In Tracker Section, Gas Chamber has been introduced for tracking purposes. Material used for the gas chamber is $XeCO_2C_4F_{10}$ in proportion $Xe(50\%)$, $CO_2(15\%)$, $C_4F_{10}(35\%)$. The material ($XeCO_2C_4F_{10}$) so chosen for the gas detector contains Xenon, which is the highest density Noble gas can be used for the proportional chambers. To increase the probability of 511 keV photons interaction with gas molecule, density of the gas should be large. Besides this one has to consider some factor, i.e. interaction cross section for the desired process, density should be enough to have the interaction probability to get the Compton interaction.

Gas detector has dimension 10 cm × 10 cm × 8.96 cm. Gaseous detector will provide the position of very first interaction of annihilation photon in the gaseous medium. Signal generated by the gaseous detector also gives information on the track of recoil electron. Time Projection Chamber (TPC) configuration can be used to get the third coordinate (depth of interaction) point. Besides that, some other methods are there to get the depth of interaction point.

Tracking of recoil electron can provide the additional benefit of getting the real direction of gamma-ray entrance within some uncertainty. This feature is very important for large diameter scanners.

136.4.1.2 Scintillator

In Scintillator Section, $CdWO_4$ crystals are used to absorb the scattered gamma ray completely. Scintillation Block of each unit have 10 × 10 arrangement of crystals of dimension 1.58 cm × 1.58 cm × 5 cm each. This section will provide position and energy of the scattered gamma ray.

136.5 Summary and Future Work

The polarization measurement technique can work as a discriminatory window for the identification of annihilation photons. Presently, working to improve the interaction probability with the analysis of the simulation data and simulation for the differential cross-sectional data needed for the polarization calculation. We will update and publish the results soon.

Acknowledgements Sushil Singh Chauhan, Sunil Kumar and Vipin Bhatnagar offer sincere gratitude to UGC, CSIR & DST for the financial support.

References

1. M.E. Phelps, PET Physics, Instrumentation and Scanner (Springer) ISBN-13: 978-0387-32302-2
2. M. Toghyani, J.E. Gillam, A.L. McNamara, Z. Kuncic, Polarization-based coincidence event discrimination, an in silico study towards a feasible scheme for Compton-PET. Phys. Med. Biol. **61**, 5803–5817 (2016)
3. D. Robley, *Evans: the Atomic Nucleus* (TATA McGRAW-HILL Publishing Company LTD, Bombay, New Delhi, 1955)
4. N. Carlin, P. Woit, Theory of angular correlation of annihilation radiation. In: The Angular Correlation of Polarization of Annihilation Radiations. Available online http://www.math.columbia.edu/~woit/angular.pdf
5. W. Hajdas, Estela Suarez-Garcia: Polarimetry at High Energies. ISSI Scientific Reports Series, ESA/ISSI. (2010), pp. 559–571 ISBN 978-92-9221-938-8

Chapter 137
Signature of Light Sterile Neutrinos at IceCube

Bhavesh Chauhan and Subhendra Mohanty

Abstract The MiniBooNE collaboration has reported evidence for a light sterile neutrino with large mixing angles which is consistent with the results by LSND collaboration approximately 20 years ago. However, any such state would be in conflict with Planck measurement of N_{eff} during Big Bang nucleosynthesis. If there is sufficient self-interaction in the sterile sector, the large effective thermal mass can suppress its production in the early universe. Our objective is to investigate if such interactions could allow for resonant absorption in the astrophysical neutrino spectrum and whether there are observable consequences for IceCube. We show that it is possible to give independent bounds on the parameter space from IceCube observations with the absorption lines corresponding to the neutrino masses.

137.1 Introduction

The MiniBooNE collaboration has recently reported an excess in the neutrino and anti-neutrino appearance channels which is consistent with the sterile neutrino hypothesis [1]. The best-fit point,

$$\Delta m_{41}^2 = 0.041 \ eV^2 \quad \text{and} \quad \sin^2(2\theta_{\mu e}) = 0.958 \tag{137.1}$$

is consistent with the earlier measurements by LSND collaboration [2]. In fact, the combined significance of the two datasets is 6.1σ. These results, however, are in tension with data from disappearance experiments like MINOS+ and IceCube. Other experiments like KARMEN and OPERA have not been able to confirm this excess,

B. Chauhan (✉) · S. Mohanty
Physical Research Laboratory, Ahmedabad 380009, India
e-mail: bhavesh@prl.res.in

S. Mohanty
e-mail: mohanty@prl.res.in

B. Chauhan
Indian Institute of Technology, Gandhinagar 382355, India

© Springer Nature Singapore Pte Ltd. 2021
P. K. Behera et al. (eds.), *XXIII DAE High Energy Physics Symposium*,
Springer Proceedings in Physics 261,
https://doi.org/10.1007/978-981-33-4408-2_137

but they do not rule it out completely either [3]. The existence of such light states with large mixing angles is also in conflict with cosmology. The Planck experiment puts severe constraints on number of thermalized relativistic degrees of freedom (N_{eff}) around the epoch of big bang nucleosynthesis (BBN) (i.e., $T_\gamma = 1$ MeV) using Cosmic Microwave Background (CMB) anisotropy [4]. A simple resolution to this puzzle is to assume self-interactions in the sterile sector [5–11]. For gauge coupling in the range 0.1–1 , one requires a gauge boson of mass 10–50 MeV to reconcile sterile neutrinos with cosmology. A consequence of this scenario is that there is a sterile neutrino background in the present universe. On the other hand, it was shown in [12] that MeV scale secret interaction of neutrinos will give rise to absorption lines in the very high-energy neutrino spectrum. Such lines can be seen by neutrino telescopes like IceCube. The IceCube HESE sample has featured a gap in the spectrum for neutrino energies in the range 400–800 TeV [13–15]. In this paper, we explain the gap as a signature of absorption by cosmic sterile neutrino background.

In this paper, we summarize the results in [16]. In Sect. 137.2, we describe the model and discuss the basics of neutrino absorption. In Sect. 137.3, we look at the allowed parameter space and the neutrino spectrum for benchmark points of the model. In Sect. 4, we conclude.

137.2 Model, Constraints, and Results

The Standard Model is extended by introducing a left-handed sterile neutrino (ν_s) which is charged under an additional gauge symmetry $U(1)_X$ with coupling strength g_X. The relevant term in the new Lagrangian is the gauge interaction of the sterile neutrino which in the mass basis as

$$- \mathcal{L}_s = \sum_{i,js} g_{ij} \bar{\nu}_i \gamma^\mu P_L \nu_j X_\mu, \tag{137.2}$$

where U is the 4×4 neutrino mixing matrix and $g_{ij} = g_X U_{si}^* U_{sj}$. The active neutrino mixing angles are taken to be the central values from the oscillation measurements [17]. In this paper, we use m_4 as the mass of the fourth (mostly sterile) mass eigenstate and M_X as the mass of new gauge boson. The details of neutrino absorption is discussed in [16] and references therein.

We examine the $m_4 - M_X$ parameter space that can explain the observed Ice-Cube spectrum. If the absorption line is between 800 and 3000 TeV, one cannot explain the three PeV events at IceCube unless exceptional circumstances are evoked. To be general, we constrain the m_3 absorption not to lie in this range. Since we wish to explain the dip in the spectrum using the fourth neutrino, we require, $E^{res} \leq 800$ TeV and $E^{res}/(1 + \langle z_s \rangle) \geq 400$ TeV. As discussed in [11], we also show the region in the parameter space that requires more than one and two extra lighter sterile neutrinos in the full theory. It can be seen from Fig. 137.1

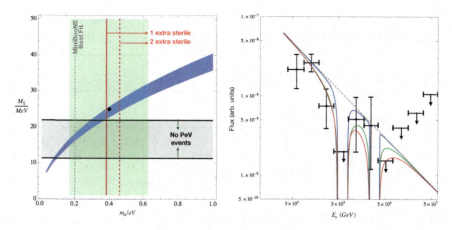

Fig. 137.1 (*Left*): The shaded blue region with solid (dashed) boundaries can explain the 400–800 TeV dip in the IceCube spectrum. The green shaded region is the allowed mass from MiniBooNE and the best-fit point is shown. The gray-shaded region denotes the range of X boson mass excluded from PeV events at IceCube. The red lines (solid, dashed) show the number of additional light particles (one, two) to be added to the theory. The black point shows the benchmark case considered in the paper. *Right*: The flux without attenuation is shown as dashed gray curve. The blue, green, and red curves is the flux with attenuation for the BM1, BM2, and BM3, respectively

that only a small portion of the parameter space is compatible with all the constraints. We chose the benchmark point in $m_4 - M_X$ plane to be (0.4 eV, 25 MeV) for our analysis. The gauge coupling is constrained from the restrictions on the recoupling temperature. We chose a benchmark point $g_X = 0.1$ which is consistent. For $m_4 = 0.4$ eV, the 68 % C.L. from MiniBooNE requires $\sin^2 \theta_{\mu e} \in (0.03, 0.075)$, whereas the 3σ range is $\sin^2 \theta_{\mu e} \in (0.0055, 0.1115)$. We use two representative values of the mixing angles for showing the neutrino spectrum with absorption. For the first benchmark case (BM1), we consider $\theta_{14} = \theta_{24} = \theta_{34} = 0.1$, for BM2, we consider $\theta_{14} = \theta_{24} = \theta_{34} = 0.3$, and for BM3, we consider $\theta_{14} = \theta_{24} = \theta_{34} = 0.5$. The spectral index is chosen to be 2.6 and the normalization is fixed from the second bin. Sources are assumed to be distributed around z = 0.6.

137.3   Conclusion

To make a light sterile neutrino observed by MiniBooNE consistent with cosmology, one must introduce self-interactions between the sterile neutrinos. Since the mediators required are at MeV scale, there will be observable effects in the spectrum of high-energy neutrinos detected by IceCube. We show that the apparent gaps in the spectrum at 400–800 TeV as well as beyond 2.6 PeV can be attributed to resonant absorption by the two heaviest mass eigenstates in the background. The features obtained not only explain the present data but may also be testable in future IceCube data. In the scenario discussed, IceCube is blind to neutrino sources beyond a certain z_{max}. Future multi-messenger astronomy should be able to confirm this.

References

1. A.A. Aguilar-Arevalo et al., MiniBooNE Collaboration. arXiv:1805.12028[hep-ex]
2. A. Aguilar-Arevalo et al., LSND Collaboration. Phys. Rev. D **64**, 112007 (2001). https://doi.org/10.1103/PhysRevD.64.112007[hep-ex/0104049]
3. M. Dentler, Hernndez-Cabezudo, J. Kopp, P. Machado, M. Maltoni, I. Martinez-Soler, T. Schwetz, JHEP **1808**, 010 (2018). https://doi.org/10.1007/JHEP08(2018)010. arXiv:803.10661[hep-ph]
4. N. Aghanim et al., Planck Collaboration. arXiv:1807.06209[astro-ph.CO]
5. S. Hannestad, R.S. Hansen, T. Tram, Phys. Rev. Lett. **112**(3), 031802 (2014). https://doi.org/10.1103/PhysRevLett.112.031802. arXiv:1310.5926[astro-ph.CO]
6. X. Chu, B. Dasgupta, J. Kopp, JCAP **1510**(10), 011 (2015). https://doi.org/10.1088/1475-7516/2015/10/011. arXiv:1505.02795[hep-ph]
7. M. Archidiacono, S. Hannestad, R.S. Hansen, T. Tram, Phys. Rev. D **91**(6), 065021 (2015). https://doi.org/10.1103/PhysRevD.91.065021. arXiv:1404.5915[astro-ph.CO]
8. M. Archidiacono, S. Hannestad, R.S. Hansen, T. Tram, Phys. Rev. D **93**(4), 045004 (2016). https://doi.org/10.1103/PhysRevD.93.045004. arXiv:1508.02504[astro-ph.CO]
9. M. Archidiacono, S. Gariazzo, C. Giunti, S. Hannestad, R. Hansen, M. Laveder, T. Tram, JCAP **1608**(08), 067 (2016). https://doi.org/10.1088/1475-7516/2016/08/067. arXiv:1606.07673[astro-ph.CO]
10. B. Dasgupta, J. Kopp, Phys. Rev. Lett. **112**(3), 031803 (2014). https://doi.org/10.1103/PhysRevLett.112.031803. arXiv:1310.6337[hep-ph]
11. X. Chu, B. Dasgupta, M. Dentler, J. Kopp, N. Saviano. arXiv:1806.10629[hep-ph]
12. M. Ibe, K. Kaneta, Phys. Rev. D **90**(5), 053011 (2014). https://doi.org/10.1103/PhysRevD.90.053011. arXiv:1407.2848[hep-ph]
13. M.G. Aartsen et al., IceCube Collaboration. Phys. Rev. Lett. **113**, 101101 (2014). https://doi.org/10.1103/PhysRevLett.113.101101. arXiv:1405.5303[astro-ph.HE]
14. M.G. Aartsen et al., IceCube Collaboration. arXiv:1510.05223[astro-ph.HE]
15. D. Williams, IceCube Collaboration, Int. J. Mod. Phys. Conf. Ser. **46**, 1860048 (2018). https://doi.org/10.1142/S2010194518600480
16. B. Chauhan, S. Mohanty, Phys. Rev. D **98**(8), 083021 (2018). https://doi.org/10.1103/PhysRevD.98.083021. arXiv:1808.04774[hep-ph]
17. I. Esteban, M.C. Gonzalez-Garcia, M. Maltoni, I. Martinez-Soler, T. Schwetz, JHEP **1701**, 087 (2017). https://doi.org/10.1007/JHEP01(2017)087. arXiv:1611.01514[hep-ph]

Chapter 138
The Measurement of the Difference Between ν and $\bar{\nu}$ Mass-Squared Splittings in Atmospheric and Long-Baseline Neutrino Experiments

Daljeet Kaur, Sanjeev Kumar, and Md Naimuddin

Abstract We present an estimated sensitivity for the differences in the mass-squared splittings of ν and $\bar{\nu}$ oscillations for the atmospheric neutrino experiment (ICAL@INO) and for long-baseline experiments (T2K & NOvA). We analysed Charged-Current (CC) ν_μ- and $\bar{\nu}_\mu$-independent interactions with the detector, assuming three flavor oscillations along with the inclusion of the Earth matter effects. The observed ν_μ and $\bar{\nu}_\mu$ events spectrum folded with realistic detector resolutions and efficiencies are separately binned to direction and energy bins, and a χ^2 is minimized with respect to each bin to find out the oscillation parameters for ν_μ and $\bar{\nu}_\mu$ independently. Assuming non-identical atmospheric oscillation parameters for ν and $\bar{\nu}$, we estimate the detectors sensitivity to confirm a non-zero difference in the mass-squared splittings ($|\Delta m^2_{32}| - |\Delta \overline{m}^2_{32}|$).

138.1 Introduction

In the phenomena of neutrino oscillations, the mass-squared splittings and mixing angles are expected to be identical for neutrinos and antineutrinos by CPT symmetry. Comparing the oscillation parameters of neutrinos and anti-neutrinos independently could, therefore, be a particular test of CPT conservation or any difference between them may indicate a sign of new physics. Here, we consider that the oscillation probability governed by a mass splitting or mixing angle for neutrinos is different compared to antineutrinos. We assume the standard three neutrino paradigm and perform realistic simulations of the current and future atmospheric experiment. We simulate the atmospheric IronCalorimeter (ICAL) detector [1] and the long-baseline

D. Kaur (✉)
S.G.T.B. Khalsa College, University of Delhi, New Delhi, India
e-mail: daljeet.kaur97@gmail.com

S. Kumar · M. Naimuddin
Department of Physics and Astrophysics, University of Delhi, New Delhi, India

© Springer Nature Singapore Pte Ltd. 2021
P. K. Behera et al. (eds.), *XXIII DAE High Energy Physics Symposium*,
Springer Proceedings in Physics 261,
https://doi.org/10.1007/978-981-33-4408-2_138

oscillation experiments such as NOvA [2] and T2K [3] in order to determine their potential in probing the difference between neutrino and anti-neutrino mass-squared splittings.

138.2   Simulation Details

We simulate neutrino events using the ICAL detector for 10 years, using NOvA for 3 years and T2K for 5 years for ν and $\bar{\nu}$ independently. Non-identical atmospheric oscillation parameters for neutrinos and anti-neutrinos are considered. All the four atmospheric oscillation parameters, i.e. $|\Delta m^2_{32}|$, $|\overline{\Delta m}^2{}_{32}|$, θ_{23} and $\overline{\theta}_{23}$ are varied in a wide range. Using the results of the four parameters analysis, we study the prospects of the scenario when the differences $(|\Delta m^2_{32}| - |\overline{\Delta m}^2{}_{32}|)$ are non-zero. Fixed true values and variation range used for oscillation parameters are listed in the below Table 138.1.

For simulating atmospheric neutrinos at the ICAL, we use the neutrino events generated through NUANCE event generation. Unoscillated neutrino/anti-neutrino events are generated for 1000 years of exposure of ICAL and it scaled to 10 years of ICAL running for estimation of sensitivity. Reconstruction and charge identification efficiencies (Obtained by INO) are implemented. Muon energy, Muon angle resolutions and hadron energy resolutions are implemented. Twenty muon energy bins ranging from 0.8 to 12.8 GeV, 20 muon direction bins ($\cos\theta$ bins ranging from -1 to $+1$)and 5 hadron bins ranging from 0.0 to 15.0 GeV. Binning is done in a similar way for neutrinos and anti-neutrinos with $20 \times 20 \times 5 = 2000$ bins with optimized bin size. A χ^2 function with method of pulls for including systematic uncertainties has been calculated for neutrino and anti-neutrino separately. Other simulation details are similar to the reference [4]. For simulating T2K and NOvA long-baseline experiments, we use the Globes Neutrino Simulator with the simulation details given in Table 138.2. Other simulation details are similar to the references [5, 6].

Table 138.1 $\nu(\bar{\nu})$ oscillation parameters used in the analysis

Neutrino/anti-neutrino oscillation parameters		
Osc. parameters	True values	Range
$\sin^2(2\theta_{12})$	0.86	Fixed
$\sin^2(\theta_{23})$	varied	0.3–0.7
$\sin^2(\theta_{13})$	0.0234	Fixed
Δm^2(sol.) (eV^2)	7.6×10^{-5}	Fixed
Δm^2(atm.) (eV^2)	varied	$(2.0$–$3.0) \times 10^{-3} (3\sigma)$
δ_{CP}	0.0	Fixed

Table 138.2   Characteristics of NOvA and T2K detectors used for simulation

Characteristics	NOvA	T2K
Baseline	812 km	295 km
Location	Fermilab-Ash river	J-PARC-Kamioka
Detector	TASD	Water Cherenkov
Target mass	14 kton	22.5 kton
Runtime	3 in ν and 3 in $\bar{\nu}$	5 in ν and 5 in $\bar{\nu}$
Beam power	0.7 MW	0.75 MW
Osc. channel	CC disappearance	CC disappearance
Signal norm. error	5%	5%
Background norm. error	10%	10%

138.3   Results and Conclusions

The independent measurements of neutrinos and anti-neutrinos for atmospheric neutrinos at the ICAL and for long-baseline experiments (T2K and NOvA) have been shown. As a preliminary result, we show the sensitivities of T2K and NOvA for lower octant, maximal mixing and for higher octant for different true values of mass-squared splittings assuming CPT is true. It is clear from the Figs. 138.1 and 138.2 that assuming CPT is a good symmetry, neutrino-only analysis gives more stringent parameters compared to the anti-neutrino only analysis, while the combined $\nu+\bar{\nu}$ analysis shows the most stringent parameter space. This study also shows that these experiments can solve octant degeneracy by taking νs and $\bar{\nu}$s alone.

Using non-identical oscillation parameters of ν_μ and $\bar{\nu}_\mu$, we measured the sensitivity of atmospheric experiment (ICAL@INO) for difference in oscillation parameters of ν_μ and $\bar{\nu}_\mu$. Figure 138.1(Right) shows the ICAL detector sensitivity to confirm a non-zero difference in the mass-squared splittings ($|\Delta m^2_{32}| - |\overline{\Delta m^2}_{32}|$). With the variation of true as well as observed values, the ICAL can rule out the null hypothesis of $|\Delta m^2_{32}| = |\overline{\Delta m^2}_{32}|$ at more than 3σ level if the difference of true values of $|\Delta m^2_{32}| - |\overline{\Delta m^2}_{32}| \geq +0.7 \times 10^{-3} eV^2$ or $|\Delta m^2_{32}| - |\overline{\Delta m^2}_{32}| \leq -0.7 \times 10^{-3} eV^2$. The study of long-baseline sensitivities for the difference of these mass-squared splitting is still in progress.

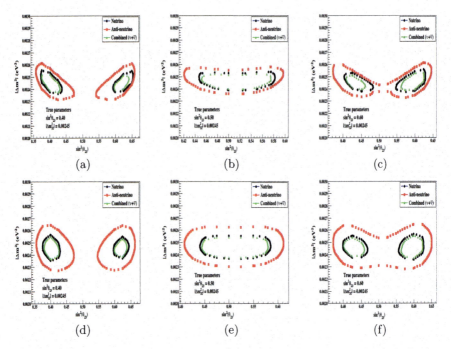

Fig. 138.1 90% C.L. from NOvA (Top) and T2K (bottom) experiments for Lower octant ($\sin^2 \theta_{23} =$ 0.40) (Left), maximal mixing ($\sin^2 \theta_{23} = 0.50$) (Middle) and for Higher octant ($\sin^2 \theta_{23} = 0.60$) (Right) with $|\Delta m^2_{32}| = 2.45 \times 10^{-3} eV^2$, assuming CPT is a true symmetry. Red, black and green contours are obtained as a results of anti-neutrino, neutrino and combined ($\nu + \bar{\nu}$) analysis, respectively

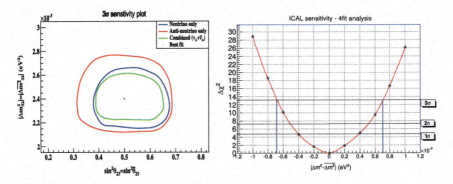

Fig. 138.2 90% C.L. expected region obtained for the ICAL at maximal mixing ($sin^2\theta_{23} = 0.50$) asuming CPT is a true symmetry (Left). The ICAL sensitivity for $(|\Delta m^2_{32}| - |\overline{\Delta m^2}_{32}|)_{True}(eV^2)$ at 1σ, 2σ and 3σ confidence levels assuming CPT is not true (Right)

References

1. The sensitivity of the ICAL detector at India-based Neutrino Observatory to neutrino oscillation parameters. arXiv:14092231v1[hep-ex]
2. P. Adamson et al., NOvA Collaboration. Phys. Rev. Lett. **118**, 231801 (2017)
3. K. Abe et al., T2K Collaboration. Phys. Rev. D **96**, 092006 (2017)
4. D. Kaur et al., Phys. Rev. D **95**, 093005 (2017)
5. https://www.mpi-hd.mpg.de/personalhomes/globes/glb/0709-nova.html
6. https://www.mpi-hd.mpg.de/personalhomes/globes/glb/0709-t2k.html

Chapter 139
eV Scale Sterile Neutrinos in A_4 Symmetric Model

Mitesh Kumar Behera and Rukmani Mohanta

Abstract The observed anomalies by several short-baseline neutrino oscillation experiments suggest the possible existence of an eV-scale sterile neutrino. Considering a modified discrete A_4 flavour symmetric model in the framework of two eV-scale sterile neutrinos, we can explain the current neutrino oscillation parameters along with the neutrinoless double beta decay (NDBD) bound on effective neutrino mass. This model also accommodates a non-zero reactor mixing angle compatible with the current observation.

139.1 Introduction

Though the Standard Model (SM) is an extremely successful theory, still it is not able to explain many observed phenomena in particle physics, astrophysics and cosmology. For instance, SM is unable to resolve the puzzles of neutrino physics. These days neutrinos are the most talked about topic in the scientific community as they are believed to contribute both in the microscopic and macroscopic world. One of the profound discoveries in neutrino physics is its flavour oscillation, which suggests that neutrinos are massive. Furthermore, the three neutrino framework is defined by six independent parameters, in which, three are mixing angles and two are mass squared differences, which is well established by the experimental data. But, the CP-violating phase is not confirmed yet due to experimental limitations and becomes the main focus to be resolved in near future. In addition to the above, neutrinos mass hierarchy is also not established which is believed to be of great importance in particle physics and cosmology. Furthermore, there are some other neutrino anomalies seen in several experiments which give rise to the existence of another type of massive neutrino called the sterile neutrino. It is a hypothetical

M. K. Behera (✉) · R. Mohanta (✉)
University of Hyderabad, Hyderabad 500 046, India
e-mail: miteshbehera1304@gmail.com

R. Mohanta
e-mail: rukmani98@gmail.com

© Springer Nature Singapore Pte Ltd. 2021
P. K. Behera et al. (eds.), *XXIII DAE High Energy Physics Symposium*,
Springer Proceedings in Physics 261,
https://doi.org/10.1007/978-981-33-4408-2_139

particle that does not interact via any fundamental interactions of the SM. Moreover, the existence of sterile neutrino can help to understand why neutrinos have a tiny mass, origin of dark matter candidate and role in baryogenesis. Sterile neutrinos are considered to have right-handed chirality, where the mass of right-handed neutrino can be between 10^{15} GeV and sub-eV. In this work, we are interested to study the neutrino phenomenology in the presence of two sterile neutrinos in the A_4 discrete flavour symmetric model.

139.2 A_4 Symmetry Group

A_4 is the symmetry group [1] of a tetrahedron, and is a discrete non-abelian group of even permutations of four objects. It has four irreducible representations: 3 one-dimensional and 1 three-dimensional which are denoted by 1, 1′, 1″ and 3, respectively, being consistent with the sum of square of the dimensions $\Sigma_i n_i^2 = 12$. Their product rules of the irreducible representations are given as

$$1 \otimes 1 = 1, \quad 1' \otimes 1' = 1'', \quad 1' \otimes 1'' = 1, \quad 1'' \otimes 1'' = 1'$$
$$3 \otimes 3 = 1 \otimes 1' \otimes 1'' \otimes 3_a \otimes 3_s$$

139.3 The Model Framework

Our proposed model contains in addition to the SM particles, two sterile neutrinos N_1 and N_2 along with the flavon fields ϕ_T, χ, ζ and ϕ and the details of their quantum numbers are listed in Table 139.1. The Yukawa interaction for charged leptons, allowed by the symmetries of the model are as follows:

$$\mathcal{L}_\ell = -H \left[\frac{y_e}{\Lambda} \left(\overline{L_L} \phi_T \right)_1 \otimes e_R + \frac{y_\mu}{\Lambda} \left(\overline{L_L} \phi_T \right)_{1'} \otimes \mu_R + \frac{y_\tau}{\Lambda} \left(\overline{L_L} \phi_T \right)_{1''} \otimes \tau_R \right]$$
$$= -\frac{v_h v_T}{\Lambda} \left[y_e (\overline{e_L} e_R) + y_\mu (\overline{\mu_L} \mu_R) + y_\tau (\overline{\tau_L} \tau_R) \right]. \tag{139.1}$$

For neutral leptons, the Lagrangian with higher dimensional mass terms is given by

Table 139.1 Field content and their corresponding charges

Field	L	e^c	μ^c	τ^c	N_1	N_2	H	ϕ_T	χ	ζ	ϕ
$SU(2)_L$	2	1	1	1	1	1	2	1	1	1	1
A_4	3	1	1″	1′	1	1	1	3	3	1′	1

$$\mathscr{L}_y{}^\nu = -\frac{y_1 v^2 v_2}{\Lambda^2}[L_1 L_1 + L_2 L_3 + L_3 L_2] - \frac{y_\chi v^2}{3\Lambda^2}[(2L_1 L_1 - L_2 L_3 - L_3 L_2)v_\chi$$
$$+ (2L_3 L_3 - L_1 L_2 - L_2 L_1)v_\chi + (2L_2 L_2 - L_1 L_3 - L_3 L_1)v_\chi]. \quad (139.2)$$

The Majorana mass terms for new fermions and their interaction with SM leptons is given by

$$L_{N_R} = -\sum_\alpha \frac{y_\alpha}{\Lambda} N_\alpha N_\alpha (\phi)^2 - \sum_\alpha \frac{y_s}{\Lambda} \overline{L} H N_\alpha \chi. \quad (139.3)$$

If one considers the case of $3 + 2$ case [2] by including two sterile neutrino along with 3 active neutrinos in this model, the mass matrix will have the form that gives a TBM mixing pattern [3] with vanishing θ_{13}. Hence, we add a perturbation term in the Lagrangian $\frac{y_p}{\Lambda^2} LLHH\zeta + $ h.c. Now the modified mass matrix can be written as

$$\mathscr{M}_\nu = \frac{v^2}{\Lambda} \begin{pmatrix} a + \frac{2d}{3} & -\frac{d}{3} & -\frac{d}{3} + b & e & f \\ -\frac{d}{3} & \frac{2d}{3} + b & a - \frac{d}{3} & e & f \\ -\frac{d}{3} + b & a - \frac{d}{3} & \frac{2d}{3} & e & f \\ e & e & e & m_{s1} & 0 \\ f & f & f & 0 & m_{s2} \end{pmatrix},$$

$$U = \begin{pmatrix} \frac{-p_+}{l_{p+}} & \frac{1}{6f}\frac{K_{5p-}}{N_{5p-}} & \frac{-p_-}{l_{p-}} & \frac{1}{6f}\frac{K_{5p+}}{N_{5p+}} & 0 \\ \frac{q_+}{l_{p+}} & \frac{1}{6f}\frac{K_{5p-}}{N_{5p-}} & \frac{q_-}{l_{p-}} & \frac{1}{6f}\frac{K_{5p+}}{N_{5p+}} & 0 \\ \frac{1}{l_{p+}} & \frac{1}{6f}\frac{K_{5p-}}{N_{5p-}} & \frac{1}{l_{p-}} & \frac{1}{6f}\frac{K_{5p+}}{N_{5p+}} & 0 \\ 0 & \frac{e}{f N_{5p-}} & 0 & \frac{e}{f N_{5p+}} & -\frac{f}{e N_5} \\ 0 & \frac{1}{N_{5p-}} & 0 & \frac{1}{N_{5p+}} & -\frac{1}{N_5} \end{pmatrix} \quad (139.4)$$

where $a = \frac{y_1 v_2 v^2}{\Lambda^2}$, $d = \frac{y_\chi v_\chi v^2}{\Lambda^2}$, $e = \frac{y_{s1} v v_4 v_\chi}{\Lambda^2}$, $f = \frac{y_{s2} v v_4 v_\chi}{\Lambda^2}$ and U is the eigenvector matrix that diagonalizes the neutrino mass matrix.

$$K_{p5\pm} = a + b - m_s \pm \sqrt{12e^2 + 12f^2 + (a + b - m_s)^2}$$

$$N_{p5\pm}^2 = 1 + \frac{\left(a + b - m_s \pm \sqrt{12e^2 + 12f^2 + (a + b - m_s)^2}\right)^2}{12e^2}$$

$$p_\pm = \frac{a \pm \sqrt{a^2 - ab + b^2}}{a - b} \qquad q_\pm = \frac{b \pm \sqrt{a^2 - ab + b^2}}{a - b}$$

$$l_{p\pm}^2 = 1 + (p_\pm)^2 + (q_\pm)^2$$

In the limit $a < m_s$, the higher order terms in the small ratio e/m_s can be neglected. The mass eigenvalues of the neutrino mass matrix in the limit of vanishing θ_{13} is given by

$$m_{\nu_1} = d + \sqrt{a^2 - ab + b^2}, \quad m_{\nu_3} = d - \sqrt{a^2 - ab + b^2}, \quad m_{\nu_5} = m_s$$
$$m_{\nu_2} = \frac{1}{2}[a + b + m_s - \sqrt{12(e^2 + f^2) + (a + b - m_s)^2}]$$
$$m_{\nu_4} = \frac{1}{2}[a + b + m_s + \sqrt{12(e^2 + f^2) + (a + b - m_s)^2}]$$

The allowed parameter space consistent with the current neutrino oscillation data is shown in the left panel of Fig. 139.1 and the correlation plot between $\sin^2 \theta_{12}$ and $\sin^2 \theta_{23}$ is shown in the right panel. The contribution to the effective mass parameter in neutrinoless double beta decay is shown in Fig. 139.2.

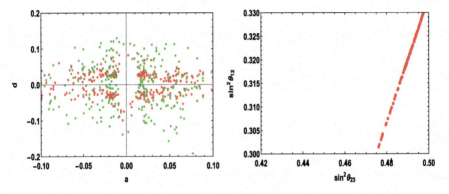

Fig. 139.1 Allowed parameter space is consistent with the neutrino oscillation data (left panel) and the correlation plot between $\sin^2 \theta_{12} - \sin^2 \theta_{23}$ (right panel)

Fig. 139.2 The variation of effective mass parameter in NDBD with the lightest neutrino mass for 3+2 neutrino case

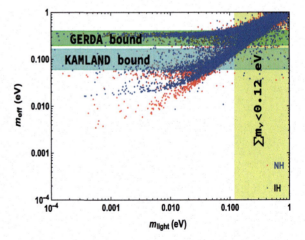

139.4   Conclusion

We discuss the neutrino masses and mixing in a A_4 flavour symmetric model with nonvanishing θ_{13} compatible with current observation. This model also contributes to NDBD within the experimental domain. The detailed study is done in [4].

Acknowledgements We acknowledge DST and SERB for their financial support.

References

1. D. Borah, B. Karmakar et al., A_4 flavour model for Dirac neutrinos: type I and inverse seesaw. Phys. Lett. B **780**, 461 (2018)
2. S. Goswami, W. Rodejohann, MiniBooNE results and neutrino schemes with 2 sterile neutrinos: possible mass orderings and observables related to neutrino masses. JHEP **10**, 073 (2007)
3. P.P. Novichkov, S.T. Petcov, M. Tanimoto, Trimaximal neutrino mixing from modular A4 invariance with residual symmetries. Phys. Lett. B **793**, 247 (2019)
4. S. Mishra, M. Kumar Behera, S. Singirala, R. Mohanta, S. Patra, (To be submitted)

Chapter 140
Active and Sterile Neutrino Phenomenology with A_4-Based Minimal Extended Seesaw

Pritam Das, Ananya Mukherjee, and Mrinal Kumar Das

Abstract Minimal extended seesaw (MES) framework is studied in this work, which plays an important role in active and sterile neutrino phenomenology in (3+1) scheme. The A_4 flavour symmetry is augmented by an additional Z_4 symmetry to constraint the Yukawa Lagrangian of the model. We use a non-trivial Dirac mass matrix, with broken $\mu - \tau$ symmetry, as the origin of leptonic mixing. Interestingly, such structure of mixing with a leading order perturbation leads to the non-zero reactor mixing angle θ_{13}. Non-degenerate mass structure for right-handed neutrino M_R is considered. A gauge singlet chiral field S is added into the picture which puts sterile neutrino in the model. Finally, this model study give us bounds on various experimental (neutrino) parameters like the atmospheric mixing angle, active-sterile mixing angles.

140.1 Introduction

The Standard Model (SM) of particle physics is the most successful and inspiring model to explain particle nature and their interactions. However, there are many unsolved phenomena that need explanation. Tiny neutrino masses were unexplored in the SM framework due to the tiny Yukawa coupling value, hence we need to go beyond the SM to explain neutrinos and their behaviours, popularly known as BSM. In standard neutrino scenario, three active neutrinos (ν_e, ν_μ, ν_τ) are involved with

P. Das (✉) · A. Mukherjee · M. K. Das
Department of Physics, Tezpur University, Tezpur 784028, India
e-mail: pritam@tezu.ernet.in

A. Mukherjee
e-mail: ananyam@tezu.ernet.in

M. K. Das
e-mail: mkdas@tezu.ernet.in

© Springer Nature Singapore Pte Ltd. 2021   969
P. K. Behera et al. (eds.), *XXIII DAE High Energy Physics Symposium*,
Springer Proceedings in Physics 261,
https://doi.org/10.1007/978-981-33-4408-2_140

two mass square differences,[1] three mixing angles (θ_{ij}; i, $j = 1, 2, 3$) and one Dirac CP phase (δ_{13}). If neutrinos are Majorana particles[2] then there are two more CP-violating phases (α and β) that come into the 3-flavour scenario. The absolute mass of individual neutrinos, the Majorana phases are still unknown. However, Planck data constrained the sum of the three neutrinos, $\Sigma m_\nu < 0.17$ eV at 95% confidence level.

Apart from these successful achievements, in the past few years, results from LSND [1], followed by MiniBooNE [2] confirms the presence of one extra flavour of neutrino which is termed as "Sterile Neutrino". Recently observed Gallium Anomaly observation [3] is also well explained by the sterile neutrino hypothesis. Results from cosmological observation also allow the existence of this fourth flavour of neutrino.

In SM language, sterile neutrinos are right-handed fermions with zero hypercharge and no colour, i.e. they are total singlets under the SM gauge group and thus perfectly neutral. These properties allow sterile neutrinos to have a mass that does not depend on the Higgs mechanism. Since RH neutrinos are SM gauge singlets, so it is possible that sterile neutrinos could fit in the canonical type-I seesaw as the RH neutrino if their masses lie in the eV regime. In order to explain eV sterile neutrino, the Yukawa Coupling relating lepton doublets and right-handed neutrinos should be of the order 10^{-12}, which implies a Dirac neutrino mass of sub-eV scale to observe the desired active-sterile mixing. These small Dirac Yukawa couplings are considered unnatural unless there is some underlying mechanism to follow. Thus, it would be more preferable to choose a framework that gives low-scale sterile neutrino masses without the need for Yukawa coupling and simultaneously explain active-sterile mixing. We have considered $(3 + 1)$ framework where the active neutrinos are in sub-eV range with the sterile neutrino is in eV range. In this work, we have extended the canonical type-I seesaw with 3 right-handed (RH) and a singlet scalar fermion, popularly known as minimal extended seesaw (MES) [4]. The beauty of the MES framework is that this extension gives rise to tiny active neutrino mass along with the sterile mass without the need for small Yukawa couplings and a wide range of sterile neutrino mass is accepted in this framework from eV to keV. In this work, we have studied the active-sterile mixing phenomenology within the MES framework based on A_4 flavour symmetry along with the discrete Z_4 and Z_3 symmetry. We have considered different flavons to construct the non-trivial Dirac mass matrix (M_D), which is responsible for generating light neutrino mass. In this context, we have also added a leading order correction to the Dirac mass matrix to accumulate non-zero reactor mixing angle (θ_{13}), in lieu of considering higher order correcting term in the Lagrangian.

---

[1]Order of 10^{-5} eV^2 and 10^{-3} eV^2 for solar (Δm_{21}^2) and atmospheric ($\Delta m_{23}^2 / \Delta m_{13}^2$) neutrino respectively.

[2]Majorana nature says that particle and anti-particle are same.

140.2   The Model

Symmetries has been playing an influential role in model building and describing phenomenology in neutrino physics. Interestingly, discrete symmetries like A_4 with Z_n is more popular in literature in explaining neutrino mass [5]. In this work, left-handed (LH) lepton doublet l to transform as A_4 triplet whereas right-handed (RH) charged leptons (e^c, μ^c, τ^c) transform as 1, $1''$ and $1'$, respectively. Triplets ζ, φ, φ', φ'' and two singlets ξ and ξ' are added in order to produce broken flavour symmetry. Besides the SM Higgs H, we have introduced two more Higgs (H', H'') which remain invariant under A_4. Non-desirable interactions while constructing the mass matrices were restricted using extra Z_4 charges to the fields. The field content with $A_4 \times Z_4$ charge assignment are shown in Table 140.1.

The leading order invariant Yukawa Lagrangian for the lepton sector is given by

$$\mathcal{L} = \mathcal{L}_{M_l} + \mathcal{L}_{M_D} + \mathcal{L}_{M_R} + \mathcal{L}_{M_S} + h.c.. \tag{140.1}$$

where

$$\mathcal{L}_{M_l} = \frac{y_e}{\Lambda}(\bar{l}H\zeta)_1 e_R + \frac{y_\mu}{\Lambda}(\bar{l}H\zeta)_{1'}\mu_R + \frac{y_\tau}{\Lambda}(\bar{l}H\zeta)_{1''}\tau_R,$$

$$\mathcal{L}_{M_D} = \frac{y_1}{\Lambda}(\bar{l}\tilde{H}\varphi)_1 \nu_{R1} + \frac{y_2}{\Lambda}(\bar{l}\tilde{H}'\varphi)_{1''}\nu_{R2} + \frac{y_3}{\Lambda}(\bar{l}\tilde{H}''\varphi)_1 \nu_{R3},$$

$$\mathcal{L}_{M_R} = \frac{1}{2}\lambda_1\xi\overline{\nu_{R1}^c}\nu_{R1} + \frac{1}{2}\lambda_2\xi'\overline{\nu_{R2}^c}\nu_{R2} + \frac{1}{2}\lambda_3\xi\overline{\nu_{R3}^c}\nu_{R3}, \tag{140.2}$$

$$\mathcal{L}_{M_S} = \frac{1}{2}\rho\chi\overline{S^c}\nu_{R1}.$$

In the Lagrangian, Λ represents the cut-off scale of the theory, $y_{\alpha,i}$, λ_i (for $\alpha = e, \mu, \tau$ and $i = 1, 2, 3$) and ρ representing the Yukawa couplings for respective interactions and all Higgs doublets are transformed as $\tilde{H} = i\tau_2 H^*$ (with τ_2 being the second Pauli's spin matrix) to keep the Lagrangian gauge invariant. Following VEV alignments of the extra flavons are required to generate the desired light neutrino mass matrix.

$$\langle \zeta \rangle = (v, 0, 0), \langle \varphi \rangle = (v, v, v), \langle \xi \rangle = \langle \xi' \rangle = v, \langle \chi \rangle = u.$$

Following the A_4 product rules and using the above-mentioned VEV alignment, one can obtain the charged lepton mass matrix as follows:

Table 140.1 Particle content and their charge assignments under $SU(2), A_4$ and Z_4 groups

Field	l	e_R	μ_R	τ_R	H	H'	H''	ζ	φ	ξ	ξ'	ν_{R1}	ν_{R2}	ν_{R3}	S	χ
SU(2)	2	1	1	1	2	2	2	1	1	1	1	1	1	1	1	1
A_4	3	1	$1''$	$1'$	1	1	1	3	3	1	$1'$	1	$1'$	1	$1''$	$1'$
Z_4	1	-1	-1	-1	1	i	-1	-1	1	1	-1	1	$-i$	-1	$-i$	i

$$M_l = \frac{\langle H \rangle v}{\Lambda} \text{diag}(y_e, y_\mu, y_\tau). \tag{140.3}$$

The Dirac[3] and Majorana neutrino mass matrices are given by

$$M_D' = \begin{pmatrix} a & b & c \\ a & b & c \\ a & b & c \end{pmatrix}, M_R = \begin{pmatrix} d & 0 & 0 \\ 0 & e & 0 \\ 0 & 0 & f \end{pmatrix}; \tag{140.4}$$

where $a = \frac{\langle H \rangle v}{\Lambda} y_1$, $b = \frac{\langle H \rangle v}{\Lambda} y_2$ and $c = \frac{\langle H \rangle v}{\Lambda} y_3$. The elements of the M_R are defined as $d = \lambda_1 v$, $e = \lambda_2 v$ and $f = \lambda_3 v$.

The structure for M_S is read as

$$M_S = \begin{pmatrix} g & 0 & 0 \end{pmatrix}, \tag{140.5}$$

The light neutrino mass matrix generated with this M_D' is a symmetric matrix (Democratic). It can produce only one mixing angle and one mass square difference. This symmetry must be broken in order to generate two mass square differences and three mixing angles. Thus we have added a perturbative matrix (M_P) to redefine the Dirac mass matrix as

$$M_P = \begin{pmatrix} 0 & 0 & p \\ 0 & p & 0 \\ p & 0 & 0 \end{pmatrix}. \tag{140.6}$$

Hence M_D from (140.4) will take new structure as

$$M_D = M_D' + M_P = \begin{pmatrix} a & b & c+p \\ a & b+p & c \\ a+p & b & c \end{pmatrix}. \tag{140.7}$$

140.3   Numerical Analysis

Following MES framework, the active mass matrix is given by

$$m_\nu \simeq M_D M_R^{-1} M_S^T (M_S M_R^{-1} M_S^T)^{-1} M_S (M_R^{-1})^T M_D^T - M_D M_R^{-1} M_D^T, \tag{140.8}$$

and the sterile neutrino mass as

$$m_s \simeq -M_S M_R^{-1} M_S^T, \tag{140.9}$$

with active-sterile mixing element as $R = M_D M_R^{-1} M_S^T (M_S M_R^{-1} M_S^T)^{-1}$.

---

[3] M_D' represents the uncorrected Dirac mass matrix which is unable to generate $\theta_{13} \neq 0$. The corrected M_D is given by (140.7).

Using the formula, we get the active mass matrix as,

$$
m_\nu = - \begin{pmatrix} \frac{b^2}{e} + \frac{(c+p)^2}{f} & \frac{b(b+p)}{e} + \frac{c(c+p)}{f} & \frac{b^2}{e} + \frac{c(c+p)}{f} \\ \frac{b(b+p)}{e} + \frac{c(c+p)}{f} & \frac{(b+p)^2}{e} + \frac{c^2}{f} & \frac{b(b+p)}{e} + \frac{c^2}{f} \\ \frac{b^2}{e} + \frac{c(c+p)}{f} & \frac{b(b+p)}{e} + \frac{c^2}{f} & \frac{b^2}{e} + \frac{c^2}{f} \end{pmatrix}.
\tag{140.10}
$$

The sterile mass and active-sterile mixing pattern emerge as

$$
m_s \simeq \frac{g^2}{10^4} \qquad R^T \simeq \left(\frac{a}{g} \quad \frac{a}{g} \quad \frac{a+p}{g} \right)^T
\tag{140.11}
$$

We have solved the light neutrino mass matrix comparing with the unitary PMNS [6] matrix, which consists of global fit light neutrino parameters. In a similar fashion we have evaluated the active-sterile mixing strength.

140.4   Results and Discussion

Canonical type-I seesaw is extended in this work to study sterile neutrino and its mixing pattern with the active neutrinos. Only Normal Hierarchy (NH) mode results are discussed in this book chapter. The required Dirac, Majorana mass matrices are generated using discrete A_4, Z_4 group symmetries. All model parameters are solved using the global 3σ bound light neutrino parameters. In Fig. 140.1, we can see the Dirac phase is constrained near to $0, \pi$ a 2π while we varied it with a model parameter (p). The addition of M_P has a great influence on producing the reactor mixing angle. Thus we have shown a variation of $\sin e$ of the reactor mixing angle with p. The value of p gets constrained near 40 GeV. A plot between $\sin e$ of the reactor and atmospheric mixing angle is shown (lower left). As we can see the upper octant of $\sin^2 \theta_{23}$ accommodating more number of data points than the lower octant. In the lower right corner plot, we have varied $|V_{e4}|^2$ [4] versus $|V_{\tau 4}|^2$ with their current 3σ bounds and our model could successfully put upper bound on both the active-sterile mixing angles.

In conclusion, the low-scale MES mechanism is analysed in normal hierarchy mode considering only one M_S structure within this work. We successfully achieved bounds on the model parameter p (which is extensively considered in this book chapter) and bounds on active-sterile mixing angles. The octant of $\sin^2 \theta_{23}$ is still unknown and as per our model, the upper octant is more preferable to the lower one. This work is a partial contribution from the original work [7].

---

[4] As predicted by our model $|V_{e4}|^2 = |V_{\mu 4}|^2$.

Fig. 140.1 In these figures, (upper left) variation of p with the Dirac phase(δ), (upper right) constrained p with sin e of reactor mixing angle, (lower left) variation of sin e of reactor mixing angle with reactor mixing angle and (lower right) allowed 3σ bound on the two active-sterile mixing elements are shown

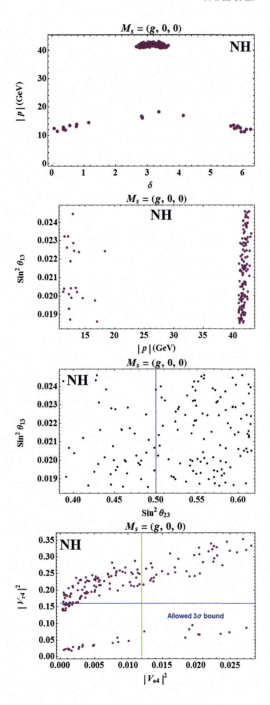

References

1. C. Athanassopoulos et al., LSND collaboration. Phys. Rev. Lett. **81**, 1774 (1998). [nucl-ex/9709006]
2. A.A. Aguilar-Arevalo et al., MiniBooNE collaboration. arXiv:1207.4809[hep-ex]
3. J.N. Abdurashitov et al., Phys. Rev. C **73**, 045805 (2006). [nucl-ex/0512041]
4. J. Barry, W. Rodejohann, H. Zhang, JHEP **1107**, 091 (2011). arXiv:1105.3911[hep-ph]
5. G. Altarelli, D. Meloni, J. Phys. G **36**, 085005 (2009). arXiv:0905.0620[hep-ph]
6. B. Kayser, AIP Conf. Proc. **1604**(1), 201 (2015). https://doi.org/10.1063/1.4883431. arXiv:1402.3028[hep-ph]
7. P. Das, A. Mukherjee, M.K. Das, Nucl. Phys. B **941**, 755 (2019). https://doi.org/10.1016/j.nuclphysb.2019.02.024. arXiv:1805.09231[hep-ph]

Chapter 141
Can New Interactions with Dark Matter Lead to Flux Change of Astrophysical Neutrinos at Icecube?

Sujata Pandey

Abstract Neutrinos can scatter off the dark matter as they travel through the cosmos and reach the Earth. These interactions can alter the neutrino spectrum at IceCube. Here, we explore the possibility of changes in the neutrino spectrum as neutrino interacts with the dark matter by considering neutrino–dark matter interaction. In this context, interactions *via* light Z' mediators are particularly interesting as they can lead to dip and cut-off like features in the neutrino spectrum at IceCube. We illustrate that various models of AGN, which predict more flux than the observed at IceCube, can be resolved through this mechanism.

141.1 Introduction

Recently, IceCube collaboration succeeded in pointing back to a specific blazar TXS 0506+056 [1] as the source of one high-energy astrophysical neutrino event observed at IceCube, located at the South Pole. With this begins a new chapter of multi-messenger astronomy. But many of the models proposed to describe the dynamics of blazars, predict large neutrino flux after a PeV [2], which is disfavored by the observed spectrum at IceCube: very few events have been observed above a PeV with no event that could correspond to the Glashow resonance. This is suggestive of a sharp cut-off in the spectrum around a PeV. Between 400 TeV and 1 PeV, very few events have been observed, implying an apparent depletion in the spectrum. In this context, it is interesting to explore neutrino–DM interactions as an explanation. Ultra-light Bose–Einstein Condensate (BEC) dark matter (DM), due to small masses, provide large number density and are interesting to explore [9] in the context of flux suppression.

S. Pandey (✉)
Discipline of Physics, Indian Institute of Technology Indore, Khandwa Road, Simrol, Indore 453552, India
e-mail: phd1501151007@iiti.ac.in

© Springer Nature Singapore Pte Ltd. 2021
P. K. Behera et al. (eds.), *XXIII DAE High Energy Physics Symposium*,
Springer Proceedings in Physics 261,
https://doi.org/10.1007/978-981-33-4408-2_141

141.2 Neutrino–Dark Matter Interaction

The interaction of scalar DM with neutrino via new gauge boson Z' is given by the Lagrangian

$$\mathcal{L} \supset f_i \bar{l}^i \gamma^\mu P_L l^i Z'_\mu + ig(\Phi^* \partial^\mu \Phi - \Phi \partial^\mu \Phi^*) Z'_\mu. \tag{141.1}$$

Here, f_i are the coupling of the $i = e, \mu, \tau$ kind of neutrinos with the new boson Z', while g is the coupling between the dark matter Φ and the mediator. Coupling f_i for $i = e, \mu$ are constrained severely from the $g - 2$ measurements, whereas τ-flavoured neutrinos are constrained from the decay width of Z, W bosons and τ. For ν_τ, the bound reads $f_i \lesssim 0.02$. Thus, the interaction with ν_τ can have larger cross section and in turn lead to flux suppression [3].

In the presence of neutrino–DM interaction, the flux of astrophysical neutrinos passing through isotropic DM background is attenuated by a factor $\sim \exp(-n\sigma L)$. Here, n, L and σ represent number density of cosmic DM, distance traversed by the neutrinos in the DM background the cross section of neutrino–DM interaction, respectively. The neutrino–DM interaction can produce appreciable flux suppression only when $n\sigma L \sim 1$. In ref. [3], many effective and renormalisable interactions were studied and it was found that light gauge boson could lead to significant flux suppression. In the case of thermal dark matter, it was shown that the bounds from relic density, collisional damping as well as N_{eff} dictates that the mass of the DM has to be greater than 10 MeV for non-negligible coupling to Z', leading to low DM number density. Further constraints from relic density do not allow thermal DM to have enough cross section, leading to no flux suppression on interaction with ν. On the other hand, ultralight BEC DM, $m_{DM} \lesssim 1\,\text{eV}$ [4], can be interesting in this context which we explore further.

141.3 Transport of Neutrinos

We consider ν–DM interaction from Eq. (141.1) and calculate the change in the neutrino spectrum if the astrophysical neutrino interacts with 200 Mpc of uniform background of ultralight scalar DM. The neutrino flux can be obtained by solving the integro-differential equation given as

$$\frac{\partial F(E, x)}{\partial x} = -n\sigma(E)F(E, x) + n \int_E^\infty dE' \frac{d\sigma(E, E')}{dE} F(E', x), \tag{141.2}$$

where $F(E, x)$ represents the flux of neutrinos of energy E after traversing a distance x from the source. The first term in the RHS of Eq. (141.2) represents the attenuation of the neutrinos, whereas the second term denotes the flux regenerated from the degradation of neutrinos of higher energies. Such a transport equation can be

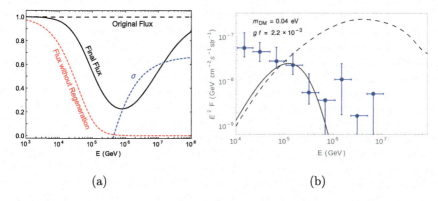

(a)   (b)

Fig. 141.1 **a** The solid and dashed black, red lines represent the final and the original flux, multiplied by E^2, in units of GeV cm^{-2} s^{-1} str^{-1} and are scaled by a factor 3×10^9. The blue-dashed line refers to the neutrino–DM scattering cross section in units of eV^{-2} and is scaled by 3×10^{21}. Here, $m_{DM} = 0.3$ eV and $gf = 7 \times 10^{-3}$. For all plots, $m_{Z'} = 10$ MeV and $m_\nu = 0.1$ eV. **b** Attenuation of the diffuse neutrino flux for the AGN core model S05 [7]. The dashed and solid black lines represent the original flux and the flux degraded by ν–DM interactions. The blue bars denote the flux from 7.5 years of neutrino flux at IceCube [8]

numerically solved in several ways: Runge–Kutta method, 'Z-factor' [6]. We have verified that the final flux obtained by these two methods match at good accuracy.

From Fig. 141.1a, it can be seen that at lower energies, when σ is not appreciable, both attenuation and regeneration are negligible. At very high energies, when σ flattens, the neutrinos lost due to attenuation get regenerated from the higher energy bins leading to no net change. In between, these two extremes attenuation occurs. This kind of feature can explain the dip ∼500 TeV at IceCube [5]. This can also lead to a cut-off kind of feature at IceCube given the dip is broader for a low mass of DM with similar coupling.

Many models of non-blazar AGNs predict a much higher flux than what is seen at IceCube. The net flux suppression due to neutrino DM interaction can reconcile such AGN and cosmogenic neutrino models. We illustrate this in Fig. 141.1b, where we have used AGN core model S05, which is used as an archetype of such AGN models.

141.4 Conclusion

Both neutrinos and dark matter are mysterious elementary particles. Their interactions are not precisely known. Cosmology suggests that DM is abundantly distributed around the cosmos and astrophysical objects produce detectable neutrino flux observed at IceCube. Therefore, neutrino-DM interactions can lead to a significant change of neutrino spectrum as ν travel through the cosmos and reach the Earth. In our work, we have found that $\nu_t au$ interacting with DM via light vector boson can lead to such effects. Due to neutrino oscillations, all the flavours of ν get depleted due to the interaction. The attenuation and regeneration of neutrinos due to ν–DM interaction can lead to paucity of events around $\sim$600 TeV as well as cut-off like feature after a PeV of neutrino energy. We show that by our mechanism the flux of AGN gets attenuated and hence a lower flux is seen at IceCube. Many such astrophysical sources which are dominant at different neutrino energies, can together explain the entire neutrino spectrum observed.

Acknowledgements This work was supported by the Department of Science and Technology, India *via* SERB grant EMR/2014/001177 and DST-DAAD grant INT/FRG/DAAD/P-22/2018.

References

1. M.G. Aartsen et al., Science **361**(6398), eaat1378 (2018)
2. K. Murase, J.F. Beacom, Phys. Rev. D **81**, 123001 (2010)
3. S. Pandey, S. Karmakar, S. Rakshit, JHEP **1901**, 095 (2019)
4. S. Das, R.K. Bhaduri, Class. Quant. Grav. **32**(10), 105003 (2015)
5. S. Karmakar, S. Pandey, S. Rakshit, arXiv:1810.04192 [hep-ph]
6. V.A. Naumov, L. Perrone, Astropart. Phys. **10**, 239 (1999)
7. F.W. Stecker, Phys. Rev. D **88**(4), 047301 (2013)
8. A. Karle, Talk presented at La Palma 2018 on behalf of IceCube Collaboration
9. W. Hu, R. Barkana, A. Gruzinov, Phys. Rev. Lett. **85**, 1158 (2000)

Chapter 142
Sensitivity of INO ICAL to Neutrino Mass Hierarchy and θ_{23} Octant in Presence of Invisible Neutrino Decay in Matter

S. M. Lakshmi, Sandhya Choubey, and Srubabati Goswami

Abstract A study of the effect of invisible decay of the third neutrino mass eigen state ν_3 on the determination of neutrino mass hierarchy and the octant of the mixing angle θ_{23} has been done. The decay is characterized by $\alpha_3 = m_3/\tau_3$, where m_3 is the mass of ν_3 and τ_3 its rest–frame lifetime. The effect of matter oscillations and decay have been taken into account. The studies are done with simulated charged current ν_μ and $\bar{\nu}_\mu$ events in the proposed 50 kt INO ICAL detector. It is found that the mass hierarchy sensitivity is not much affected if the values of α_3 are smaller than 2.35×10^{-4} eV^2 (which is the 90% C.L. limit on α_3 from the analysis of MINOS data). A significant reduction in hierarchy sensitivity is observed for values of α_3 as large as 2.35×10^{-4} eV^2. The dependence of sensitivity to mass ordering on θ_{23} is also studied. Octant sensitivity is found to increase (worsen) with increase in α_3 if θ_{23} is in the first (second) octant, for non-zero values of $\alpha_3 < 2.35 \times 10^{-4}$ eV^2, as compared to the no decay case. For $\alpha_3 = 2.35 \times 10^{-4}$ eV^2, the octant sensitivity is found to increase for both octants of θ_{23} .

S. M. Lakshmi (✉)
Indian Institute of Technology Madras, Chennai 600036, India
e-mail: ph17ipf06@smail.iitm.ac.in; slakshmi@physics.iitm.ac.in

S. Choubey
Harish-Chandra Research Institute, Chhatnag Road, Jhunsi, Allahabad 211019, India
e-mail: sandhya@hri.res.in

Department of Physics, School of Engineering Sciences, KTH Royal Institute of Technology, AlbaNova University Center, 10691 Stockholm, Sweden

Homi Bhabha National Institute, Training School Complex, Anushakti Nagar, Mumbai 400085, India

S. Goswami
Physical Research Laboratory, Navrangpura, Ahmedabad 380009, India
e-mail: sruba@prl.res.in

© Springer Nature Singapore Pte Ltd. 2021   981
P. K. Behera et al. (eds.), *XXIII DAE High Energy Physics Symposium*,
Springer Proceedings in Physics 261,
https://doi.org/10.1007/978-981-33-4408-2_142

142.1 Introduction

Neutrinos can decay either visibly or invisibly, since they have a tiny but finite mass. Non-radiative decay of a neutrino into a neutral fermion and a (pseudo)scalar is possible. If the final state neutral fermion is an active neutrino, the decay is visible; if it is a sterile state, the decay is *invisible*. INO ICAL [1] is a proposed magnetised iron calorimeter detector mainly sensitive to atmospheric ν_μ and $\overline{\nu}_\mu$. In addition to neutrino oscillations, this detector is sensitive to subdominant effects like invisible neutrino decay of the mass eigen state ν_3. The sensitivity of ICAL to ν_3 decay life time, when it is the heaviest mass eigen state was studied in [2]. The main goal of ICAL, i.e., to determine neutrino mass hierarchy can be affected by the presence of invisible decay of ν_3. A study of the effects of invisible decay of ν_3 and matter oscillations on the measurement of hierarchy and octant of θ_{23} is presented here.

142.2 Oscillation Probabilities for Different Mass Hierarchies and Octants in Presence of Invisible Decay of ν_3

The 3-flavour evolution equation in presence of Earth matter and invisible decay are given in [2]. The effects of invisible decay and oscillations on the matter oscillation probabilities of the dominant channels $P_{\mu\mu}$ and $\overline{P}_{\mu\mu}$ for different hierarchies and octants of θ_{23} are shown in Fig. 142.1. The oscillograms in the figure are plotted as a funciton of the energy and direction of incident neutrinos. The central values of oscillation probabilities used to generate these are given in [2].

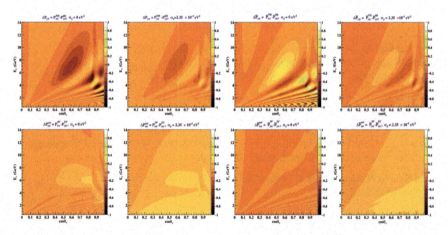

Fig. 142.1 As a function of $(E_\nu, \cos\theta_\nu)$, (top) $\Delta P_{\mu\mu} = P_{\mu\mu}^{NH} - P_{\mu\mu}^{IH}$ (with $\theta_{23} = 45°$) (bottom) $\Delta P_{\mu\mu}^{oct} = P_{\mu\mu}^{39°} - P_{\mu\mu}^{52°}$ (with NH), for $\alpha_3 = 0$ (no decay), 2.35×10^{-4} eV^2 (MINOS limit [3])

With increasing α_3, the difference between the NH and IH probabilities, especially in the matter resonance region $E_\nu = 4.0$–11.0 GeV and $\cos\theta_\nu = 0.4$–0.8 decreases. The ability to identify whether θ_{23} belongs to the first $(\theta_{23} < 45°)$ or the second $(\theta_{23} > 45°)$ octant is also affected by invisible decay. The difference of probabilities with $\theta_{23} = 39°$ and $52°$ get enhanced in the presence of α_3. The event spectra are driven by $P_{\mu\mu}$ and $\overline{P}_{\mu\mu}$. The presence of α_3 results in a decrease in hierarchy sensitivity. The octant sensitivity increases with increasing α_3 for θ_{23}^{true} in the lower octant but decreases with θ_{23}^{true} in the higher octant.

142.3 Results

The details of event generation, central values of parameters used for applying oscillations, binning scheme, marginalisation range, systematic uncertainties and χ^2 analysis can be found in Ref. [2].

142.3.1 Hierarchy Sensitivity

The hierarchy sensitivity as a function of exposure time in years for 50 kt ICAL in the presence of α_3 is shown in Fig. 142.2. It is found that for both fixed and marginalised parameter cases, the hierarchy sensitivity decreases with increase in α_3. This is because the hierarchy is determined using the difference between the amplitudes of the event spectra with NH and IH; both NH and IH amplitudes decrease and the difference between them also decrease in presence of invisible decay. Table 142.1 summarises the mass hierarchy sensitivity χ^2s with different true α_3 and $\sin^2\theta_{23} = 0.5$. With marginalised parameters, if the true hierarchy is NH, there is no significant reduction of sensitivity for $\alpha_3 < 2.35 \times 10^{-4}$ eV^2. For true IH with marginalised parameters, the sensitivity decreases with increase in α_3 for all values of α_3.

142.3.2 Sensitivity to Octant of θ_{23}

The ability to distinguish between the lower $(\theta_{23} < 45°)$ and higher $(\theta_{23} > 45°)$ octants varies in the presence of invisible neutrino decay. Octant sensitivity for $\theta_{23}^{true} = 39°$ and $52°$ and four different values of α_3 including no decay is shown in Fig. 142.3. For θ_{23}^{true} in the lower octant, the rejection of wrong octant increases with increase in the value of α_3. For θ_{23}^{true} in the higher octant, the rejection capability decreases with increase in α_3. Table 142.2 summarises the rejection χ^2 for the wrong octant for 500 kton year exposure of ICAL with fixed parameters and α_3 increasing from $0 - 1 \times 10^{-5}$ eV^2 and $\theta_{23}^{true} = 39°, 52°$ with fixed parameters.

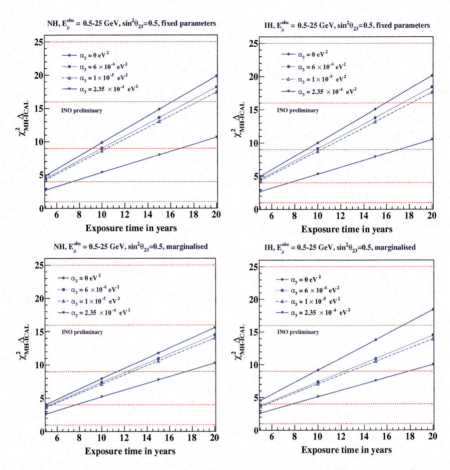

Fig. 142.2 Sensitivity to neutrino mass hierarchy as a function of exposure time of a 50 kton ICAL, and $\sin^2 \theta_{23}^{true} = 0.5$ with true (left) NH (right) IH. The top panels are for the fixed parameter cases and the bottom ones with marginalization. The values of α_3^{true} used are $0, 6 \times 10^{-6}, 1 \times 10^{-5}$ and 2.35×10^{-4} eV^2

Table 142.1 Comparison of fixed parameter and marginalised $\Delta\chi^2_{MH-ICAL}$ obtained with 500 kton year exposure of ICAL with true NH and IH, $\sin^2 \theta_{23}^{true} = 0.5$ and different α_3^{true} values

$\sin^2 \theta_{23}$	α_3 eV^2	$\Delta\chi^2_{MH-ICAL}$ (true NH - fp)	$\Delta\chi^2_{MH-ICAL}$ (true IH - fp)	$\Delta\chi^2_{MH-ICAL}$ (true NH - marg)	$\Delta\chi^2_{MH-ICAL}$ (true IH - marg)
	0	9.86	9.99	7.92	9.17
0.5	6×10^{-6}	9.02	9.13	7.37	7.36
	1×10^{-5}	8.64	8.74	7.11	7.07
	2.35×10^{-4}	5.38	5.29	5.19	5.10

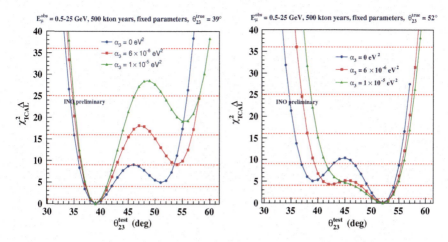

Fig. 142.3 Rejection of wrong octant with fixed parameters and marginalisation in the presence and absence of decay for θ_{23}^{true} ($\sin^2\theta_{23}^{true}$) 39° (0.396) (left) & 52° (0.621) (right)

Table 142.2 Comparison of $\Delta\chi^2_{ICAL-OCT}$ obtained with 500 kton year exposure of ICAL with true NH, $\theta_{23}^{true} = 39°, 52°$ and different α_3^{true} values for fixed parameters

α_3^{true} (eV^2)	θ_{23}^{true}	$\Delta\chi^2_{ICAL-OCT}$	θ_{23}^{true}	$\Delta\chi^2_{ICAL-OCT}$
0		4.83		5.04
6×10^{-6}	39°	9.00	52°	4.00
1×10^{-5}		19.35		4.69

142.4   Summary and Conclusions

The effect of invisible decay of ν_3 and matter oscillations in for the 3-flavour paradigm, on the determination of mass hierarchy and octant of θ_{23} is discussed. While invisible decay does not affect the determination of mass hierarchy significantly if the value of α_3 is much less than 2.35×10^{-4} eV^2, decay has a significant effect on the determination of θ_{23} octant. For values of the order of 2.35×10^{-4} eV^2, decay takes over entirely and results in a significant deterioration of the mass hierarchy sensitivity. But the limits from [2, 4] α_3 are atleast two orders magnitude smaller than 2.35×10^{-4} eV^2, the MINOS 90% C.L. [3]. So there will not be too much loss of hierarchy sensitivity. While a large enough α_3 value would diminish the hierarchy sensitivity, it would have increased (decreased) the octant sensitivity for true θ_{23} in the lower (higher) octant as compared to the no decay case in the fixed parameter case.

Acknowledgements The authors acknowledge Prof. Amol Dighe, Tata Institute of Fundamental Research, Mumbai, Prof. D. Indumathi, Institute of Mathematical Sciences (IMSc), Chennai for the discussions. S.M. Lakshmi acknowledges Prof. James Libby, Indian Institute of Technology Madras. She also acknowledges Nandadevi and Satpura clusters belonging to the computing facility of IMSc Chennai, with which this analysis was made possible.

References

1. A. Kumar, A.M. Vinod Kumar, A. Jash et al., Invited review: physics potential of the ICAL detector at the India-based Neutrino Observatory (INO). Pramana - J. Phys. **88**(79) (2017). https://doi.org/10.1007/s12043-017-1373-4; **1505.07380** (2015); M.S. Athar et al., India-based Neutrino Observatory: Project Report. Vol. I. INO-2006-01 (2006)
2. S. Choubey, S. Goswami, C. Gupta, S.M. Lakshmi, T. Thakore, Sensitivity to neutrino decay with atmospheric neutrinos at the INO-ICAL detector. Phys. Rev. D **97**, 033005 (2018)
3. R.A. Gomes, A.L.G. Gomes, O.L.G. Peres, Constraints on neutrino decay lifetime using long-baseline charged and neutral current data. Phys. Lett. B **740** (2015); M.C. Gonzalez-Garcia, Michele Maltoni, Status of oscillation plus decay of atmospheric and long-baseline neutrinos. Phys. Lett. B **663**, 405–409 (2008). arXiv: hep-ph/0802.3699
4. P.F. de Salas et al., Constraining the invisible neutrino decay with KM3NeT–ORCA. Phys. Lett. B **789**, 472–479 (2019)

Chapter 143
Icecube Spectrum: Neutrino Splitting and Neutrino Absorption

Soumya Sadhukhan, Ashish Narang, and Subhendra Mohanty

Abstract A single power law flux of high energy neutrinos does not adequately explain the entire event spectrum observed at the IceCube, specially the lack of Glashow resonance (GR) events expected around 6.3 PeV. To remedy this we consider two new scenarios (1) ν to 3 ν splitting of neutrino and (2) CRν absorption by CνB, both of which happen over cosmological distances. In the first scenario we show that for the neutrino splitting, the flavor ratios of the daughter neutrinos are different from the standard oscillation or invisible decay cases and can be used as a test of this scenario. In the second scenario we show that capture of CRν by CνB produces a dip in the flux of neutrino which then translates into the absence of GR in the IceCube spectrum.

143.1 Introduction

After six years of its operation, a clear 6σ excess of events is observed at IceCube for energies above 60 TeV and these events cannot be explained by the atmospheric neutrinos [1]. The initial choices to explain the ultra high energetic (UHE) neutrino events were different astrophysical sources [2–5]. But some puzzles remain, The astrophysical neutrino flux modeled by a single power law does not give a good fit to the IceCube event distribution at all energy bins upto $\sim$2 PeV. With a fit at lower energy ($\sim$100 TeV) bins, this flux predicts an excess of neutrino events at higher energy bins, but IceCube has not observed that effect till now. Even if the decrease of flux amplitude can fit neutrino event observation at some bin, there remains huge

S. Sadhukhan (✉)
Department of Physics and Astrophysics, University of Delhi, Delhi 110007, India
e-mail: physicsoumya@gmail.com

A. Narang · S. Mohanty
Physical Research Laboratory, Navrangpura, Ahmedabad 380009, India
e-mail: ashish@prl.res.in

S. Mohanty
e-mail: mohanty@prl.res.in

© Springer Nature Singapore Pte Ltd. 2021
P. K. Behera et al. (eds.), *XXIII DAE High Energy Physics Symposium*,
Springer Proceedings in Physics 261,
https://doi.org/10.1007/978-981-33-4408-2_143

mismatch with predicted events at other energies. We explore a two phenomenon in the context of IceCube observation. First we consider a BSM model where the neutrino splitting $\nu \to 3\nu$ occurs through a one-loop diagram with a light mediator [6]. Then we discuss a scenario where an ultra high energetic (UHE) neutrino originating from an astrophysical source interacts with the cosmic neutrino background (CνB) and get absorbed through a t-channel (and also u-channel) process mediated by a scalar causing a suppression of the neutrino flux [7].

143.2 The ν2HDM

The ν2HDM theory [8, 9] is based on the symmetry group $SU(3)_c \times SU(2)_L \times U((1)_Y \times Z_2$. The model has two Higgs doublets, Φ_1 and Φ_2. The Higgs sector is considered to be CP invariant here. The potential with Z_2 symmetry is given as [9]

$$V = -\mu_1^2 \, \Phi_1^\dagger \Phi_1 - \mu_2^2 \, \Phi_2^\dagger \Phi_2 + m_{12}^2 \, (\Phi_1^\dagger \Phi_2 + h.c.) + \lambda_1 (\Phi_1^\dagger \Phi_1)^2 + \lambda_2 (\Phi_2^\dagger \Phi_2)^2$$
$$+ \lambda_3 (\Phi_1^\dagger \Phi_1)(\Phi_2^\dagger \Phi_2) + \lambda_4 |\Phi_1^\dagger \Phi_2|^2 + \frac{1}{2}\lambda_5 [(\Phi_1^\dagger \Phi_2)^2 + (\Phi_2^\dagger \Phi_1)^2]. \quad (143.1)$$

In the above potential, $\lambda_{6,7} = 0$. After the EW symmetry breaking, the two doublets can be written as follows in the unitary gauge

$$\Phi_1 = \frac{1}{\sqrt{2}} \begin{pmatrix} \sqrt{2}(v_2/v)H^+ \\ h_0 + i(v_2/v)A + v_1 \end{pmatrix}, \quad (143.2)$$

$$\Phi_2 = \frac{1}{\sqrt{2}} \begin{pmatrix} -\sqrt{2}(v_1/v)H^+ \\ H_0 - i(v_1/v)A + v_2 \end{pmatrix} \quad (143.3)$$

where charged fields $H^\pm$, two neutral CP even scalar fields h and H, and a neutral CP odd field A are the physical Higgs fields and $v_1 = \langle \Phi_1 \rangle$, $v_2 = \langle \Phi_2 \rangle$, and $v^2 = v_1^2 + v_2^2$.

For the fermionic sector we consider two different variants of the ν2HDM model:

- **Variant 1**: We add a right handed neutrino ν_R, which is odd under Z_2 symmetry. This RH neutrino along with Φ_2, which is also odd under Z_2 forms the Yukawa interaction as

$$\mathscr{L}_Y = y_e L_e^\dagger \tilde{\Phi}_2 \nu_R + y_\mu L_\mu^\dagger \tilde{\Phi}_2 \nu_R + y_\tau L_\tau^\dagger \tilde{\Phi}_2 \nu_R + h.c. \quad (143.4)$$

$$= \sum_{e,\mu,\tau} \frac{y_e v_2}{\sqrt{2}} U_{ei} \nu_i \nu_R + \sum_{e,\mu,\tau} \frac{y_e}{\sqrt{2}} U_{ei} H(iA)\nu_i \nu_R + \text{otherterms} \quad (143.5)$$

$$= m_{\nu_i} \nu_i \nu_R + \frac{m_{\nu_i}}{v_2} H(iA)\nu_i \nu_R + \text{otherterms} \quad (143.6)$$

where $m_{\nu_i} = \sum_{e,\mu,\tau} \frac{y_\ell v_2}{\sqrt{2}} U_{\ell i}$ are neutrino masses and $U_{\ell i}$ is the PMNS matrix.

- **Variant 2**: In this variant, we have three EW singlet right-handed (RH) neutrinos, N_{R_i}, for each flavor of SM lepton. With all the SM fermions being Z_2 even and the RH neutrinos and the Higgs doublet Φ_2 being odd under Z_2, the Yukawa interaction in this model in the flavor basis takes the form,

$$\mathscr{L}_Y = Y^d_{\alpha\beta}\bar{Q}_{L,\alpha}\Phi_1 d_{R,\beta} + Y^u_{\alpha\beta}\bar{Q}_{L,\alpha}\tilde{\Phi}_1 u_{R,\beta} + Y^l_{\alpha\beta}\bar{L}_{L,\alpha}\Phi_1 l_{R,\beta} + Y^\nu_{\alpha\beta}\bar{L}_{L,\alpha}\tilde{\Phi}_2 N_{R,\beta} + \text{h.c.}$$
(143.7)

If we restrict our model to only one right handed Majorana neutrino N_R, then the relevant Yukawa and mass terms of the right handed neutrino in the mass basis of the SM neutrinos are written as,

$$\mathscr{L} = y_i \bar{L}_i \tilde{\Phi}_2 N_R + \frac{m_R}{2} N_R N_R.$$
(143.8)

With a Yukawa coupling $y_i \sim O(0.1)$, we get the Majorana neutrino mass of 0.1 eV for $v_2 \approx 10$ keV with right handed neutrino mass $m_R \sim 10$ MeV. This type of low scale seesaw mechanism was first proposed in the Ref. [10].

143.3 Solutions to IceCube Anomalies

The all flavor initial astrophysical neutrino spectrum with single component is parametrized as

$$\left(\frac{d\Phi}{dE_\nu}\right) = \Phi^0_{\text{astro}} \left(\frac{E_\nu}{100 \text{ TeV}}\right)^{-\gamma}$$
(143.9)

$$\left(\frac{d\Phi}{dE_\nu}\right)_{BSM} = \Phi^0_{\text{astro}} \left(\frac{E_\nu}{100 \text{ TeV}}\right)^{-\gamma} A(E_\nu),$$
(143.10)

if the neutrinos with energy E_ν have any BSM interaction while traveling a distance L from the source then the flux of the neutrino is corrected by a factor $A(E_\nu)$ which depends on the BSM interaction the neutrinos go through.

143.3.1 Splitting of Neutrinos

Here we use the variant 1 of ν2HDM in which the active neutrinos do not decay to the invisible final state particles. Instead, one heavier neutrino mass eigenstate splits to three lighter neutrino mass eigenstates with a process like $\nu_3 \to 3\nu_1$, creating three daughter neutrinos for each initial neutrinos (Fig. 143.1).

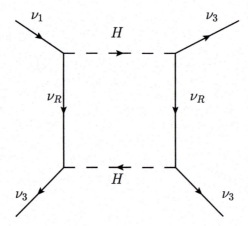

Fig. 143.1 Visible decay of active neutrinos to other active neutrinos in the ν2HDM set up through a box diagram: $\nu_1 \rightarrow 3\nu_3$

In this case the flux is modified by the multiplication of a the decay factor. The amount of the initial flux that remains unchanged after the decay is of the form

$$\left(\frac{d\Phi}{dE_\nu}\right)_{nd} = \Phi^0_{astro}\left(\frac{E_\nu}{100\,\text{TeV}}\right)^{-\gamma} e^{-\frac{\beta}{E_\nu}}. \tag{143.11}$$

$$\left(\frac{d\Phi}{dE_\nu}\right)_{d} = \Phi^0_{astro}\left(\frac{E_\nu}{100\,\text{TeV}}\right)^{-\gamma}\left(1 - e^{-\frac{\beta}{E_\nu}}\right). \tag{143.12}$$

We assume all three daughter neutrinos will have same energy i.e. $E_d \approx E_\nu/3$ with which they interact at the IceCube. So, the part of the flux that is decayed to three neutrinos each, can be written as It is shown in Fig. 143.2 that a single power law flux cannot fit the full IceCube event spectrum. This issue can be resolved introducing

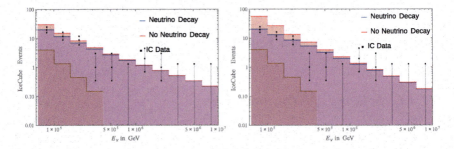

Fig. 143.2 IceCube event distribution with comparison of the cases of no neutrino decay for a. $\Phi^0_{astro} = 2 \times 10^{-18}(\text{GeV cm}^2\text{s sr})^{-1}$, $\gamma = 2.8$ (left) b. $\Phi^0_{astro} = 5 \times 10^{-18}(\text{GeV cm}^2\text{s sr})^{-1}$, $\gamma = 3$ (right) with cases with a visible neutrino decay a. $\beta = 5 \times 10^4$ GeV (left) and b. $\beta = 1.25 \times 10^5$ GeV (right) respectively. Atmospheric neutrino background is shown as brown shaded region

a split of active neutrinos and therefore suppressing the energy spectrum with the factor $e^{-\frac{\beta}{E_\nu}}$ with greater suppression at lower energy bins, which therefore can help expected neutrino events to match the observed ones. Figure 143.2 present effects of the introduction of a splitting of neutrinos for different benchmark points comparing with the case where there is no neutrino splitting.

143.3.2 Neutrino Absorption

Variant 2 of the ν2HDM allows us to have a t-channel process through which UHE neutrino gets absorbed by the cosmic neutrino background and releases two right handed neutrinos. These two right handed neutrinos, unlike their left handed partners, do not have charged current and neutral current interaction with IceCube matter. Therefore those will not be detected in the IceCube, which results in vanishing one astrophysical neutrino in this process. The t-channel diagram cross matrix element is computed as:

$$M^2 = \frac{4y_i^2 y_j^2}{(t - m_h^2)^2}\left(-\frac{1}{2}(t - m_R^2) + m_{\nu_i} m_R\right)^2 \tag{143.13}$$

where t represents the energy transfer to the final state right handed neutrinos. Here m_h and m_R are the ultralight scalar mass and the right handed neutrino mass respectively, with y being the neutrino-scalar Yukawa coupling.

The variation of t-channel process cross section with incident neutrino energy is shown in Fig. 143.3. The absorption of CRν produces a dip in the CRν flux starting at the threshold energy of the t-channel process. The mean free path is,

$$\lambda_i(E_i, z) = \left(\sum_j \int \frac{d^3\mathbf{p}}{(2\pi)^3} f_j(p, z)\sigma_{ij}(p, E_i, z)\right)^{-1} \approx \left(n_\nu(z)\sum_j \sigma_{ij}(p, E_i, z)\right)^{-1} \tag{143.14}$$

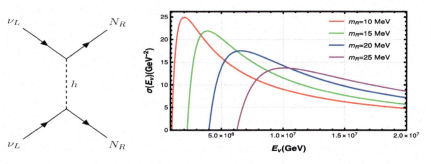

Fig. 143.3 Feynman diagram and the cross section for t-channel absorption of UHE neutrino by CνB

where f_i is the distribution function for the neutrinos and $T_i = 1.95\ K$. Away from the sources, due to the mixing, flavor ratio of neutrino in the cosmic ray flux is (1:1:1). The modified flux due to the absorption is given as

$$\left(\frac{d\Phi}{dE_\nu}\right)_{cap} = \exp\left[-\int_0^{z_s} \frac{1}{\lambda_i}\frac{dL}{dz}dz\right]\frac{d\Phi}{dE_\nu} \tag{143.15}$$

where z_s denotes the redshift of source. The modified flux and the IceCube spectrum due to the t-channel resonant absorption in this model, for the Benchmark Point(BP1) $\Phi_{astro}^0 = 1.1 \times 10^{-18}$ (GeV cm^{2}s sr)$^{-1}$, $\gamma = 2.5$ are shown in Fig. 143.4 and Fig. 143.5 respectively. The χ^2 value for the IceCube best fit is 21.9 and the same for our benchmark points is 7.23 for NH and 7.17 for the IH.

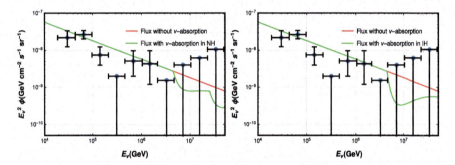

Fig. 143.4 Comparison of $E_\nu^2 \times$ flux for the incoming cosmic neutrinos after they got absorbed by the CνB (green) with that when there is no cosmic neutrino absorption (red) for a. normal mass hierarchy with $(m_1, m_2, m_3) = 2 \times 10^{-3}, 8.8 \times 10^{-3}, 5 \times 10^{-2}$ eV (left) b. inverted mass hierarchy with $(m_1, m_2, m_3) = 4.9 \times 10^{-2}, 5 \times 10^{-2}, 2 \times 10^{-3}$ eV (right). The data points obtained from the IceCube measurement are given in black. Yukawa couplings here are taken to be 0.1 for the representation purpose

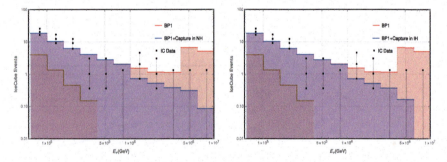

Fig. 143.5 IceCube event spectrum with (violet) and without (red) neutrino capture for Benchmark Point are shown here. Effect on the event spectrum due to neutrino absorption is shown for normal hierarchy (left) and inverted hierarchy (right). The atmospheric background is given in brown. Here we have taken $m_R \sim 15$ MeV and $y \sim 1$

143.4 Discussion and Conclusion

The anomalies in the IceCube data have been addressed in terms of excess at PeV energies or a cutoff at 6 PeV Glashow resonance energies. We discuss neutrino splitting which transfers neutrino energies to roughly 1/3 after decay and due to the Lorentz factor there is a larger depletion of lower energy neutrinos. The benchmark points that explain the IceCube spectrum put the decay parameter in the range, $\beta = L/(\tau/m) = (5 \times 10^4 - 1.25 \times 10^5)$ GeV with the correlated spectral index being in the range $\gamma = (2.8 - 3)$. We also discussed a scenario where cosmic ray neutrinos are absorbed by cosmic neutrino background through a t-channel resonant process, therefore causing multiple dips in the cosmic ray neutrino flux. The occurrence of dips in the neutrino flux is tuned to the energy of the expected GR events, explaining their absence at the IceCube.

References

1. **IceCube** Collaboration, M.G. Aartsen et al., Evidence for astrophysical muon neutrinos from the northern sky with IceCube. Phys. Rev. Lett. **115**(8), 081102 (2015). arXiv:1507.04005
2. I. Cholis, D. Hooper, On the origin of IceCube's PeV neutrinos. JCAP **1306**, 030 (2013). arXiv:1211.1974
3. L.A. Anchordoqui et al., Cosmic neutrino pevatrons: a brand new pathway to astronomy, astrophysics, and particle physics. JHEAP **1–2**, 1–30 (2014). arXiv:1312.6587
4. K. Murase, On the origin of high-energy cosmic neutrinos. AIP Conf. Proc. **1666**, 040006 (2015). arXiv:1410.3680
5. S. Sahu, L.S. Miranda, Some possible sources of IceCube TeVpeV neutrino events. Eur. Phys. J. **C75**, 273 (2015). arXiv:1408.3664
6. S. Mohanty, S. Sadhukhan, Explanation of IceCube spectrum with $\nu \rightarrow 3\nu$ neutrino splitting in a ν2HDM model. JHEP **1810**, 111 (2018). https://doi.org/10.1007/JHEP10(2018)111, arXiv:1802.09498 [hep-ph]
7. S. Mohanty, A. Narang, S. Sadhukhan, Cutoff of IceCube neutrino spectrum due to t-channel resonant absorption by CνB. JCAP **1903**(03), 041 (2019). https://doi.org/10.1088/1475-7516/2019/03/041, arXiv:1808.01272 [hep-ph]
8. S. Gabriel, S. Nandi, A New two Higgs doublet model. Phys. Lett. **B655**, 141–147 (2007). arXiv:hep-ph/0610253
9. P.A.N. Machado, Y.F. Perez, O. Sumensari, Z. Tabrizi, R.Z. Funchal, On the viability of minimal neutrinophilic two-higgs-doublet models. JHEP **12**, 160 (2015). arXiv:1507.07550
10. E. Ma, Naturally small seesaw neutrino mass with no new physics beyond the TeV scale. Phys. Rev. Lett. **86**, 2502–2504 (2001). arXiv:hep-ph/0011121

Chapter 144
Majorana Unitarity Triangle in Two-Texture Zero Neutrino Mass Model and Associated Phenomenology

Surender Verma, Shankita Bhardwaj, and Monal Kashav

Abstract Non-zero value of θ_{13} has, now, shifted the focus of the neutrino oscillation experiments to measure leptonic CP violation. If neutrinos are Majorana particles, the mixing matrix contains two Majorana-type CP-violating phases along with Dirac phase. In the present work, we have established a possible relation between Majorana CP phases and geometric parameters of Majorana unitarity triangle (MUT) in two-texture zero neutrino mass model. Similar relations can, also, be derived for other theoretical and phenomenological neutrino mass models. For two-texture zero models, we find that Majorana CP phases depend on one of the interior angles of MUT. The non-trivial orientation of MUT in complex plane and its non-vanishing area suggests that CP violation is inherent in two-texture models.

144.1 Introduction

In three active flavor neutrino framework, flavor eigenstates and mass eigenstates are connected through 3×3 unitary mixing matrix V parametrized in terms of three angles (θ_{12}, θ_{23}, and θ_{13}) and one phase (three phases, for Majorana neutrino). The information about CP violation is embodied in the mixing matrix V. In charged lepton basis, this 3×3 mixing matrix is $V \equiv UP$, U being Pontecorvo-Maki-Nakagawa–Sakata (PMNS) matrix in standard PDG representation and $P = diag(1, e^{i\rho}, e^{i(\sigma+\delta)})$, where δ is Dirac phase and ρ, σ are Majorana phases. The unitarity condition $VV^{\dagger} = V^{\dagger}V = 1$, imposes six orthogonality conditions and will lead to two sets of unitarity triangle in the complex plane, viz.

S. Verma · S. Bhardwaj · M. Kashav (✉)
Department of Physics and Astronomical Science, Central University of Himachal Pradesh, Dharamshala 176215, India
e-mail: monalkashav@gmail.com

S. Verma
e-mail: s_7verma@yahoo.co.in

S. Bhardwaj
e-mail: shankita.bhardwaj982@gmail.com

© Springer Nature Singapore Pte Ltd. 2021
P. K. Behera et al. (eds.), *XXIII DAE High Energy Physics Symposium*,
Springer Proceedings in Physics 261,
https://doi.org/10.1007/978-981-33-4408-2_144

$$\Delta_{e\mu} \equiv V_{e1}V_{\mu1}^* + V_{e2}V_{\mu2}^* + V_{e3}V_{\mu3}^* = 0,$$
$$\Delta_{\mu\tau} \equiv V_{\mu1}V_{\tau1}^* + V_{\mu2}V_{\tau2}^* + V_{\mu3}V_{\tau3}^* = 0,$$
$$\Delta_{\tau e} \equiv V_{\tau1}V_{e1}^* + V_{\tau2}V_{e2}^* + V_{\tau3}V_{e3}^* = 0, \tag{144.1}$$

$$\Delta_{12} \equiv V_{e1}V_{e2}^* + V_{\mu1}V_{\mu2}^* + V_{\tau1}V_{\tau2}^* = 0,$$
$$\Delta_{23} \equiv V_{e2}V_{e3}^* + V_{\mu2}V_{\mu3}^* + V_{\tau2}V_{\tau3}^* = 0,$$
$$\Delta_{31} \equiv V_{e3}V_{e1}^* + V_{\mu3}V_{\mu1}^* + V_{\tau3}V_{\tau1}^* = 0, \tag{144.2}$$

where $\Delta_{ij}(i, j = e, \mu, \tau; i \neq j)$ represents Dirac unitarity triangle (DUT) obtained from multiplication of any two rows of V and $\Delta_{pq}(p, q = 1, 2, 3; p \neq q)$ are Majorana unitarity triangles (MUT) obtained from multiplication of any two columns of V. Under the leptonic field transformation, the elements of unitary mixing matrix V transforms as $V_{li} \rightarrow e^{i\phi_l}V_{li}$. In complex plane, DUT will rotate under these transformation and their orientation do not have any physical significance. Non-zero area of the triangles in complex plane imply CP-violation. However, if neutrinos are Majorana then zero-area triangle does not necessarily mean that CP is conserved because of presence of Majorana phases. MUT orientations, on the other hand, are physical as they contain bilinear rephasing invariant terms. Thus, MUTs provide complete elucidation of the CP-violation phenomena. In the present work, we have established a possible connection of geometric parameters of MUT with CP-violating phases in two-texture zero neutrino mass model [1].

144.2   Two-Texture Zero Neutrino Mass Model and Geometric Parameters of MUT

In the charged lepton basis, the Majorana neutrino mass matrix M is given by $M = VM_\nu V^T$, where $M_\nu = Diag\{m_1, m_2, m_3\}$. Two-texture zeros have seven phenomenologically allowed patterns $A_1, A_2, B_1, B_2, B_3, B_4$ and C. Any two vanishing elements of M provide two constraining equations, i.e., $M_{st} = 0, M_{xy} = 0$, where s, t, x, and $y = e, \mu, \tau$, viz $\sum_{i=1}^3 V_{si}V_{ti}m_i = 0$, $\sum_{i=1}^3 V_{xi}V_{yi}m_i = 0$. We have used these constraining equations to derive mass ratios$(\frac{m_1}{m_2}, \frac{m_1}{m_3})$ and Majorana phases(ρ, σ) as

$$\frac{m_1}{m_2} = \left| \frac{U_{x2}U_{y2}U_{s3}U_{t3} - U_{s2}U_{t2}U_{x3}U_{y3}}{U_{s1}U_{t1}U_{x3}U_{y3} - U_{s3}U_{t3}U_{x1}U_{y1}} \right|, \tag{144.3}$$

$$\frac{m_1}{m_3} = \left| \frac{U_{x3}U_{y3}U_{s2}U_{t2} - U_{s3}U_{t3}U_{x2}U_{y2}}{U_{s1}U_{t1}U_{x2}U_{y2} - U_{s2}U_{t2}U_{x1}U_{y1}} \right|, \tag{144.4}$$

Table 144.1 Mass ratios for A_1, B_1, and C-type textures up to first order in s_{13}

Type of texture	Mass ratios
$A_1(M_{ee} = 0; M_{e\mu} = 0)$	$\frac{m_1}{m_2} \approx \tan^2\theta_{12}\left(1 - \frac{\cot\theta_{23}}{s_{12}c_{12}}s_{13}\cos\delta\right)$ $\frac{m_1}{m_3} \approx \tan\theta_{12}\tan\theta_{23}s_{13}$
$B_1(M_{e\tau} = 0; M_{\mu\mu} = 0)$	$\frac{m_1}{m_2} \approx 1 + \frac{c_{23}}{c_{12}s_{12}s_{23}^3}s_{13}\cos\delta$ $\frac{m_1}{m_3} \approx \tan^2\theta_{23}\left(1 + \frac{\tan\theta_{23}\cot\theta_{12}}{s_{23}^2}s_{13}\cos\delta\right)$
$C(M_{\mu\mu} = 0; M_{\tau\tau} = 0)$	$\frac{m_1}{m_2} \approx \frac{1}{\tan^2\theta_{12}}\left(1 - \frac{\tan\theta_{23}}{s_{12}c_{12}}s_{13}\cos\delta\right)$ $\frac{m_1}{m_3} \approx \frac{1}{\tan\theta_{12}\tan 2\theta_{23}s_{13}}\left(1 + \frac{4(-s_{12}^2 + c_{12}^2\cos^2 2\theta_{23})}{\sin 4\theta_{23}\sin\theta_{12}}s_{13}\cos\delta\right)$

and

$$\rho = -\frac{1}{2}Arg\left(\frac{U_{x2}U_{y2}U_{s3}U_{t3} - U_{s2}U_{t2}U_{x3}U_{y3}}{U_{s1}U_{t1}U_{x3}U_{y3} - U_{s3}U_{t3}U_{x1}U_{y1}}\right), \qquad (144.5)$$

$$\sigma = -\frac{1}{2}Arg\left(\frac{U_{x3}U_{y3}U_{s2}U_{t2} - U_{s3}U_{t3}U_{x2}U_{y2}}{U_{s1}U_{t1}U_{x2}U_{y2} - U_{s2}U_{t2}U_{x1}U_{y1}}\right) - \delta, \qquad (144.6)$$

respectively. By using Eqs. (144.3) and (144.4), we have obtained these mass ratios up to first order in s_{13} and for the sake of simplicity, we have shown these mass ratios for only A_1, B_1 and C types in Table 144.1. In similar way, the relations can, also, be obtained for A_2, B_2, B_3, B_4. We have calculated the allowed parameter space using Eqs. (144.3) and (144.4). In numerical analysis, the known parameters(mixing angles and mass-squared differences) are randomly generated (10^7 points) with Gaussian distribution, whereas unknown parameters are uniformly generated in their full range. Two-texture zero patterns A_1, A_2, B_1, B_2, B_3, B_4 and C satisfies the neutrino oscillation data [2] and cosmological bound on sum of neutrino masses [3]. The best-fit point (bfp) and 1σ range of these parameters for A_1, B_1, and C types are tabulated in Table 144.2.

MUTs Δ_{12} and Δ_{31} are sensitive to one of the Majorana phase, ρ and σ respectively, in contrast to Δ_{23} which is sensitive to both. In general, the sides and angles of MUT can be expressed as,

$$(S_1, S_2, S_3) = \left(|V_{ef}V_{ef'}|, |V_{\mu f}V_{\mu f'}|, |V_{\tau f}V_{\tau f'}|\right), \qquad (144.7)$$

$$\alpha_{ff'} = Arg\left(-\frac{V_{\mu f}V_{\mu f'}^*}{V_{\tau f}V_{\tau f'}^*}\right), \beta_{ff'} = Arg\left(-\frac{V_{\tau f}V_{\tau f'}^*}{V_{ef}V_{ef'}^*}\right), \gamma_{ff'} = Arg\left(-\frac{V_{ef}V_{ef'}^*}{V_{\mu f}V_{\mu f'}^*}\right),$$
$$\qquad (144.8)$$

where, S_1, S_2, S_3 represent three sides and α, β, γ are three angles with subscript $f, f' = (1, 2, 3)$ and $f \neq f'$. The obtained relations of Majorana phases with interior angles of MUT for different mass matrices patterns A_1, B_1, and C of two-texture zero model are tabulated in Table 144.3. Using scanned neutrino oscillation parameter

Table 144.2 The neutrino mixing parameters for A_1, B_1 and C in two-texture zero neutrino mass model

Type of texture	bfp $\pm 1\sigma$ in degrees(^{o}) normal hierarchy	Inverted hierarchy
A_1	$\theta_{12} = 33.52^{+0.74}_{-0.74}$, $\theta_{13} = 8.45^{+0.14}_{-0.14}$, $\theta_{23} = 41.63^{+1.38}_{-1.38}$, $\rho = -77.42^{+34.71}_{-34.71}$, $\sigma = -102.60^{+48.37}_{-48.37}$, $\delta = 67.58^{+31.99}_{-31.99}$.	–
B_1	$\theta_{12} = 33.75^{+0.57}_{-0.57}$, $\theta_{13} = 8.45^{+0.14}_{-0.14}$, $\theta_{23} = 41.10^{+1.07}_{-1.07}$, $\rho = -3.67^{+1.73}_{-1.73}$, $\sigma = -0.36^{+0.25}_{-0.25}$, $\delta = 267.10^{+15.21}_{-15.21}$.	$\theta_{12} = 33.52^{+0.77}_{-0.77}$, $\theta_{13} = 8.47^{+0.13}_{-0.13}$, $\theta_{23} = 45.76^{+0.52}_{-0.52}$, $\rho = 0.47^{+0.32}_{-0.32}$, $\sigma = -177.36^{+28.84}_{-28.84}$, $\delta = 268.87^{+14.47}_{-14.47}$.
C	–	$\theta_{12} = 33.54^{+0.75}_{-0.75}$, $\theta_{13} = 8.44^{+0.14}_{-0.14}$, $\theta_{23} = 41.08^{+1.26}_{-1.26}$, $\rho = 46.65^{+12.33}_{-12.33}$, $\sigma = -165.06^{+19.07}_{-19.07}$, $\delta = 292.80^{+15.17}_{-15.17}$.

Table 144.3 Relation between Majorana phases and angles of MUT for A_1, B_1, and C classes of two-texture zero neutrino mass model

Type of texture	Majorana phases in terms of angles of Majorana unitarity triangle
A_1	$\rho = -\frac{1}{2}\left(\gamma_{12} - \pi + Arg\left(\frac{U_{e3}U_{\mu2}U_{\mu1}U_{e2}^2 - U_{e2}^3U_{\mu3}U_{\mu1}}{U_{\mu2}U_{e1}^3U_{\mu3} - U_{\mu2}U_{e3}U_{\mu1}U_{e1}^2}\right)\right)$, $\sigma = -\frac{1}{2}\left(\gamma_{12} - \pi + Arg\left(\frac{U_{e2}^3U_{e3}U_{\mu3}U_{\mu1} - U_{e3}^2U_{e2}^2U_{\mu2}U_{\mu1}}{U_{e1}^3U_{e2}U_{\mu2}^2 - U_{e2}^2U_{e1}^2U_{\mu1}U_{\mu2}}\right)\right) - \delta$.
B_1	$\rho = -\frac{1}{2}\left(\beta_{23} - \pi + Arg\left(\frac{U_{\mu2}^2U_{e3}U_{\tau3}^2U_{e2} - U_{\mu3}^2U_{e2}^2U_{\tau2}U_{\tau3}}{U_{\mu3}^2U_{e1}U_{\tau1}U_{\tau2}U_{e3} - U_{\mu1}^2U_{e3}^2U_{\tau2}U_{\tau3}}\right)\right)$, $\sigma = -\frac{1}{2}\left(\beta_{23} - \pi + Arg\left(\frac{U_{\mu3}^2U_{e2}^2U_{\tau2}U_{\tau3} - U_{\mu2}^2U_{e2}U_{e3}U_{\tau3}^2}{U_{\mu2}^2U_{e1}U_{\tau1}U_{\tau2}U_{e3} - U_{\mu1}^2U_{e2}U_{\tau2}^2U_{e3}}\right)\right) - \delta$.
C	$\rho = \frac{1}{2}\left(\alpha_{31} - \pi + Arg\left(\frac{U_{\tau2}^2U_{\mu3}^3U_{\tau3}^3 - U_{\mu2}^2U_{\tau3}^3U_{\mu3}}{U_{\mu1}^3U_{\tau3}^2U_{\tau1}^3 - U_{\mu3}^2U_{\tau1}^3U_{\mu1}}\right)\right)$, $\sigma = \frac{1}{2}\left(\alpha_{31} - \pi + Arg\left(\frac{U_{\tau3}^3U_{\mu1}U_{\tau2}^2 - U_{\mu3}^2U_{\tau2}^2U_{\mu1}U_{\tau3}}{U_{\mu1}^3U_{\tau2}^2U_{\mu3}U_{\tau1} - U_{\mu2}^2U_{\tau1}^3U_{\mu3}}\right)\right) - \delta$.

space in Table 144.2, we have constructed the Majorana triangles shown with solid line in Fig. 144.1. The dashed triangles in Fig. 144.1 are obtained for the case in which Majorana phases are zero.

144.3 Results and Discussion

We know, from neutrino oscillation experiments $\Delta m_{12}^2 > 0$ i.e. $\frac{m_1}{m_2} < 1$ holds, implies that $\cos\delta$ will be positive lying in I or IV quadrant for A_1 type pattern. For pattern B_1, $\Delta m_{12}^2 > 0$ implies $\cos\delta$ will be negative lying in II or III quadrant which can also

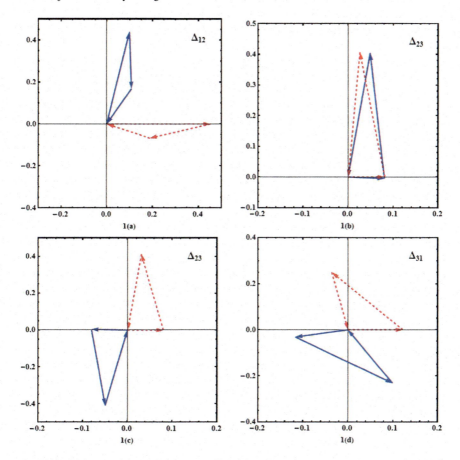

Fig. 144.1 Majorana triangles for type A_1(NH) in Fig. 144.1a, for type B_1(NH and IH) Fig. 144.1b and 144.1c and for type C_1(IH) in Fig. 144.1d

be verified from Table 144.1. We have obtained bfp value $\delta = 267.10^o (268.87^o)$, for B_1 with NH(IH). For the class B, we obtain a quasi-degenerate spectrum of neutrino masses. In case of pattern C, $\Delta m_{12}^2 > 0$ satisfies if factor $\tan 2\theta_{23} \cos \delta$ is positive, lying in I or IV quadrant. The theoretical predictions on the basis of mass ratios obtained in Table 144.1 are in agreement with the bfp values of δ obtained in Table 144.2. The key point here is that the bfp values for δ lies near to $\frac{3\pi}{2}$ for class B and C. The connecting relations for Majorana CP violating phase and interior angle of MUT are obtained individually for class A_1, B_1, and C of two-texture zero model as shown in Table 144.3. The generic feature of two-texture zero model is the dependence of Majorana CP violating phase on one of the independent interior angles of MUT. In particular, for our representative case of two-texture zero model, MUT's construction and obtained relations signals that CP violation is essentially non-zero as no side is parallel to the axis and MUT's have non-vanishing area.

144.4   Conclusions

In this work, we have studied the phenomena of *CP* violation in two-texture zero neutrino mass model which can, also, be studied in other various theoretical mass models. In particular, we have established a possible relation of geometric parameters of MUT with Majorana *CP* violating phases(ρ, σ). Since, upcoming neutrino oscillation experiments are hopeful for measuring Dirac *CP* phase δ, making this study more significant. During the analysis of different mass models in lepton number violating (LNV) processes such as $\nu - \bar{\nu}$ oscillations, these relation becomes vital and much informative. For two-texture zero model, we have obtained different neutrino oscillation parameters (bfp$\pm$ $\pm 1\sigma$). It is found that for class *B*, $\delta \approx \frac{3\pi}{2}$ is in good agreement with hits obtained from T2K and Daya Bay experiments [4]. Also, two-texture models are necessarily *CP* violating due to non-zero area and non-trivial orientations of MUT's.

Acknowledgements M. K. acknowledges the financial support provided by DST, Government of India under DST-INSPIRE fellowship program vide Grant No. DST/INSPIRE Fellowship/2018/IF180327.

References

1. S. Verma, S. Bhardwaj, Phys. Rev. D **97**, 095022 (2018)
2. I. Esteban, M.C. Gonzalez-Garcia, M. Maltoni, I. Martinez-Soler, T. Schwetz, JHEP **087**, 1701 (2017)
3. F. Couchot et al., Astron. Astrophys. **606**, A104 (2017)
4. F. Capozzi, E. Lisi, A. Marrone, D. Montanino, A. Palazzo, Nucl. Phys. B **908**, 218–234 (2016)

Chapter 145
Implications of Non-unitarity on θ_{23}, Neutrino Mass Hierarchy and *CP*-Violation Discovery Reach in Neutrino Oscillation Experiments

Surender Verma and Shankita Bhardwaj

Abstract We have studied the effect of θ_{23} octant, neutrino mass hierarchy and *CP* violation discovery reach on the sensitivity of non-unitarity in short baseline (SBL) experiments using minimal unitarity violation (MUV) scheme. We find that the θ_{23} octant has distinguishing implications towards the sensitivities of non-unitary parameter $|\rho_{\mu\tau}|$ with normal and inverted mass hierarchies. Using *CP*-fraction formalism, we find the possibilities to distinguish the *CP*-violating effects due to unitary (δ) and non-unitary ($\omega_{\mu\tau}$) *CP* phases.

145.1 Introduction

The oscillation parameters in the neutrino sector have been deciphered in the oscillation experiments under the perspective that the neutrino mixing matrix is unitary. However, the unitarity of mixing matrix has not been established yet and there exist new physics scenarios which prompt the non-unitarity effects. In the foreseeable future, the observation of non-unitarity is not an uphill task. For example, in case of one additional sterile neutrino, the complete mixing matrix U^4 is of dimension 4×4:

$$U^4 = \begin{pmatrix} N_{3\times3} & U_{es} \\ U_{s1} & U_{s2} & U_{s3} & U_{s4} \end{pmatrix}, \tag{145.1}$$

where the 'active' 3×3 part ($N_{3\times3}$) is, in general, non-unitary. This type of situation occurs in scenarios in which the LSND [1] and MiniBooNE [2] results are being considered responsible for the presence of one or two additional sterile neutrinos. In the present work, we have focused on $\nu_\mu \to \nu_\tau$ channel at short baseline (SBL)

S. Verma · S. Bhardwaj (✉)
Department of Physics and Astronomical Science, Central University of Himachal Pradesh, Dharamshala 176215, India
e-mail: shankita.bhardwaj982@gmail.com

S. Verma
e-mail: s_7verma@yahoo.co.in

© Springer Nature Singapore Pte Ltd. 2021
P. K. Behera et al. (eds.), *XXIII DAE High Energy Physics Symposium*,
Springer Proceedings in Physics 261,
https://doi.org/10.1007/978-981-33-4408-2_145

to study the non-unitary neutrino mixing [3]. We have used the minimal unitarity violation (MUV) scheme for parametrizing the non-unitary neutrino mixing and to investigate the sensitivities of θ_{23} octant to non-unitary parameters $|\rho_{\mu\tau}|$ and $\omega_{\mu\tau}$ with normal hierarchical (NH) and inverted hierarchical (IH) neutrino masses. We have investigated the CP violation discovery reach of unitary (δ) and non-unitary ($\omega_{\mu\tau}$) CP phases for a wide range of L/E ratio.

145.2 Non-unitary Mixing Matrix with MUV Scheme

Non-unitary mixing matrix under MUV scheme can be expressed as $N = HU$ where U is a unitary matrix and H is a Hermitian matrix [4, 5]. Also, $H \equiv (1 + \rho)$ with $\rho = |\rho_{ff'}|e^{-i\omega_{ff'}}$ for $f \neq f'$ and $\rho_{ff'}$, $\omega_{ff'}$ are non-unitary parameters. In this case, the flavour and mass eigenstates are connected via a non-unitary mixing matrix, i.e. $\nu_f = N_{fi}\nu_i$, where (f, i) represents flavour and mass indices, respectively. At SBL, the non-unitary CP-violating effects become dominant over the unitary one, the total oscillation amplitude can be expressed as $< \nu_{f'}|\nu_f(L) > = A_{ff'}^{SM}(L) + 2\rho_{ff'}^* + \mathcal{O}(\rho A)$, where all the ρ components having flavour indices other than ff' are considered up to terms $\mathcal{O}(\rho A)$ [5].

$$P_{ff'} = |\sum_{i=1}^{3}(U_{f'i}e^{-iE_i t}U_{fi}^*) + 2\rho_{ff'}^*|^2, \tag{145.2}$$

where $E_i - E_j \simeq \Delta m_{ji}^2/2E$ and $\Delta m_{ji}^2 = m_i^2 - m_j^2$. We find the oscillation probability for $\nu_\mu \to \nu_\tau$ channel, in vacuum, is given by

$$
\begin{aligned}
P_{\mu\tau} = \ & 4|\rho_{\mu\tau}|^2 + 4c_{12}^2 c_{23}^4 s_{12}^2 s_{13}^2 \sin^2(\Delta_{21}/2) + 4c_{12}^2 s_{12}^2 s_{13}^2 s_{23}^4 \sin^2(\Delta_{21}/2) \\
& + 4c_{23}^2 s_{12}^4 s_{13}^4 s_{23}^2 \sin^2(\Delta_{21}/2) - 8c_{12}^2 c_{23}^2 s_{12}^2 s_{13}^2 s_{23}^2 \sin^2(\Delta_{21}/2) \\
& - 8c_{12}^2 c_{23}|\rho_{\mu\tau}|s_{23} \sin(\Delta_{21}/2) \cos(\omega_{\mu\tau} - \Delta_{21}/2) + 4c_{12}^4 c_{23}^2 s_{23}^2 \sin^2(\Delta_{21}/2) \\
& - 8c_{12}^2 c_{23}^2 s_{12}^2 s_{13}^2 s_{23}^2 \sin^2(\Delta_{21}/2) \cos 2\delta + 4c_{13}^4 c_{23}^2 s_{23}^2 \sin^2(\Delta_{31}/2) \\
& + 8c_{23}|\rho_{\mu\tau}|s_{12}^2 s_{13}^2 s_{23} \sin(\Delta_{21}/2) \cos(\omega_{\mu\tau} - \Delta_{21}/2) \\
& - 8c_{12} c_{23}^2 |\rho_{\mu\tau}|s_{12} s_{13} \cos\delta \sin(\Delta_{21}/2) \cos(\omega_{\mu\tau} - \Delta_{21}/2) \\
& + 8c_{12}|\rho_{\mu\tau}|s_{12} s_{13} s_{23}^2 \cos(\Delta_{21}/2) \sin(\Delta_{21}/2) \cos(\delta + \omega_{\mu\tau}) \\
& + 8c_{12}|\rho_{\mu\tau}|s_{12} s_{13} s_{23}^2 \sin^2(\Delta_{21}/2) \sin(\delta + \omega_{\mu\tau}) \\
& + 8c_{12}^3 c_{23}^3 s_{12} s_{13} s_{23} \cos\delta \sin^2(\Delta_{21}/2) - 8c_{12}^3 c_{23} s_{12} s_{13} s_{23}^3 \cos\delta \sin^2(\Delta_{21}/2) \\
& - 8c_{12} c_{23}^3 s_{12}^3 s_{13}^3 s_{23} \cos\delta \sin^2(\Delta_{21}/2) + 8c_{12} c_{23} s_{12}^3 s_{13}^3 s_{23}^3 \cos\delta \sin^2(\Delta_{21}/2) \\
& + 8c_{13}^2 c_{23}|\rho_{\mu\tau}|s_{23} \sin(\Delta_{31}/2) \cos(\omega_{\mu\tau} - \Delta_{31}/2) - 8c_{12}^2 c_{13}^2 c_{23}^2 s_{23}^2 A \\
& + 8c_{13}^2 c_{23}^2 s_{12}^2 s_{13}^2 s_{23}^2 A - 8c_{12} c_{13}^2 c_{23}^3 s_{12} s_{13} s_{23} \cos\delta A + 8c_{12} c_{13}^2 c_{23} s_{12} s_{13} s_{23}^3 \cos\delta A \\
& - 8c_{12} c_{13}^2 c_{23}^3 s_{12} s_{13} s_{23} \sin\delta B - 8c_{12} c_{13}^2 c_{23} s_{12} s_{13} s_{23}^3 \sin\delta B \\
& - 8c_{12} c_{23}^2 |\rho_{\mu\tau}|s_{12} s_{13} \sin\delta \sin(\Delta_{21}/2) \sin(\omega_{\mu\tau} - \Delta_{21}/2), \tag{145.3}
\end{aligned}
$$

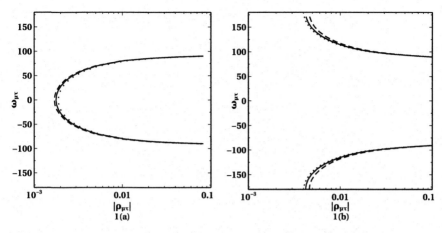

Fig. 145.1 3σ contour plots (for $\delta = \pi/2$) for the NH (left) and IH (right) neutrino mass spectrum. The solid and dashed (dotted) lines represent 3σ best fit and upper (lower) values of neutrino mixing parameters. The considered best fit value of θ_{23} is 42.3^o and 3σ upper (lower) bound is $53.3^o(38.2^o)$

where $c_{ij} = \cos\theta_{ij}$, $s_{ij} = \sin\theta_{ij}$, $\Delta_{ji} = 1.27\frac{\Delta m_{ji}^2 L}{E}$, L is the source-detector distance in km, E is energy of neutrino beam in GeV, $A = \sin(\Delta_{21}/2)\sin(\Delta_{31}/2)\cos(\Delta_{32}/2)$ and $B = \sin(\Delta_{21}/2)\sin(\Delta_{31}/2)\sin(\Delta_{32}/2)$.

We have used GLoBES software [6] with neutrino mixing parameters [7] to study SBL experiments having baseline $L = 130$ km similar to the distance between CERN and FREJUS. Figure 145.1 shows that the sensitivity to $|\rho_{\mu\tau}|$ is at its peak value for $\omega_{\mu\tau} = 0$ and at lowest value for $\omega_{\mu\tau} = \pm\pi$ for NH neutrino masses. For IH neutrino masses, the sensitivity gets reversed, i.e. $|\rho_{\mu\tau}|$ is at its peak value for $\omega_{\mu\tau} = \pm\pi$ and lowest value at $\omega_{\mu\tau} = 0$. These contours also show reflection symmetry about $\omega_{\mu\tau} = 0$ in both NH and IH neutrino masses due to cosine dependence of $\omega_{\mu\tau}$. The maximum sensitivity to $|\rho_{\mu\tau}|$ is $10^{-2.5}$ at 3σ, which is better than the limit provided by CHORUS and NOMAD($|\rho_{\mu\tau}| = 10^{-1.7}$ at 3σ) [8].

145.3 Unitary and Non-unitary *CP*-Violation Effects

To distinguish the *CP*-violating effects coming from unitary and non-unitary *CP* phases in oscillation probability $P_{\mu\tau}$ (Eq. 145.3) can be divided into four parts

$$P_{\mu\tau} = P_0 + P(\delta) + P(\omega_{\mu\tau}) + P(\delta, \omega_{\mu\tau}). \qquad (145.4)$$

First term P_0 is independent of both unitary and non-unitary *CP* phases δ.

$$P_0 = 4|\rho_{\mu\tau}|^2 + s_{13}^2 \left(c_{23}^4 + s_{23}^4\right) \sin^2 2\theta_{12} \sin^2 \left(\Delta_{21}/2\right) \tag{145.5}$$
$$-(1/2)s_{13}^2 \sin^2 2\theta_{23} \sin^2 2\theta_{12} \sin^2 \left(\Delta_{21}/2\right)$$
$$+ \left(s_{12}^4 s_{13}^4 + c_{12}^4\right) \sin^2 2\theta_{23} \sin^2 \left(\Delta_{21}/2\right)$$
$$-2c_{13}^2 \left(c_{12}^2 - s_{12}^2 s_{13}^2\right) A \sin^2 2\theta_{23}$$
$$+c_{13}^4 \sin^2 2\theta_{23} \sin^2 \left(\Delta_{31}/2\right).$$

Second (third) term $P(\delta)$ $(P(\omega_{\mu\tau}))$ depends on $\delta(\omega_{\mu\tau})$. The last term $P(\delta, \omega_{\mu\tau})$ has dependence on both unitary and non-unitary CP phases.

$$P(\delta) = -(1/2)s_{13}^2 \sin^2 2\theta_{12} \sin^2 2\theta_{23} \cos 2\delta \sin^2 \left(\Delta_{21}/2\right) \tag{145.6}$$
$$+s_{13} \left(c_{12}^2 - s_{12}^2 s_{13}^2\right) \sin 4\theta_{23} \sin 2\theta_{12} \cos \delta \sin^2 \left(\Delta_{21}/2\right)$$
$$-c_{13} \sin 2\theta_{23} \sin 2\theta_{12} \sin 2\theta_{13} \left(A \cos 2\theta_{23} \cos \delta + B \sin \delta\right),$$
$$P(\omega_{\mu\tau}) = -4|\rho_{\mu\tau}| \left(c_{12}^2 - s_{12}^2 s_{13}^2\right) \sin 2\theta_{23} \sin \left(\Delta_{21}/2\right) \cos \left(\omega_{\mu\tau} - \Delta_{21}/2\right) \tag{145.7}$$
$$+4c_{13}^2 |\rho_{\mu\tau}| \sin 2\theta_{23} \sin \left(\Delta_{31}/2\right) \cos \left(\omega_{\mu\tau} - \Delta_{31}/2\right),$$
$$P(\delta, \omega_{\mu\tau}) = -4s_{13}c_{23}^2 |\rho_{\mu\tau}| \sin 2\theta_{12} \sin \left(\Delta_{21}/2\right) \cos \left(\Delta_{21}/2 + \delta - \omega_{\mu\tau}\right) \tag{145.8}$$
$$+4s_{13}s_{23}^2 |\rho_{\mu\tau}| \sin 2\theta_{12} \sin \left(\Delta_{21}/2\right) \cos \left(\Delta_{21}/2 - \delta - \omega_{\mu\tau}\right).$$

It is evident from Fig. 145.2a that for $L/E < 200$ km/GeV the contribution from $P(\delta)$ is negligibly small, whereas the contribution from $P(\omega_{\mu\tau})$ and $P(\delta, \omega_{\mu\tau})$ is predominantly large for this range. Therefore, the non-unitary CP-violating effects will be more prominent in the oscillation experiments possessing $L/E < 200$ km/GeV. From Fig. 145.2b, we can interpret that the CP fraction [9] $F(\delta)$ for SBL experiments

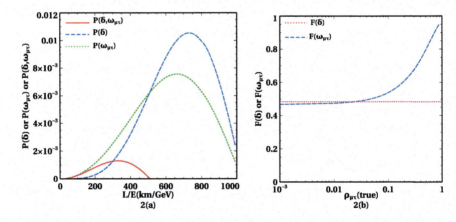

Fig. 145.2 $P(\delta)$ or $P(\omega_{\mu\tau})$ or $P(\delta, \omega_{\mu\tau})$ versus L/E(km/GeV) ratio (left) and CP fractions $F(\delta)$ or $F(\omega_{\mu\tau})$ versus $\rho_{\mu\tau}(true)$ (right) for $L = 130$ km. The value of $|\rho_{\mu\tau}| = 0.02$ is considered in Fig. 145.2a

remains constant, i.e. $F(\delta) = 0.49$, at 3σ and shows no variation for the entire range of $\rho_{\mu\tau}$. For the case, when $\delta = 0$, and the CP fraction $F(\omega_{\mu\tau})$ is the dominant one, we find that $F(\omega_{\mu\tau})$ show similar behaviour (remain constant) as that of $F(\delta)$ up to $\rho_{\mu\tau} \leq 10^{-2}$ after which it starts increasing with increase in $\rho_{\mu\tau}$.

145.4 Conclusions

We investigated the sensitivities to non-unitary parameter $|\rho_{\mu\tau}|$ for θ_{23} octant with both the hierarchies. We found that θ_{23} octant has distinguishing implications towards the $|\rho_{\mu\tau}|$ sensitivities with NH and IH of neutrino masses. The maximum sensitivity to $\rho_{\mu\tau}$, at 3σ, is $10^{-2.5}$ which is better than the limit provided by CHORUS and NOMAD. To differentiate the CP-violating effects from the unitary and non-unitary sources would be very difficult; however, we have worked upon the effects of non-unitary parameter ($\rho_{\mu\tau}$) on CP-violation discovery reach. We found that the non-unitary effects will be more prominent in the oscillation experiments having $L/E \leq 200$ km/GeV. Moreover, in SBL experiments, CP fraction $F(\omega_{\mu\tau})$ plays the significant role above $\rho_{\mu\tau} \geq 10^{-2}$, whereas the CP fraction $F(\delta)$ corresponding to unitary phase δ remains insignificant.

Acknowledgements S. V. acknowledges the financial support provided by University Grants Commission (UGC)-Basic Science Research (BSR), Government of India vide Grant No. F.20-2(03)/2013(BSR).

References

1. C. Athanassopoulos et al., Phys. Rev. Lett. **75**, 2650 (1995)
2. A.A. Aguilar-Arevalo et al., Phys. Rev. Lett. **121**, 221801 (2018)
3. S. Verma, S. Bhardwaj, Indian J. Phys. **92**, 1161–1167 (2018)
4. S. Antusch, C. Biggio, E. Fernandez-Martinez, M.B. Gavela, J. Lopez-Pavon, JHEP **10**, 084 (2006)
5. E. Fernandez-Martinez, M.B. Gavela, J. Lopez-Pavon, O. Yasuda, Phys. Lett. B **649**, 427 (2007)
6. P. Huber, J. Kopp, M. Lindner, M. Rolinec, W. Winder, Comput. Phys. Commun. **177**, 432 (2007)
7. M.C. Gonzalez-Garcia, M. Maltoni, T. Schwetz, JHEP **052**, 1411 (2014)
8. F.J. Escrihuela, D.V. Forero, O.G. Miranda, M. Tortola, J.W.F. Valle, New J. Phys. **19**, 093005 (2017)
9. Z. Rahman, Arnab Dasgupta, R. Adhikari, J. Phys. G: Nucl. Part. Phys. **42**, 065001 (2015)

Chapter 146
General Structure of a Gauge Boson Propagator and Pressure of Deconfined QCD Matter in a Weakly Magnetized Medium

Bithika Karmakar, Aritra Bandyopadhyay, Najmul Haque, and Munshi G. Mustafa

Abstract We have systematically constructed the general structure of the gauge boson self-energy and the effective propagator in presence of a nontrivial background like a hot magnetized material medium. Based on this as well as the general structure of fermion propagator in the weakly magnetized medium, we have calculated pressure of deconfined QCD matter within HTL approximation.

146.1 Introduction

Quark-Gluon Plasma is a thermalized color deconfined state of nuclear matter in the regime of Quantum Chromo Dynamics (QCD) under extreme conditions such as very high temperature and/or density. For the past couple of decades, different high energy Heavy-Ion-Collisions (HIC) experiments are underway, e.g., RHIC @ BNL, LHC @ CERN and upcoming FAIR @ GSI, to study this novel state of QCD matter. In recent years the non-central HIC is also being studied, where a very strong magnetic field can be created in the direction perpendicular to the reaction plane due

B. Karmakar (✉) · M. G. Mustafa
Theory Division, Saha Institute of Nuclear Physics, HBNI, 1/AF, Bidhannagar, Kolkata 700064, India
e-mail: bithika.karmakar@saha.ac.in; karmakarbithika.93@gmail.com

M. G. Mustafa
e-mail: munshigolam.mustafa@saha.ac.in

A. Bandyopadhyay
Departamento de Física, Universidade Federal de Santa Maria, Santa Maria, RS 97105-900, Brazil
e-mail: aritrabanerjee.444@gmail.com

N. Haque
School of Physical Sciences, National Institute of Science Education and Research, HBNI, Jatni 752050, Khurda, India
e-mail: nhaque@niser.ac.in

© Springer Nature Singapore Pte Ltd. 2021
P. K. Behera et al. (eds.), *XXIII DAE High Energy Physics Symposium*, Springer Proceedings in Physics 261, https://doi.org/10.1007/978-981-33-4408-2_146

to the spectator particles that are not participating in the collisions [1–3]. Also, some studies have showed that the strong magnetic field generated during the non-central HIC is time dependent and rapidly decreases with time [4, 5]. At the time of the non-central HIC, the value of the created magnetic field B is very high compared to the temperature T ($T^2 < q_f B$, where q_f is the absolute charge of the quark with flavor f) associated with the system, whereas after few fm/c, the magnetic field is shown to decrease to a very low value ($q_f B < T^2$). In this regime one usually works in the weak magnetic field approximation.

The presence of an external anisotropic field in the medium calls for the appropriate modification of the present theoretical tools to investigate various properties of QGP and a numerous activity is in progress. The EoS is a generic quantity and of phenomenological importance for studying the hot and dense QCD matter, QGP, created in HIC.

146.2 General Structure of Gauge Boson Propagator

Finite temperature breaks the boost symmetry of a system, whereas magnetic field or anisotropy breaks the rotational symmetry. We consider the momentum of gluon as $P_\mu = (p_0, p_1, 0, p_3)$. We work in the rest frame of the heat bath, $i.e$, $u^\mu = (1, 0, 0, 0)$ and represent the background magnetic field as $n_\mu \equiv \frac{1}{2B}\epsilon_{\mu\nu\rho\lambda} u^\nu F^{\rho\lambda} = \frac{1}{B} u^\nu \tilde{F}_{\mu\nu} = (0, 0, 0, 1)$. The general structure of the gauge boson self-energy in presence of magnetic field can be written as [6]

$$\Pi^{\mu\nu} = bB^{\mu\nu} + cR^{\mu\nu} + dQ^{\mu\nu} + aN^{\mu\nu}, \tag{146.1}$$

where the form factors b, c, d, and a can be calculated as

$$b = B^{\mu\nu}\Pi_{\mu\nu}, \quad c = R^{\mu\nu}\Pi_{\mu\nu}, \quad d = Q^{\mu\nu}\Pi_{\mu\nu}, \quad a = \frac{1}{2}N^{\mu\nu}\Pi_{\mu\nu}. \tag{146.2}$$

Using Dyson–Schwinger equation, one can write the general structure of gluon propagator as

$$\mathcal{D}_{\mu\nu} = \frac{\xi P_\mu P_\nu}{P^4} + \frac{(P^2 - d)B_{\mu\nu}}{(P^2 - b)(P^2 - d) - a^2} + \frac{R_{\mu\nu}}{P^2 - c} + \frac{(P^2 - b)Q_{\mu\nu}}{(P^2 - b)(P^2 - d) - a^2}$$
$$+ \frac{aN_{\mu\nu}}{(P^2 - b)(P^2 - d) - a^2}. \tag{146.3}$$

It is found from the poles of Eq. (146.3) that the gluon in hot magnetized medium has three dispersive modes which are given as

$$P^2 - c = 0, \tag{146.4}$$

$$(P^2 - b)(P^2 - d) - a^2 = (P^2 - \omega_n^+)(P^2 - \omega_n^-) = 0, \tag{146.5}$$

where $\omega_{n^+} = \frac{b+d+\sqrt{(b-d)^2+4a^2}}{2}$ and $\omega_{n^-} = \frac{b+d-\sqrt{(b-d)^2+4a^2}}{2}$.

We consider small magnetic field approximation and calculate all the quantities up to $\mathcal{O}[(eB)^2]$. Within this approximation (146.5) becomes

$$\left(P^2 - b\right)\left(P^2 - d\right) = 0. \tag{146.6}$$

The form factors b, c, and d are calculated [6] from (146.2) using HTL approximation.

146.3   Free Energy and Pressure in Weak Field Approximation

The total one-loop free energy of deconfined QCD matter in a weakly magnetized hot medium reads as [7]

$$F = F_q + F_g + F_0 + \Delta\mathcal{E}_0, \tag{146.7}$$

where F_q, F_g are quark and gluon free energy in weak magnetized medium which are calculated [7] using the form factors corresponding to quark [8] and gluon self-energy [6]. $F_0 = \frac{1}{2}B^2$ is the tree-level contribution due to the constant magnetic field and the $\Delta\mathcal{E}_0$ is the HTL counter term given as

$$\Delta\mathcal{E}_0 = \frac{d_A}{128\pi^2\epsilon}m_D^4, \tag{146.8}$$

with $d_A = N_c^2 - 1$, N_c is the number of color in fundamental representation and m_D is the Debye screening mass in HTL approximation. The divergences present in the total free energy are removed by redefining the magnetic field in F_0 and by adding counter terms [7].

The pressure of the deconfined QCD matter in weakly magnetized medium is given by

$$P(T, \mu, B, \Lambda) = -F(T, \mu, B, \Lambda), \tag{146.9}$$

where Λ is the renormalization scale.

146.4 Results

The variation of scaled pressure with temperature is shown in Fig. 146.1 for $\mu = 0$ and $\mu = 300$ MeV. It can be seen from the figure that the magnetic field dependence of the scaled pressure decreases with temperature because temperature is the dominant scale in the weak field approximation ($|eB| < T^2$). At high temperatures, all the plots for different magnetic fields asympotically reach the one-loop HTL pressure. The scaled pressure is plotted with magnetic field strength in Fig. 146.2. We note

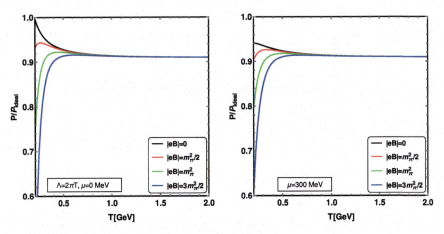

Fig. 146.1 Variation of the scaled one-loop pressure with temperature for $N_f = 3$ with $\mu = 0$ (left panel) for $\mu = 300$ MeV (right panel) in presence of weak magnetic field. Renormalization scales are chosen as $\Lambda_g = 2\pi T$ for gluon and $\Lambda_q = 2\pi\sqrt{T^2 + \mu^2/\pi^2}$ for quark

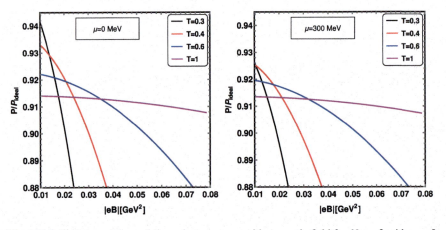

Fig. 146.2 Variation of the scaled one-loop pressure with magnetic field for $N_f = 3$ with $\mu = 0$ (left panel) and $\mu = 300$ MeV (right panel) for $T = (0.3, 0.4, 0.6$ and $1)$ GeV

from Fig. 146.2 that the slopes of the plots decrease with an increase of temperature reflecting the reduced magnetic field dependence.

Acknowledgements BK and MGM were funded by the Department of Atomic Energy, India via the project TPAES. BK gratefully acknowledges the organizers of XXIII DAE-BRNS High Energy Physics Symposium 2018 for the invitation.

References

1. I.A. Shovkovy, Magnetic catalysis: a review. Lect. Notes Phys. **871**, 13 (2013). arXiv:1207.5081 [hep-ph]
2. M. D'Elia, Lattice QCD simulations in external background fields. Lect. Notes Phys. **871**, 181 (2013). arXiv:1209.0374 [hep-lat]
3. K. Fukushima, Views of the chiral magnetic effect. Lect. Notes Phys. **871**, 241 (2013). arXiv:1209.5064 [hep-ph]
4. A. Bzdak, V. Skokov, Anisotropy of photon production: initial eccentricity or magnetic field. Phys. Rev. Lett. **110**, 192301 (2013). arXiv:1208.5502 [hep-ph]
5. L. McLerran, V. Skokov, Comments about the electromagnetic field in heavy-ion collisions. Nucl. Phys. A **929**, 184 (2014). arXiv:1305.0774 [hep-ph]
6. B. Karmakar, A. Bandyopadhyay, N. Haque and M.G. Mustafa, General structure of gauge boson propagator and its spectra in a hot magnetized medium. arXiv:1805.11336 [hep-ph]
7. A. Bandyopadhyay, B. Karmakar, N. Haque and M.G. Mustafa, The pressure of a weakly magnetized hot and dense deconfined QCD matter in one-loop Hard-Thermal-Loop perturbation theory. arXiv:1702.02875 [hep-ph]
8. A. Das, A. Bandyopadhyay, P.K. Roy, M.G. Mustafa, General structure of fermion two-point function and its spectral representation in a hot magnetized medium. Phys. Rev. D **97**, 034024 (2018). arXiv:1709.08365 [hep-ph]

Chapter 147
Transport Phenomena in the Hot Magnetized Quark–Gluon Plasma

Manu Kurian and Vinod Chandra

Abstract The transport coefficients such as bulk viscosity and electrical conductivity of the hot quark–gluon plasma (QGP) have been estimated in the presence of a strong magnetic field. A quasiparticle description of the hot QCD equation of state (EoS) has been adopted to encode the collective excitations of the magnetized nuclear matter. Transport coefficients have been estimated by setting up a $1 + 1-$dimensional effective covariant kinetic theory to describe the Landau level dynamics of the quarks, with thermal relaxation time as the dynamical input. The effective kinetic theory enables one to include the mean field corrections in terms of nontrivial modified quasiparticle energy dispersion to the transport parameters.

147.1 Introduction

The quantitative estimation of the experimental observables in highly energetic heavy-ion collision experiments from the dissipative hydrodynamic simulations involves the dependence upon the transport coefficients of the hot QCD medium [1, 2]. Recent studies on the QGP suggest the presence of intense magnetic field in the non-central asymmetric heavy-ion collisions [3–5]. The effect of magnetic field on the thermodynamic and transport properties of the QGP is investigated in Refs. [6–12]. The effective description of the transport coefficients requires the quasiparticle modelling of the system followed by the setting up of the $1 + 1-$ Boltzmann equation for the hot magnetized QGP medium [13]. We employed relaxation time approximation (RTA) to incorporate the proper collision integral for the processes in the presence of a magnetic field. It is important to note that quark–antiquark pair production and annihilation ($1 \rightarrow 2$ processes) are dominant in the presence of strong magnetic field [10]. The prime focus of the current contribution is to investigate the

M. Kurian (✉) · V. Chandra
Indian Institute of Technology Gandhinagar, Gandhinagar 382355, Gujarat, India
e-mail: manu.kurian@iitgn.ac.in

V. Chandra
e-mail: vchandra@iitgn.ac.in

© Springer Nature Singapore Pte Ltd. 2021
P. K. Behera et al. (eds.), *XXIII DAE High Energy Physics Symposium*,
Springer Proceedings in Physics 261,
https://doi.org/10.1007/978-981-33-4408-2_147

temperature behaviour of the bulk viscosity and electrical conductivity, incorporating the effects of hot QCD medium and higher Landau levels in the presence of the magnetic field. The thermal QCD medium interaction effects are embedded in the analysis by adopting the effective fugacity quasiparticle model (EQPM) [14]. By setting up a covariant effective kinetic theory, one can obtain the mean field correction to the transport coefficients from the local conservations of number current and stress-energy tensor [15].

147.2 Formalism

Covariant effective kinetic theory

The description of the system away from equilibrium can be obtained from the effective Boltzmann equation, which quantifies the rate of change of momentum distribution function in terms of collision kernel. Here, we are primarily focusing on the quark and antiquark Landau level dynamics of the magnetized medium. Within RTA, we have [13, 15]

$$\frac{1}{\omega_{l_q}} \bar{p}_q^\mu \partial_\mu f_q^0(x, \bar{p}_{z_q}) + F_z \frac{\partial f_q^0}{\partial p_{z_q}} = -\frac{\delta f_q}{\tau_{eff}}, \tag{147.1}$$

where $F_z = -\partial_\mu(\delta \omega u^\mu u_z)$ is the force term from the basic conservation laws and τ_{eff} is the thermal relaxation time for the dominant $1 \to 2$ process in the presence of the magnetic field. The EQPM quark/antiquark distribution function in the presence of magnetic field $\mathbf{B} = B\hat{z}$

$$f_q^0 = \frac{z_q \exp\left[-\beta(u^\mu p_\mu)\right]}{1 + z_q \exp\left[-\beta(u^\mu p_\mu)\right]}, \tag{147.2}$$

where z_q is the effective fugacity for the quarks/antiquarks which encodes the thermal medium effects, and $p_q^\mu = (E_{l_q}, 0, 0, p_z)$, with $E_{l_q} = \sqrt{\bar{p}_z^2 + m^2 + 2l \mid q_f e B \mid}$ for the Landau level (LL) l. Quasiparticle momenta and bare particle four momenta can be related as $\bar{p}_q^\mu = p_q^\mu + \delta \omega u^\mu$, where $\delta \omega = T^2 \partial_T \ln(z_q)$ and modifies the zeroth component of the four momenta as $\bar{p}_q^0 \equiv \omega_{l_q} = E_{l_q} + \delta \omega$. The local distribution function of quarks can expand as $f_q = f_q^0(p_z) + f_q^0(1 \pm f_q^0)\phi_q$. Here, ϕ_q defines the deviation of the quark/antiquark distribution function from its equilibrium and can be obtained by solving the Boltzmann equation as described in Ref. [13]. The analysis is done by employing the Chapman–Enskog method.

Bulk viscosity and longitudinal conductivity

The non-equilibrium component of the pressure tensor can be decomposed from the energy–momentum tensor as $P^{\mu\nu} = -P\Delta_\parallel^{\mu\nu} + \Pi^{\mu\nu}$, where $\Pi^{\mu\nu}$ is the viscous pressure tensor which takes the form $\Pi^{\mu\nu} = 2\eta\langle\langle\partial^\mu u^\nu\rangle\rangle + \zeta\Delta_\parallel^{\mu\nu}\partial.u$. Here, u^μ is the hydrodynamic four velocity and $\Delta_\parallel^{\mu\nu} = g_\parallel^{\mu\nu} - u^\mu u^\nu = -b^\mu b^\nu$, with $g_\parallel^{\mu\nu} = (1, 0, 0, -1)$. Comparing the macroscopic definition with the microscopic definition which is obtained by solving the Boltzmann equation as in Ref. [13], we have

$$\zeta = \sum_l \sum_{k=q,\bar{q}} \mu_l \frac{|q_{f_k}eB|}{2\pi} \frac{v_k}{3T} \int \frac{d\bar{p}_{z_k}}{(2\pi)} \frac{1}{\omega_{l_k}^2} \{\bar{p}_{z_k}^2 - \omega_{l_k}^2 c_s^2\}^2 \tau_{eff} f_k^0(1 - f_k^0)$$
$$+ \sum_l \sum_{k=q,\bar{q}} \delta\omega\mu_l \frac{|q_{f_k}eB|}{2\pi} \frac{v_k}{3T} \int \frac{d\bar{p}_{z_k}}{(2\pi)} \frac{1}{\omega_{l_k}^2} \{\bar{p}_{z_k}^2 - \omega_{l_k}^2 c_s^2\}^2 \frac{1}{E_{l_k}} \tau_{eff} f_k^0(1 - f_k^0).$$

(147.3)

The second term represents the mean field corrections to the bulk viscosity due to the quasiparton excitations. To analyse the induced longitudinal current density of the hot magnetized nuclear matter in the presence of the electric field $\mathbf{E} = E(X)\hat{z}$, the quantity δf_q need to be obtained from the transport theory as described in Eq. (147.1). The current density with the mean field contribution takes the form [16]

$$J_z = -\sum_{l=0}^\infty \frac{q_f^2}{2\pi} \frac{|q_f B|}{2\pi} \mu_l N_c \int_{-\infty}^\infty dp_z v_z^2 \tau_{eff}(f_q^{(0)'} + f_{\bar{q}}^{(0)'})E$$
$$+ \delta\omega \sum_{l=0}^\infty \frac{q_f^2}{2\pi} \frac{|q_f B|}{2\pi} \mu_l N_c \int_{-\infty}^\infty dp_z v_z^2 \frac{1}{E_l} \tau_{eff}(f_q^{(0)'} + f_{\bar{q}}^{(0)'})E.$$

(147.4)

The longitudinal conductivity σ_{zz} could be read off from Ohm's, law $J_z = \sigma_{zz}E$. The linear and additional components due to the inhomogeneity of the electric field of the current density can also be obtained with this formalism [16]. Thermal relaxation time for the $1 \rightarrow 2$ process in the strong magnetic field with higher Landau levels is estimated in Refs. [7, 13] (Figs. 147.1 and 147.2).

147.3 Summary

We have estimated the temperature dependence of the bulk viscosity and conductivity within the effective kinetic theory. The key observation is that the transport coefficients of the magnetized QGP are significantly affected by the higher Landau level contributions and mean field corrections.

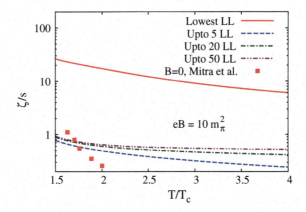

Fig. 147.1 Effects of higher Landau levels in the temperature behaviour of the ζ/s at $|q_f B|= 10m_\pi^2$ within the RTA. The results are compared with that of $B = 0$ case [15]

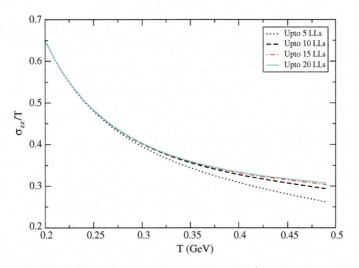

Fig. 147.2 Temperature behaviour of the σ_{zz}/T at $|q_f B| = 10m_\pi^2$ with different Landau levels

Acknowledgements VC would like to acknowledge the Department of Science and Technology (DST), Govt. of India for INSPIRE-Faculty Fellowship and SERB, Govt. of India for the Early Career Research Award (ECRA/2016).

References

1. M. Luzum, P. Romatschke, Phys. Rev. C **78**, 034915 (2008)
2. J. Adam et al., ALICE Collaboration. Phys. Rev. Lett. **117**, 182301 (2016)
3. V. Skokov, A.Y. Illarionov, V. Toneev, Int. J. Mod. Phys. A **24**, 5925 (2009)

4. Y. Zhong, C.B. Yang, X. Cai, S.Q. Feng, Adv. High Energy Phys. **2014**, 193039 (2014)
5. S.K. Das, S. Plumari, S. Chatterjee, J. Alam, F. Scardina, V. Greco, Phys. Lett. B **768**, 260 (2017)
6. M. Kurian, V. Chandra, Phys. Rev. D **96**(11), 114026 (2017)
7. K. Fukushima, Y. Hidaka, Phys. Rev. Lett. **120**(16), 162301 (2018)
8. M. Kurian, V. Chandra, Phys. Rev. D **97**(11), 116008 (2018)
9. A. Bandyopadhyay, B. Karmakar, N. Haque, M.G. Mustafa, arXiv:1702.02875 [hep-ph]
10. K. Hattori, S. Li, D. Satow, H.U. Yee, Phys. Rev. D **95**(7), 076008 (2017)
11. K. Hattori, X.G. Huang, D.H. Rischke, D. Satow, Phys. Rev. D **96**(9), 094009 (2017)
12. S. Koothottil, V.M. Bannur, Phys. Rev. C **99**(3), 035210 (2019)
13. M. Kurian, S. Mitra, S. Ghosh, V. Chandra, Eur. Phys. J. C **79**(2), 134 (2019)
14. V. Chandra, V. Ravishankar, Phys. Rev. D **84**, 074013 (2011)
15. S. Mitra, V. Chandra, Phys. Rev. D **97**(3), 034032 (2018)
16. M. Kurian, V. Chandra, arXiv:1902.09200 [nucl-th]

Chapter 148
Strangeness Production in p-Pb Collisions at 8.16 TeV

Meenakshi Sharma

Abstract The analysis status of strange hadrons (K_s^0 and Λ) in p-Pb collisions in multiplicity bins is presented as a function of p_T for $-0.5 < y_{CMS} < 0$. The excellent tracking and particle identification capabilities of ALICE can be used to reconstruct the strange hadrons using the tracks produced by their weak decays. The yield of strange hadrons is one of the various observables sensitive to the evolution of the system after nuclear collisions. It is now confirmed that strange quarks would be produced with higher probability in a QGP scenario with respect to that expected in a pure hadron gas scenario. Therefore, studies of strangeness production can help to determine the properties of the created system.

148.1   Introduction

Strangeness production plays a key role in the study of hot and dense systems created in nuclear collision. Rafelski and Muller [1] reported for the first time that the enhancement of the relative strangeness production could be one of the signatures of a phase transition from hadronic matter to the new phase consisting of almost free quarks and gluons (QGP). Strangeness enhancement was observed in several experiments [2–4] as a function of number of participating nucleons $<N_{part}>$. The enhancement is relative to the production of strangeness in small systems (p-p or p-Be) where the enhanced production was not expected. Recent studies by the ALICE collaboration show that the strange particle yield with respect to pion yield increases smoothly with increase in $\langle dN_{ch}/d\eta \rangle$ [5, 6].

Meenakshi Sharma, for the ALICE Collaboration

M. Sharma (✉)
Department of Physics, University of Jammu, Jammu and Kashmir, Jammu 180006, India
e-mail: meenakshi.sharma@cern.ch

© Springer Nature Singapore Pte Ltd. 2021   1019
P. K. Behera et al. (eds.), *XXIII DAE High Energy Physics Symposium*,
Springer Proceedings in Physics 261,
https://doi.org/10.1007/978-981-33-4408-2_148

148.2 Detection of Strange Hadrons in P-Pb Collisions with ALICE

With an overall branching ratio of 69.2% (63.9%), K_s^0 (Λ) hadrons decay weakly into $\pi^+\pi^-$ ($p\pi^-$) pairs. The tracks formed by the daughter particles of the decays are reconstructed in Inner Tracking System (ITS) and in the large Time Projection Chamber (TPC). The particle identity is determined through the measured ionisation energy loss and momentum measured in the TPC. The secondary vertex reconstruction from particle decays is performed as shown in Fig. 148.1.

To reduce the combinatorial background in the selection of the weakly decaying particles, a set of geometrical cuts based on the decay topologies were applied. These cuts were imposed on the minimum Distance of Closest Approach (DCA) between the V0 daughter tracks and the primary vertex. Another cut was applied on the DCA between charged daughter tracks.

148.3 Analysis

A detailed description of the ALICE detector can be found in [7]. The data used for the analysis was recorded in 2016. The particle identification is done using following ALICE sub-detectors: the Inner Tracking System (ITS), the Time Projection Chamber (TPC), by using different PID techniques [8]. The multiplicity bins are defined based on the signal amplitude measured in the V0A scintillator.

Fig. 148.1 Secondary vertex reconstruction principle, with K_s^0 and Ξ^- decays shown as examples. For clarity, the decay points are placed between the first two ITS layers (radii are not to scale). The solid lines represent the reconstructed charged particle tracks, extrapolated to secondary vertex. Extrapolations to the primary and secondary vertices are shown in dashed lines

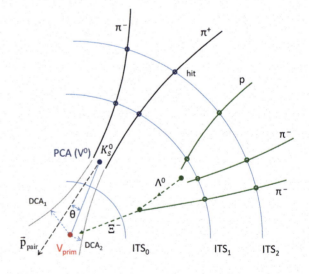

The signal would be extracted from the invariant mass distribution of the decay daughters. In every p_T bin, a Gaussian function will be used to fit to the mass peak and a central area will be defined as -4σ and 4σ for that peak.

148.4   Outlook

The invariant mass distributions have been presented in Fig. 148.2 in four different p_T bins for the 0–100% multiplicity class. Figure 148.3 shows the comparison of the yields from previous analysis of various hadrons relative to pions in p-p collisions at $\sqrt{s}$ = 13 TeV, Pb-Pb collisions at $\sqrt{s_{NN}}$ = 5.02 TeV, p-p collisions at $\sqrt{s}$ = 7 TeV, p-Pb collisions at $\sqrt{s_{NN}}$ = 5.02 TeV and Xe-Xe collisions at $\sqrt{s_{NN}}$ = 5.44 TeV.

From this comparison, one can see that the points from different systems but with similar value of $\langle dN_{ch}/d\eta \rangle$ overlap. This suggests that relative strangeness production does not depend on the initial stage parameters such as identity of the colliding ions or collision energy, but depends on the final state charged particle density. The $\langle dN_{ch}/d\eta \rangle$ covers three orders of magnitude starting with no strangeness enhancement for multistrange particles at low value of $\langle dN_{ch}/d\eta \rangle$ (possible canonical suppression scenario) with constant relative production of strangeness for high values of $\langle dN_{ch}/d\eta \rangle$ (possible grand canonical saturation scenario). The present analysis is

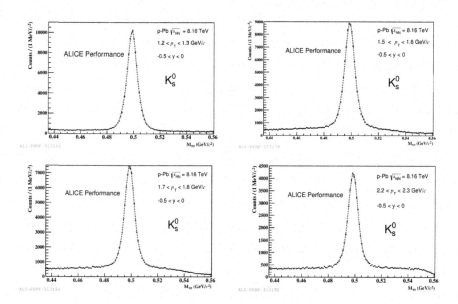

Fig. 148.2   Invariant mass distributions of K_s^0 in four p_T bins 1.2–1.3 GeV/c, 1.5–1.6 GeV/c, 1.7–1.8 GeV/c and 2.2–2.3 GeV/c, fitted with Gaussian peaks (dashed red curves) and linear backgrounds. The distributions are for the 0–100% multiplicity class

Fig. 148.3 Comparison of yields of protons, K_s^0, $(\Lambda + \overline{\Lambda})$, ϕ, $(\Xi^- + \Xi^+)$ and $(\Omega^- + \overline{\Omega}^+)$ relative to pion yields in p-p collisions at $\sqrt{s}$ = 13 TeV, Pb-Pb collisions at $\sqrt{s_{NN}}$ = 5.02 TeV, p-p collisions at $\sqrt{s}$ = 7 TeV, p-Pb collisions at $\sqrt{s_{NN}}$ = 5.02 TeV, Xe-Xe collisions at $\sqrt{s_{NN}}$ = 5.44 TeV and p-Pb collisions at $\sqrt{s_{NN}}$ = 8.16 TeV

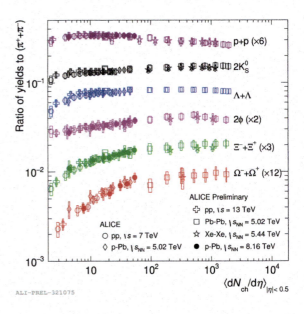

ALI-PREL-321075

a preliminary work and measurements of K_s^0 and Λ in p-Pb collisions at $\sqrt{s_{NN}}$ = 8.16 TeV will be added to Fig. 148.3. This will allow us to see whether p-Pb system at 8.16 TeV shows the same trend as the other mentioned systems or not.

References

1. J. Rafelski, B. Muller, Phys. Rev. Lett. **48**, 1066–1069 (1982)
2. The WA97 Collaboration, Phys. Lett. B **449**(3–4), 401–406 (1999)
3. B.I. Alelev et al., (STAR Collaboration) Phys. Rev. C **77**, 044908. Accessed 24 April 2008
4. D. Elia (for the ALICE Collaboratio), J. Phys.: Conf. Ser. **455**, 012005 (2013)
5. ALICE Collaboration, B. Abelev et al., Multiplicity dependence of pion, kaon, proton, and lambda production in p-Pb collisions at $\sqrt{s_{NN}}$ = 5.02 TeV. Phys. Lett. B **728**, 25–28 (2014). arXiv:1307. 6796 [nucl-ex]
6. ALICE Collaboration, J. Adam et al., Multi-strange baryon production in p-Pb collisions at $\sqrt{s_{NN}}$ = 5.02 TeV. Phys. Lett. B **758**, 389–401 (2016). arXiv:1512. 07227 [nucl-ex]
7. K. Aamodt et al., ALICE collaboration. J. Inst. **3**, S08002 (2008)
8. J. Adam et al., ALICE collaboration. Eur. Phys. J. C **75**, 226 (2015)

Chapter 149
Open Heavy Flavor Production in Hot QCD Matter at CERN LHC

Captain R. Singh, Nikhil Hatwar, and M. Mishra

Abstract Quarkonia suppression has been looked up as a prominent signature of Quark–Gluon Plasma (QGP). Charmonium and bottomonium suppression has been extensively studied under the QGP scenario with various hot and cold nuclear matter effects using Unified Model of Quarkonia Suppression (UMQS) in our earlier work. Here we have modified our model to explain suppression data of D^0 mesons at 2.76 TeV. For space-time evolution of temperature, we used $(1 + 1)$-dimensional Bjorken's hydrodynamics. We are using quasi-particle model as the equation of state for QGP. We calculate the survival probability for color screening, gluonic dissociation, and collision damping under QGP scenario. We also consider recombination of mesons through gluonic de-excitation approach. Among cold nuclear matter effects, we include nuclear shadowing. We finally calculate the survival probability of D^0 meson as a function of transverse momentum and centrality and compare it with the nuclear modification factor data available at LHC.

149.1 Introduction

Quarkonium suppression in heavy-ion collisions at the Relativistic Heavy-ion Collider (RHIC) and Large Hadron Collider (LHC) experiments is a prominent signature of the Quark–Gluon Plasma (QGP) formation [1]. There had been many attempts to theoretically reproduce the suppression of quarkonia that has been observed in heavy-ion collisions [2]. At energies greater than twice the rest masses of heavy quarks, pair production happens due to hard scattering at the point of collision. Some of these heavy quarks (antiquarks) forms quarkonia immediately and the rest

C. R. Singh · N. Hatwar (✉) · M. Mishra
Department of Physics, Birla Institute of Technology and Science, Pilani 333031, India
e-mail: nikhil.hatwar@gmail.com

C. R. Singh
e-mail: captainriturajsingh@gmail.com

M. Mishra
e-mail: madhukar.12@gmail.com

P. K. Behera et al. (eds.), *XXIII DAE High Energy Physics Symposium*,
Springer Proceedings in Physics 261,
https://doi.org/10.1007/978-981-33-4408-2_149

drifts apart along with the medium. Most of these drifting heavy quark (antiquark) combine with lighter quark (antiquarks) to form heavy-light mesons like D^0, B^0, etc. The UMQS model has been developed over a period of time by our group to explain quarkonia suppression. Here we employ the same model to reproduce open flavor suppression data with minor modifications. We usually consider the simplified picture where the complete QGP hadronizes at a single instance. This is not necessarily true. We could have different species of hadrons forming at different times due to the crossover nature of phase transition. So, we make an assumption here that D^0 like heavy-light mesons can form during the evolution of QGP. Our present formulation is based on a model consisting of suppression due to color screening, gluonic dissociation along with the collisional damping. Regeneration due to correlated $q\bar{q}$ pairs has also been taken into account in the current work. We estimate here the net D^0 meson suppression in terms of survival probability, which is a theoretically equivalent quantity for experimentally measured Nuclear Modification Factor (R_{AA}). We mainly concentrate here on the centrality and transverse momentum dependence of R_{AA} for D^0 meson's suppression in Pb−Pb collisions at mid-rapidity. We compare the survival probability thus obtained for D^0 suppression with the corresponding experimentally measured, R_{AA}, at the LHC center of mass energy of 2.76 TeV.

In the section below, we have given a very brief overview of the UMQS model followed by results. The in-depth details about the model can be found in our latest paper [3].

149.2   Brief Overview of Unified Model of Quarkonia Suppression

The UMQS employed here has been modified over the years and has been used to analyze suppression data of quarkonia and its excited states from RHIC to LHC, for A−A and p−A collision at various energies [3–6]. Considering heavy-quark-like charm being pair produced at the hard scattering of heavy-ion collision, we model the quarkonia evolution in this system using the following rate equation first proposed by Thews [7]:

$$\frac{N_{D^0}}{d\tau} = \frac{\Gamma_{F,nl} N_c N_{\bar{u}}}{V(\tau)} - \Gamma_{D,nl} N_{D^0}, \tag{149.1}$$

where N_{D^0} is the number of D^0 at a given time. First term on right-hand side of above equation is the formation term for new meson. $\Gamma_{F,nl}$ is the formation rate calculated in Sub-section 149.2.5; N_c and $N_{\bar{u}}$ are the number of charm quark and up antiquark, respectively, available at the initial time. $V(\tau)$ is the volume of fireball. The second term on right-hand side is the dissociation term for D^0 and $\Gamma_{D,nl}$ is the corresponding dissociation rate.

This differential equation could be solved under the approximation that $N_{D^0} < N_c, N_{\bar{u}}$. The solution is expressed as follows:

$$
N_{D^0}(\tau_{QGP}, b, p_T) = \epsilon(\tau_{QGP}, b, p_T)\left[N_{D^0}(\tau_0, b) \right.
$$

$$
\left. + N_c N_{\bar{u}} \int_{\tau_0}^{\tau_{QGP}} \frac{\Gamma_{F,nl}(\tau, b, p_T)}{V(\tau, b)\epsilon(\tau, b, p_T)} d\tau \right], \tag{149.2}
$$

where τ_{QGP} is the lifetime of QGP medium, $N_{D^0}(\tau_{QGP}, b, p_T)$ is the final number of D^0 produced. The initial number of D^0 is calculated as follows:

$$
N_{D^0}(\tau_0, b) = \sigma_{D^0}^{NN} T_{AA}(b), \tag{149.3}
$$

where σ is the production cross section of D^0 at 2.76 TeV [8]. $T_{AA}(b)$ is the nuclear overlap function whose value is obtained from [9–11]. $\epsilon(\tau_{QGP}, b, p_T)$ and $\epsilon(\tau, b, p_T)$ in (149.2) are the decay factors calculated as follows:

$$
\epsilon(\tau_{QGP}, b, p_T) = exp\left(-\int_{\tau_{nl}'}^{\tau_{QGP}} \Gamma_{D,nl}(\tau, b, p_T)d\tau \right) \tag{149.4}
$$

and

$$
\epsilon(\tau, b, p_T) = exp\left(-\int_{\tau_{nl}'}^{\tau} \Gamma_{D,nl}(\tau', b, p_T)d\tau' \right). \tag{149.5}
$$

Following is the temperature cooling equation from the Quasi-particle Model (QPM) EOS that we used in our calculation [12]:

$$
T(\tau, b) = T_c \left(\frac{N_\beta}{N_{\beta_0}} \right)^{\frac{1}{3}} \left[\left(\frac{\tau}{\tau_{QGP}} \right)^{\frac{1}{R}-1} \left(1 + \frac{a}{b'T_c^3} \right) - \frac{a}{b'T_c^3} \right]^{\frac{1}{3}}. \tag{149.6}
$$

The D^0 dissociation and recombination mechanisms used in our formalism are described briefly below.

149.2.1   Color Screening

The screening of color charge just as in case of electromagnetic plasma has been long thought of as a reason for dissociation of heavy meson [1]. The color screening model used in the present work assumes that the pressure abruptly falls at freeze-out. Within QGP, color charges are free where screening can happen. So we have a screening region where effective temperature is greater than dissociation temperature of the meson of interest (here, D^0). We have the pressure profile for cooling as

$$p(\tau, r) = -c_1 + c_2 \frac{c_s^2}{\tau^{(c_s^2+1)}} + \frac{4\eta}{3\tau}\left(\frac{c_s^2+1}{c_s^2-1}\right) + \frac{c_3}{\tau^{c_s^2}}, \qquad (149.7)$$

where c_s is the speed of sound and c_1, c_2, c_3 are the constants determined using following boundary conditions:

$$p(\tau_i, r) = p(\tau_i, 0)h(r); \quad p(\tau_s, r) = p_{QGP}, \qquad (149.8)$$

where τ_i is the initial thermalization time, τ_s is the screening time, p_{QGP} is the QGP pressure inside the screening region, and $h(r)$ is the radial distribution function in transverse direction given by

$$h(r) = \left(1 - \frac{r^2}{R_T^2}\right)^\beta \theta(R_T - r). \qquad (149.9)$$

The $c - \bar{u}$ pair present inside the screening region may escape this region provided they have enough kinetic energy and are near to the boundary of this region, which itself evolves with time. This restricts the allowed values of azimuthal angle $\phi_{max}(r)$ for survival of D^0. We find the survival probability, $S_c^{D^0}(p_T, b)$ of D^0 by integrating over $\phi_{max}(r)$, whose expression is given by equation (15) of [4].

149.2.2 Gluonic Dissociation

On absorption of a soft gluon, a singlet state of D^0 could excite to an octet state. This is the principle behind suppression of a meson due to gluonic dissociation. The cross section for dissociation is given by [13]

$$\sigma_{diss,nl}(E_g) = \frac{\pi^2 \alpha_{nl} E_g}{N_c N_{\bar{u}}} \sqrt{\frac{m_b}{E_g + E_{nl}}} \frac{(l+1)|J_{nl}^{q,l+1}|^2 + l|J_{nl}^{q,l-1}|^2}{2l+1}, \qquad (149.10)$$

where $J_{nl}^{q,l'} = \int_0^\infty dr r g_{nl}^*(r) h_{qi'}(r)$ is the probability density. Here, g_{nl}^* and $h_{qi'}(r)$ are the singlet and octet wave functions of D^0, respectively, obtained by numerically solving Schrodinger's equation.

We take the thermal average of above cross section over a modified Bose–Einstein distribution to get the decay width $\Gamma_{gd,nl}$ corresponding to gluonic dissociation.

149.2.3 Collisional Damping

Collisional damping is the dissociation of bound state of quarks due to collision with medium particles. We hence find the associated decay width given by the expectation value of imaginary part of effective quark–antiquark potential in QGP [14]:

$$\Gamma_{damp,nl}(\tau, p_T, b) = \int g_{nl}(r)^{\dagger} Im(V) g_{nl}(r) dr, \qquad (149.11)$$

where $g_{nl}(r)$ is the D^0 singlet wave function.

149.2.4 Shadowing

Shadowing is a cold nuclear matter effect caused by multiple scattering of patrons. We have used *EPS09* parametrization to calculate shadowing for nuclei [15]. We find suppression due to shadowing as

$$S_{sh}(p_T, b) = \frac{d\sigma_{AA}/dy}{T_{AA}d\sigma_{pp}/dy}. \qquad (149.12)$$

Shadowing effect influence the initial production of D^0, and hence we replace (149.3) by shadowing corrected initial number of D^0 given by

$$N^i_{D^0}(\tau_0, b) = N_{D^0}(\tau_0, b)S_{sh}(p_T, b). \qquad (149.13)$$

149.2.5 Recombination Mechanisms

We have considered the possibility of recombination of $c - \bar{u}$ due to de-excitation of octet to singlet state by emission of a gluon. We find the recombination cross section in QGP using detailed balance from gluonic dissociation cross section as

$$\sigma_{f,nl} = \frac{48}{36}\sigma_{d,nl}\frac{(s - M_{nl}^2)^2}{s(s - 4m_b m_{\bar{u}})}, \qquad (149.14)$$

where s is the Mandelstam variable; M_{nl}, m_c, and $m_{\bar{u}}$ are the masses of D^0, charm quark, and up antiquark, respectively. We then define the recombination factor, $\Gamma_{F,nl}$, as the thermal average of product of the above cross section and relative velocity between $c - \bar{u}$ [3].

149.2.6 Survival Probability

We combine the two decay width obtained in Sub-section 149.2.5 and 149.2.3 as follows:

$$\Gamma_{D,nl} = \Gamma_{damp,nl} + \Gamma_{gd,nl}. \tag{149.15}$$

This $\Gamma_{D,nl}$ is used to calculate the decay factors (ϵ) in equations (149.4) and (149.5). On numerically solving equation (149.2), we get the final number of D^0 mesons. Using this we calculate the survival probability due to shadowing, gluonic dissociation, and collisional damping as

$$S_{sgc}^{D^0} = \frac{N_{D^0,nl}^f(p_T, b)}{N_{D^0,nl}(\tau_0, b)}. \tag{149.16}$$

We now combine this with the survival probability due to color screening of Sub-section 149.2.1, as we have introduced it independently. Therefore, we write the final D^0 survival probability as

$$S_P(p_T, b) = S_{sgc}^{D^0}(p_T, b) S_c^{D^0}(p_T, b). \tag{149.17}$$

149.3 Results and Discussion

We have calculated survival probability of D^0 for two values of dissociation temperature, T_D as $1.5\,T_c$ and $2\,T_c$, where T_c is 170 MeV. The experimental value of prompt D^0 nuclear modification factor (R_{AA}) as a function of transverse momentum and centrality at $\sqrt{s_{NN}} = 2.76\,TeV$ is obtained from [8]. The model requires cross section for quark–antiquark pair formation as an input, which we have calculated using the cross-section formula for the process ($gg \rightarrow q\bar{q}$) [16]. The formation time of D^0 meson is also an unknown parameter. Therefore, we have plotted the results for some specific values of formation times.

In Figs. 149.1 and 149.2, we have plotted R_{AA} versus N_{PART} for $T_D = 255$ MeV and 340 MeV respectively. The curve corresponding to $\tau_{form} = 1$ fm is showing reasonable agreement with the suppression data. We see that curves with higher values of τ_{form} stay mostly unchanged with increase in dissociation temperature, T_D. The curves differ with dissociation temperature significantly only for $\tau_{form} = 1.5$ fm and $\tau_{form} = 1$ fm. It suggests that if dissociation temperature is lower, then the meson is relatively weakly bound which results in more D^0 suppression.

For R_{AA} versus p_T plots in Figs. 149.3 and 149.4, our model predictions are showing close agreement with the data points but it fails to reproduce the pattern at $p_T < 10 GeV/c$.

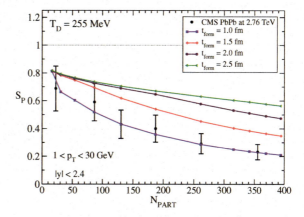

Fig. 149.1 The survival probability (S_P) of D^0 meson is plotted and compared with R_{AA} as a function of centrality at mid-rapidity at $T_D = 1.5T_c$

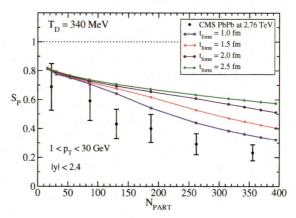

Fig. 149.2 The survival probability (S_P) of D^0 meson is plotted and compared with R_{AA} as a function of centrality at mid-rapidity at $T_D = 2T_c$

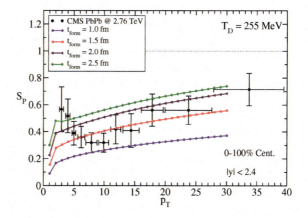

Fig. 149.3 The survival probability (S_P) of D^0 meson R_{AA} as a function of traverse momentum at mid-rapidity at $T_D = 1.5T_c$

Fig. 149.4 The survival probability (S_P) of D^0 meson R_{AA} as a function of traverse momentum at mid-rapidity at $T_D = 2T_c$

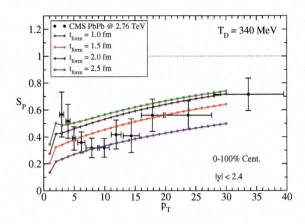

149.3.1   Conclusion

We found that our predicted values for centrality-dependent suppression are following the trend of data points. For R_{AA} versus p_T, our model shows reasonable agreement with the data, but only for transverse momentum greater than 10 GeV. These are our preliminary results for open flavor D^0 using UMQS model. We further need to refine the parameters used and test the UMQS model with more open flavor mesons like B^0, D^+, $D_s^\pm$, B_s^0, and B_c^+.

References

1. T. Matsui, H. Satz, Phys. Lett. B **178**, 416 (1986)
2. N. Brambilla, M.A. Escobedo, J. Soto, A. Vairo, Phys. Rev. D **96**, 034021 (2017)
3. C.R. Singh, S. Ganesh, M. Mishra, Eur. Phys. J. C **79**, 147 (2019)
4. P.K. Srivastava, M. Mishra, C.P. Singh, Phys. Rev. C **87**, 034903 (2013)
5. S. Ganesh, M. Mishra, Phys. Rev. C **88**, 044908 (2013)
6. C.R. Singh, P. Srivastava, S. Ganesh, M. Mishra, Phys. Rev. C **92**, 034916 (2015)
7. R.L. Thews, M. Schroedter, J. Rafelski, Phys. Rev. C **63**, 054905 (2001)
8. C. CMS, Collaboration (2015)
9. ALICE Collaboration, J. e. a. Adam, Phys. Rev. C **91**, 064905 (2015)
10. ATLAS Collaboration, G. e. a. Aad, Phys. Rev. Lett. **114**, 072302 (2015)
11. T. C. c. Khachatryan, V. et al., Journal of High Energy Physics **2017**, 39 (2017)
12. P.K. Srivastava, S.K. Tiwari, C.P. Singh, Phys. Rev. D **82**, 014023 (2010)
13. F. Nendzig, G. Wolschin, Phys. Rev. C **87**, 024911 (2013)
14. M. Laine, O. Philipsen, M. Tassler, P. Romatschke, J. High Energy Phys. **2007**, 054 (2007)
15. R. Vogt, Phys. Rev. C **81**, 044903 (2010)
16. Particle Data Group, M. e. a. Tanabashi, Phys. Rev. D **98**, 030001 (2018)

Chapter 150
Quark and Gluon Condensates in Strongly Magnetized Nuclear Matter

Rajesh Kumar and Arvind Kumar

Abstract Medium modification of the quark and gluon condensates in the strongly magnetized symmetric nuclear medium has been calculated using chiral $SU(3)$ model. The explicit symmetry breaking term and the broken scale invariance part of Lagrangian density of the model are exploited to express the quark and gluon condensates in terms of σ, ζ, δ, and χ fields. The density and temperature dependence of these scalar fields, for different magnetic field strength, are evaluated first and subsequently used as input in the expressions of quark and gluon condensates. Our present study on medium modification of quark and gluon condensates can be used further as input in different QCD sum rules. Consequently, this may help to understand the experimental observables arising from various heavy-ion collision experiments whose one of the aim is to explore the in-medium properties of hadrons.

150.1 Introduction

The medium modification of hadrons under the effect of a strong magnetic field is a challenging area of research. In Heavy-Ion Collisions (HICs) such as RHIC and LHC, strong magnetic field (estimated to the order of $eB \sim 2 - 15 m_\pi^2$ ($1m_\pi^2 = 2.818 \times 10^{18}$ gauss) is believed to exist. To value how strong is the magnetic field, compare it with the strongest magnetic field generated on earth by electromagnetic shock wave ($\sim 10^7$ gauss) [1], and magnetic field of a compact stars (10^{14} gauss) [2]. It may be the strongest magnetic field that has ever existed in Universe. In the presence of finite magnetic field, density, and temperature, the chiral condensates such as gluon and quark condensates get modified which further affects the critical temperature T_c and this effect is known as (inverse) magnetic catalysis [3]. In the presence of

R. Kumar (✉) · A. Kumar
Dr. B R Ambedkar National Institute of Technology, Jalandhar 144011, Punjab, India
e-mail: rajesh.sism@gmail.com

A. Kumar
e-mail: kumara@nitj.ac.in

© Springer Nature Singapore Pte Ltd. 2021
P. K. Behera et al. (eds.), *XXIII DAE High Energy Physics Symposium*,
Springer Proceedings in Physics 261,
https://doi.org/10.1007/978-981-33-4408-2_150

external magnetic field, Landau quantization (ν) occurs which results in unequal values of scalar densities of proton and neutron [2]. The non-perturbative QCD sum rule techniques are used to evaluate the effective mass of different mesons (J/ψ, η_c, ρ, ω, and ϕ) through the in-medium behavior of quark and gluon condensates [2, 4]. In the present work, we evaluate the effect of external magnetic field at finite temperature and density on quark and gluon condensates.

150.2 Methodology

As discussed above, in Chiral SU(3) model, these condensates are expressed in terms of the σ, ζ, δ, and χ field which are solved in the presence of external magnetic field at finite density and temperature from the coupled equations of motion [2]. Here, the δ field is introduced to incorporate the effect of isospin asymmetry $\eta = (\rho_n^v - \rho_p^v)/2\rho_N$ in nuclear medium. The effect of magnetic field and temperature is introduced through the scalar ρ_i^s and vector densities ρ_i^v of nucleons [2] which are given as

$$\rho_p^v = \frac{|q_p|B}{2\pi^2}\left[\sum_{\nu=0}^{\nu_{max}^{(s=1)}}\int_0^\infty dk_\parallel^p \left(f_{k,\nu,s}^p - \bar{f}_{k,\nu,s}^p\right) + \sum_{\nu=1}^{\nu_{max}^{(s=-1)}}\int_0^\infty dk_\parallel^p \left(f_{k,\nu,s}^p - \bar{f}_{k,\nu,s}^p\right)\right],$$

(150.1)

$$\rho_p^s = \frac{|q_p|Bm_p^*}{2\pi^2}\left[\sum_{\nu=0}^{\nu_{max}^{(s=1)}}\int_0^\infty \frac{dk_\parallel^p}{\sqrt{(k_\parallel^p)^2 + (\bar{m}_p)^2}}\left(f_{k,\nu,s}^p + \bar{f}_{k,\nu,s}^p\right)\right.$$

$$\left. + \sum_{\nu=1}^{\nu_{max}^{(s=-1)}}\int_0^\infty \frac{dk_\parallel^p}{\sqrt{(k_\parallel^p)^2 + (\bar{m}_p)^2}}\left(f_{k,\nu,s}^p + \bar{f}_{k,\nu,s}^p\right)\right],$$

(150.2)

$$\rho_n^v = \frac{1}{2\pi^2}\sum_{s=\pm1}\int_0^\infty k_\perp^n dk_\perp^n \int_0^\infty dk_\parallel^n \left(f_{k,s}^n - \bar{f}_{k,s}^n\right),$$

(150.3)

and

$$\rho_n^s = \frac{1}{2\pi^2}\sum_{s=\pm1}\int_0^\infty k_\perp^n dk_\perp^n \left(1 - \frac{s\mu_N\kappa_n B}{\sqrt{m_n^{*2} + (k_\perp^n)^2}}\right)\int_0^\infty dk_\parallel^n \frac{m_n^*}{\tilde{E}_s^n}\left(f_{k,s}^n + \bar{f}_{k,s}^n\right).$$

(150.4)

In above, $f_{k,\nu,s}^p$, $\bar{f}_{k,\nu,s}^p$, $f_{k,s}^n$, and $\bar{f}_{k,s}^n$ represent the finite temperature distribution functions for particles and antiparticles for proton and neutron. Also, μ_N, k_i, and m_i^* represent the nuclear magneton, anomalous magnetic moment, and effective mass of nucleons, respectively. In this model, the up $\langle \bar{u}u \rangle$, down $\langle \bar{d}d \rangle$, and strange $\langle \bar{s}s \rangle$ quark condensates can be expressed as [4]

$$\langle \bar{u}u \rangle = \frac{1}{m_u}\left(\frac{\chi}{\chi_0}\right)^2 \left[\frac{1}{2}m_\pi^2 f_\pi (\sigma + \delta)\right], \langle \bar{d}d \rangle = \frac{1}{m_d}\left(\frac{\chi}{\chi_0}\right)^2 \left[\frac{1}{2}m_\pi^2 f_\pi (\sigma - \delta)\right],$$

(150.5)

and

$$\langle \bar{s}s \rangle = \frac{1}{m_s}\left(\frac{\chi}{\chi_0}\right)^2 \left(\sqrt{2}m_K^2 f_K - \frac{1}{\sqrt{2}}m_\pi^2 f_\pi \right)\zeta,$$

(150.6)

respectively. In above, m_u, m_d, m_s, m_π, and f_π are the mass of up quark, down quark, strange quark, π meson, and decay constant of π meson, respectively. Moreover, the scalar $\langle \frac{\alpha_s}{\pi} G_{\mu\nu}^a G^{a\mu\nu} \rangle$ and tensorial gluon condensates $\langle \frac{\alpha_s}{\pi} G_{\mu\sigma}^a G^a{}_\nu{}^\sigma \rangle$ are expressed as [2]

$$G_0 = \frac{8}{9}\left[(1-d)\chi^4 + \left(\frac{\chi}{\chi_0}\right)^2 \left(m_\pi^2 f_\pi \sigma + \left(\sqrt{2}m_K^2 f_K - \frac{1}{\sqrt{2}}m_\pi^2 f_\pi\right)\zeta\right)\right],$$

(150.7)

and

$$G_2 = \frac{\alpha_s}{\pi}\left[(1 - d + 4k_4)(\chi_0^4 - \chi^4) - \chi^4\ln\left(\frac{\chi^4}{\chi_0^4}\right) + \frac{4}{3}d\chi^4\ln\left(\left(\frac{(\sigma^2 - \delta^2)\zeta}{\sigma_0^2\zeta_0}\right)\left(\frac{\chi}{\chi_0}\right)^3\right)\right],$$

(150.8)

respectively. Here m_K and f_K are the mass and decay constant of K meson and α_s is the running strong coupling constant.

150.3   Results and Discussion

In this section, results are shown for strong external magnetic field, $eB = 3m_\pi^2$ and $6m_\pi^2$ at nucleon density $\rho_N = \rho_0$ and $4\rho_0$, and isospin asymmetry parameter $\eta = 0, 0.3$ and 0.5 as a function of temperature. Various parameters used in the present work are taken from [2]. In Fig. 150.1, we have shown the variation of gluon condensates G_0 and G_2 as a function of temperature. As can be seen from (150.7) and (150.8), the scalar and tensorial condensate depends on the scalar fields and both of them strongly depend on the χ field due to its fourth power dependence. In low-temperature regime, it is observed that the magnitude of scalar gluon condensate G_0 decreases with the increase in magnetic field whereas the magnitude of twist-2 gluon operator G_2 increases. Also, the values of both G_0 and G_2 increase up to certain

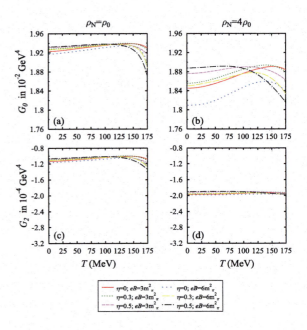

Fig. 150.1 Scalar gluon condensate G_0 and tensorial gluon condensates G_2

value of temperature and then start decreasing with further increase in temperature. Moreover, the scalar gluon condensate G_0 varies appreciably in high density whereas tensorial gluon condensate G_2 shows fewer variations due to the cancellation effect of its second term. In Fig. 150.2, the non-strange ($\langle \bar{u}u \rangle$, $\langle \bar{d}d \rangle$) and strange ($\langle \bar{s}s \rangle$) quark condensates have been plotted as a function of temperature. The magnitude of quark condensates decreases with the increase in magnetic field at low temperature. Also, on contrary to gluon condensates, the magnitude of quark condensates first increases up to a particular value of temperature and then decreases with the further increase in temperature. Moreover, it is observed that the quark condensates vary appreciably in the nuclear medium as compared to gluon condensates. In both plots, at finite asymmetry of the nuclear medium, the crossover behavior is observed which reflects the inequality between scalar densities of proton and neutron (see (150.1)–(150.4)).

150.4 Conclusions

At $4\rho_0$, the quark and gluon condensates vary appreciably as a function of temperature in the presence of magnetic field except for tensorial gluon operator G_2, which shows very less variation. These condensates can be used in QCD sum rules to calculate the in-medium mass-shift and decay width of mesons [2, 4], which may be further used to understand the experimental observables in non-central HICs planned at Compressed Baryonic Matter (CBM) at FAIR.

Fig. 150.2 Up $\langle \bar{u}u \rangle$, down $\langle \bar{d}d \rangle$ and strange $\langle \bar{s}s \rangle$ quark condensates

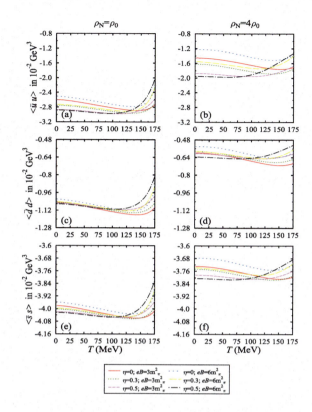

Acknowledgements One of the authors, Rajesh Kumar, sincerely acknowledges the support toward this work from the Ministry of Science and Human Resources Development (MHRD), Government of India.

References

1. B.A. Boyko et al., *Proceedings of 12th IEEE International Pulsed Power Conference*, 764 (1999)
2. Rajesh Kumar, Arvind Kumar, Eur. Phys. J C **79**, 403 (2019)
3. D. Kharzeev et al., *Strongly Interacting Matter in Magnetic Fields* (Springer, Berlin, 2013)
4. A. Mishra, Phys. Rev. C **91**, 035201 (2014)

Chapter 151
Measurement of Heavy-Flavour Correlations and Jets with ALICE at the LHC

Samrangy Sadhu

Abstract In this article, we report the latest results on heavy-flavour correlations and jets measured with the ALICE detector in pp, p–Pb and Pb–Pb collisions from the LHC Run 2. The results of azimuthal correlations between D mesons and charged particles in pp collisions at $\sqrt{s}$ = 7 TeV and 13 TeV and in p–Pb collisions at $\sqrt{s_{NN}}$ = 5.02 TeV are presented. The centrality-dependent study on azimuthal correlations between electrons from open heavy-flavour hadron decays and charged particles in Pb–Pb collisions at $\sqrt{s_{NN}}$ = 5.02 TeV are performed. In addition, measurements of D-meson tagged jet production in pp collisions at $\sqrt{s}$ = 7 TeV and in p–Pb collisions at $\sqrt{s_{NN}}$ = 5.02 TeV are presented.

151.1 Introduction

ALICE is the dedicated experiment at the Large Hadron Collider (LHC) to study nuclear matter under extreme conditions of high temperature and high energy density which are obtained in relativistic heavy-ion collisions and where a new state of deconfined matter known as the Quark-Gluon Plasma (QGP) is expected to be formed. Due to their large masses, heavy quarks (charm and beauty), are produced in the early stages of the collision via hard partonic scattering processes and they are expected to experience the full evolution of the system propagating through the medium produced in such collisions [1]. Therefore, they are considered to be the ideal probes to study the properties of the medium formed in heavy-ion collisions. The angular correlations between open heavy-flavour particles and charged particles are sensitive to the charm quark production and fragmentation mechanisms [2]. Thus, this study allows us to characterize the heavy-quark fragmentation process and the possible modification

Samrangy Sadhu on behalf of ALICE Collaboration.

S. Sadhu (✉)
Variable Energy Cyclotron Centre, Kolkata, India
e-mail: samrangy.sadhu@cern.ch

Homi Bhabha National Institute, Mumbai, India

© Springer Nature Singapore Pte Ltd. 2021
P. K. Behera et al. (eds.), *XXIII DAE High Energy Physics Symposium*,
Springer Proceedings in Physics 261,
https://doi.org/10.1007/978-981-33-4408-2_151

inside the medium [3, 4]. The measurement of heavy-flavour jets gives more direct access to the initial parton kinematics and provides further constraints for heavy-quark energy loss models, in particular adding information on the in-medium energy loss mechanisms. It also provides input to the possible flavour dependence of jet quenching in the medium.

Besides constituting the necessary baseline for nucleus–nucleus measurements, the studies in pp collisions are important for testing expectations from pQCD-inspired Monte Carlo generators. Comparisons between results from pp and p–Pb collisions can give insight into the cold nuclear matter effects on the heavy-quark production and hadronization into jets.

151.2   Results

151.2.1   Azimuthal Correlations Between D Mesons and Charged Particles

Measurements of azimuthal correlations between D mesons and charged particles have been performed at mid-rapidity in pp collisions at $\sqrt{s} = 7$ TeV [5] and 13 TeV and in p–Pb collisions at $\sqrt{s_{NN}} = 5.02$ TeV. In order to study the properties of the measured correlations, the $\Delta\varphi$ (the difference in azimuthal angle between the trigger and associated particle) projections of azimuthal-correlation distributions are fitted with a function composed of two Gaussian terms to describe the near- and away-side peak properties and a constant term to describe the pedestal of the distributions, determined mainly by correlations with particles from the underlying event. The associated-particle yields and the widths ($\sigma_{fit,NS}$) of the near-side peak are shown in Fig. 151.1, as a function of the D-meson p_T for different associated particle p_T intervals. The results in p–Pb collisions show compatibility with that in pp collisions at two energies within uncertainties. Also, a qualitative agreement with Monte Carlo PYTHIA6 tunes (Perugia0, Perugia2010 and Perugia2011) [6], PYTHIA8 [7] and POWHEG+PYTHIA6 [8, 9] simulations is found in most of the studied kinematic ranges, as presented in [5, 10]. The correlation distributions in p–Pb collisions at $\sqrt{s_{NN}} = 5.02$ TeV are also studied as a function of the event multiplicity, for three centrality classes defined by the energy deposited in the ZNA neutron calorimeter [11]: 0–20, 20–60 and 60–100%. No modification of the near-side peak properties is found within the current uncertainties.

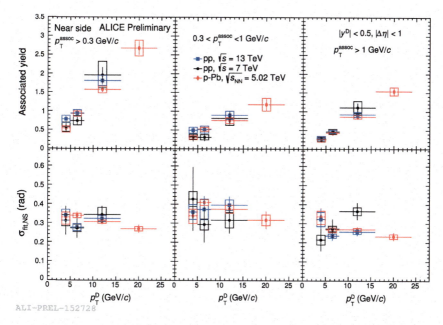

ALI-PREL-152728

Fig. 151.1 Near-side associated yields and widths extracted from azimuthal correlations between D mesons and charged particles in pp collisions at $\sqrt{s} = 7$ TeV and 13 TeV and p–Pb collisions at $\sqrt{s_{NN}} = 5.02$ TeV as a function of D-meson p_T in different associated p_T region

151.2.2 Azimuthal Correlations Between Heavy-Flavour Decay Electrons and Charged Particles

The two-particle azimuthal correlations between heavy-flavour hadron decay electrons (e^{HF}) and charged particles have been measured in p–Pb and Pb–Pb collisions at $\sqrt{s_{NN}} = 5.02$ TeV. The near-side associated yields are extracted for the e^{HF} interval of $4 < p_T^e < 12$ GeV/c in different associated particle p_T region. The results in p–Pb collisions and central (0–20%) and semi-central (20–50%) Pb–Pb collisions, presented in Fig. 151.2 (left panel), are consistent with each other, with a hint of an enhancement in the central Pb–Pb collisions at low associated particles p_T. In addition, studies in high-multiplicity p–Pb collisions show evidence of a positive elliptic flow (v_2) of heavy-flavour hadron decay electrons in an interval $1.5 < p_T^e < 4$ GeV/c [12].

151.2.3 D-Meson Tagged Jets

The D-meson tagged jets are reconstructed with the anti-k_T algorithm [14] implemented in the FastJet package [13]. The production cross-sections of prompt D^0 and D^{*+} tagged charged jets have been measured by ALICE in pp collisions at $\sqrt{s} = 7$

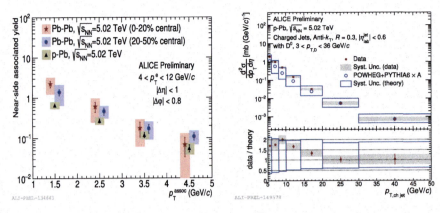

Fig. 151.2 Near-side associated yields extracted from e^{HF}-h correlations in p–Pb and 0–20%, 20–50% central Pb–Pb collisions at $\sqrt{s_{NN}} = 5.02$ TeV as a function of associated particles p_T and for $4 < p_T^e < 12$ GeV/c (left). The p_T-differential production cross-section of D^0-meson tagged charged jets in p–Pb collisions at $\sqrt{s_{NN}} = 5.02$ TeV compared to POWHEG+PYTHIA6 predictions (right)

TeV [10] and p–Pb collisions at $\sqrt{s_{NN}} = 5.02$ TeV, respectively. The covered charged jet transverse momentum $p_{T,ch,jet}$ ranges from 5 to 30 GeV/c with the jet resolution parameter R = 0.4 and a requirement of the D-meson transverse momentum $p_{T,D} > 3$ GeV/c. Figure 151.2 (right) presents a new measurement of the D^0-tagged jet p_T-differential cross-section in p–Pb collisions using R = 0.3 with a $p_{T,ch,jet}$ extended up to 50 GeV/c showing an agreement with NLO pQCD POWHEG+PYTHIA6 simulation.

151.3 Summary and Outlook

In this contribution, the results of azimuthal correlations between D mesons and charged particles in pp and p–Pb collisions have been reported. The near-side observables are found to be compatible with Monte Carlo predictions. The results in p–Pb collisions suggest that the modification of jets from charm is not significantly affected by the cold nuclear matter effects as well as by possible final-state effects. The analysis of azimuthal correlations between heavy-flavour hadron decay electrons and charged particles shows a hint of near-side yield enhancement in central Pb–Pb collisions. The p_T-differential production cross-sections of D-meson tagged jets have been measured in pp and p–Pb collisions and the jet momentum fraction carried by the D-meson is extracted in pp collisions. The measurements are in agreement with POWHEG+PYTHIA6 predictions. More precise and differential measurements are expected with data taken by ALICE in pp collisions in 2017 and 2018 and in Pb–Pb collisions in 2018.

References

1. R. Rapp, H. van Hees, arXiv:0903.1096 [hep-ph]
2. S. Acharya et al., [ALICE Collaboration], Phys. Rev. Lett. **122**(7), 072301 (2019). arXiv:1805.04367 [nucl-ex]
3. G.C. Fox, S. Wolfram, Phys. Rev. Lett. **41**, 1581 (1978). https://doi.org/10.1103/PhysRevLett. 41.1581
4. M. Cacciari, E. Gardi, Nucl. Phys. B **664**, 299 (2003). [hep-ph/0301047]
5. J. Adam et al., Measurement of azimuthal correlations of D mesons and charged particles in pp collisions at $\sqrt{s}$ = 7 TeV and p-Pb collisions at $\sqrt{s_{NN}}$ = 5.02 TeV. Eur. Phys. J. **C77**(4), 245 (2017)
6. T. Sjostrand, S. Mrenna, P.Z. Skands, PYTHIA 6.4 physics and manual. JHEP **05**, 026 (2006)
7. T. Sjostrand, S. Mrenna, P.Z. Skands, A brief introduction to PYTHIA 8.1. Comput. Phys. Commun. **178**, 852–867 (2008)
8. P. Nason, A new method for combining NLO QCD with shower Monte Carlo algorithms. JHEP **11**, 040 (2004)
9. S. Frixione, P. Nason, C. Oleari, Matching NLO QCD computations with parton shower simulations: the POWHEG method. JHEP **11**, 070 (2007)
10. S. Sakai, Measurement of heavy-flavour production, correlations and jets with ALICE. Eur. Phys. J. Web Conf. **171**, 18008 (2018). https://doi.org/10.1051/epjconf/201817118008
11. J. Adam et al., Centrality dependence of particle production in p-Pb collisions at $\sqrt{s_{NN}}$ = 5.02 TeV. Phys. Rev. **C91**(6), 064905 (2015)
12. S. Acharya et al., Azimuthal anisotropy of heavy-flavour decay electrons in p-Pb collisions at $\sqrt{s_{NN}}$ = 5.02 TeV
13. M. Cacciari, G.P. Salam, G. Soyez, FastJet user manual. Eur. Phys. J. C **72**, 1896 (2012)
14. M. Cacciari, G.P. Salam, G. Soyez, The Anti-k(t) jet clustering algorithm. JHEP **04**, 063 (2008)

Chapter 152
Enhanced Production of Multi-strange Hadrons in Proton-Proton Collisions

Sarita Sahoo, R. C. Baral, P. K. Sahu, and M. K. Parida

Abstract Strangeness enhancement is proposed as a signature of QGP formation in high-energy nuclear collisions. The observed relative yield enhancements of hadrons increase with the strangeness content but not with the mass and baryon number of the hadron. To understand this behavior, we have studied on the simulation models. We have found the models EPOS and AMPT are not able to explain simultaneously the effect of strangeness canonical suppression in low-multiplicity events and QGP-like effect in high-multiplicity pp collisions at LHC energies.

152.1 Introduction

The production of strange hadrons in high-energy hadronic interactions provides one of the tools to investigate properties of Quark-Gluon Plasma (QGP). Strangeness is produced in hard partonic scattering processes by flavor creation, flavor excitation, and gluon splitting. These productions dominate in higher p_T region. The non-perturbative processes like string fragmentation dominate the productions at low p_T region. The relative yields of strange particles to π in heavy-ion collisions from top RHIC to LHC energy are found to be compatible with those of a hadron gas in thermal and chemical equilibrium, and this behavior can be described using grand-canonical statistical models [1]. In peripheral collisions, the relative yields of strange particles to π decrease and tend toward those observed in pp collisions. This behavior can be described by statistical mechanics approach. The statistical models by implementing strangeness canonical suppression can predict a suppression of

S. Sahoo (✉) · M. K. Parida
CETMS, SOA University, Bhubaneswar, India
e-mail: ssarita.2006@gmail.com

R. C. Baral
NISER, HBNI, Jatni, Bhubaneswar, Odisha, India

S. Sahoo · P. K. Sahu
Institute of Physics, HBNI, Bhubaneswar, India

© Springer Nature Singapore Pte Ltd. 2021
P. K. Behera et al. (eds.), *XXIII DAE High Energy Physics Symposium*,
Springer Proceedings in Physics 261,
https://doi.org/10.1007/978-981-33-4408-2_152

strangeness production in small systems. ALICE has shown in pp collisions at $\sqrt{s} = 7$ TeV that the p_T integrated yields of strange and multi-strange particles relative to π increase significantly with multiplicity [2]. The observed relative yield enhancements increase with the strangeness contained in hadrons. To understand this behavior, we have studied on the simulations like AMPT (A Multi-Phase Transport Model) [3] and EPOS (3 + 1 Hydrodynamics Model) [4].

152.2 Model Description

The AMPT model (version 1.26t5/2.26t5) mainly provides the initial conditions, partonic interactions, conversion from the partonic to hadronic matter, and hadronic interactions. The initial conditions, which include the spatial and momentum distributions of minijet partons and string excitations, are obtained from the HIJING model [5–8]. The partonic interactions or the scatterings among partons are modeled by Zhang's parton cascade (ZPC) [9]. Here, we have studied both the versions of AMPT model, i.e., default version of AMPT and string melting version of AMPT. In the default AMPT model, partons are recombined with their parent strings when they stop interacting, and the resulting strings are converted to hadrons using the Lund string fragmentation model [10, 11]. In the AMPT model with string melting, a quark coalescence model is used instead to combine partons into hadrons. The dynamics of the subsequent hadronic matter is described by a hadronic cascade, which is based on ART model [12, 13] and extended to include additional reaction channels those are important in high energies, like formation and decay of resonances and antibaryon resonances, baryon and antibaryon production from mesons and their inverse reactions of annihilation. Final results from the AMPT model are obtained after hadronic interactions are terminated at a cutoff time ($t_{cut} = 20$ fm/c) when observables under study are considered to be stable, i.e., when further hadronic interactions after t_{cut} will not significantly affect these observables. As per recommended by the model, default set of parameters are used for this study.

The model EPOS (version 3.107) describes the full evolution of heavy-ion collisions. The initial stage is treated via a multiple scattering approach based on pomerons and strings. The reaction volume is divided into a core and corona part. The core is taken as the initial condition for the QGP evolution, for which one employs viscous hydrodynamics. The corona part is simply composed of hadrons from string decays. After hydronization of the fluid (core part), these hadrons, as well the corona hadrons, are fed into UrQMD [14], which describes hadronic interactions in microscopic approach. The chemical and kinetic freeze-out occur within this phase. Details of EPOS model can be found in [4] and references therein.

152.3   Results and Discussion

Figure 152.1 shows the charged particle event-multiplicity density ($\langle dN_{ch}/d\eta\rangle_{|\eta|<0.5}$) dependence of primary yield ratios of strange hadrons (Λ, Ξ, Ω) to π ($\langle h/\pi\rangle$) divided by the same values measured in the inclusive inelastic pp collisions ($\langle h/\pi\rangle^{pp}_{INEL>0}$). Here a primary particle is defined as a particle created in the collision, but not coming from a weak decay. The measurements were performed for events having at least one charged particle produced in the pseudorapidity interval $|\eta| < 1.0$ (represented as INEL > 0). The experimental data points are shown by markers in the figure. The measurements were performed at mid-rapidity ($|y| < 0.5$). A detailed information about data can be found in [2] and the references therein.

We compare the AMPT and the EPOS results for the scaled strange hadrons to π ratios with the measurements from the ALICE experiment. The AMPT default version shows no apparent variation in the ratios with event charged particle multiplicity. The string melting version of AMPT explains qualitatively both Λ/β and Ξ/β ratios except at the lowest event multiplicity. It fails in explaining the enhancement in Ω/β ratios toward higher event multiplicity. The EPOS describes Λ/β ratios well whereas it fails to explain Ξ/β and Ω/β ratios. This means EPOS fails to describe the production rate for multi-strange hadrons.

Fig. 152.1 Yield ratios of hadrons to π normalized to the values measured in the inclusive (INEL > 0) pp collisions at $\sqrt{s}$ = 7 TeV. The markers show ALICE data [2] and the bands show model results

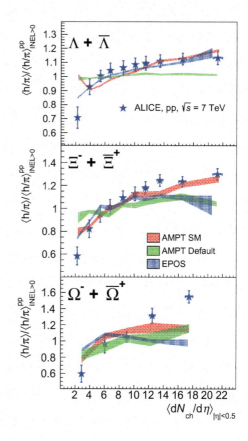

152.4 Summary

Models are insufficient to explain the relative yield enhancement of strange hadrons to π, which are shown by ALICE. The AMPT and the EPOS models are not able to explain simultaneously the effect of strangeness canonical suppression in low-multiplicity events and QGP-like effect in high-multiplicity pp collisions at this LHC energy.

References

1. J. Cleymans, I. Kraus, H. Oeschler, K. Redlich, S. Wheaton, Phys. Rev. C **74**, 034903 (2006)
2. A. Andronic, P. Braun-Munzinger, J. Stachel, J. Phys. Lett. B **673**, 142 (2009)
3. J. Adam et al., ALICE Collaboration. Nat. Phys. **13**, 535–539 (2017)
4. Z.-W. Lin, C.M. Ko, B.-A. Li, B. Zhang, S. Pal, Phys. Rev. C **72**, 064901 (2005)
5. K. Werner, I. Karpenko, T. Pierog, M. Bleicher, K. Mikhailov, Phys. Rev. C **82**, 044904 (2010)
6. K. Werner, B. Guiot, I. Karpenko, T. Pierog, Phys. Rev. C **89**, 064903 (2014)
7. X.-N. Wang, Phys. Rev. D **43**, 104 (1991)
8. X.-N. Wang, M. Gyulassy, Phys. Rev. D **44**, 3501 (1991)
9. X.-N. Wang, M. Gyulassy, Phys. Rev. D **45**, 844 (1992)
10. M. Gyulassy, X.-N. Wang, Comput. Phys. Commun. **83**, 307 (1994)
11. B. Zhang, Comput. Phys. Commun. **109**, 193 (1998)
12. B. Andersson, G. Gustafson, B. Söderberg, Z. Phys. C **20**, 317 (1983)
13. B. Andersson, G. Gustafson, G. Ingelman, T. Sjostrand (Lund U., Dept. Theor. Phys.), Phys. Rept. **97**, 31 (1983)
14. B.-A. Li, C.M. Ko, Phys. Rev. C **52**, 2037 (1995)

Chapter 153
$\eta N N$ and $\eta' N N$ Coupling Constants in QCD Sum Rules

Janardan P. Singh and Shesha D. Patel

Abstract We have analyzed $\eta N N$ and $\eta' N N$ coupling constants by considering correlation functions of the interpolating fields of two nucleons between vacuum and one-meson states. By taking matrix elements of these correlation functions with respect to nucleon spinors, unwanted pole contributions have been avoided and dispersion integrals kept well defined in chiral limit. A characteristically important contribution to these coupling constants comes from pseudoscalar gluonic operator. The results obtained have been compared with other evaluations of these coupling constants from literature.

153.1 Introduction

Study of η-η' mesons reveals interesting connection between chiral symmetry breaking and gluon dynamics. Higher masses of η-η' are due to their association with non-perturbative gluon dynamics and QCD axial anomaly [1]. $\eta^{(\prime)}$ phenomenology is characterized by large OZI violations [2]. Goldberger–Treiman relation relates $\eta^{(\prime)}$N coupling constant to the flavor singlet-axial charge of the nucleon $g_A^{(0)}$ extracted from polarized deep inelastic scattering [3, 4]. A reliable determination of $g_{\eta N N}$ and $g_{\eta' N N}$ is necessary to understand $U(1)_A$ dynamics of QCD [4]. It has essential roles in construction of realistic NN potential [5, 6], in estimation of electric dipole moment of the neutron [7] and in analysis of photoproduction of these mesons [8–10].

QCD sum rules have been used in past to calculate η-nucleon coupling constant $g_{\eta N N}$ in SU(3) symmetry limit [11] as well as with SU(3)-flavor-violating effects taken into account [12]. It has also been used to calculate singlet-axial vector coupling constant of the nucleon without resorting to any instanton contribution [13]. In the

J. P. Singh · S. D. Patel (✉)
Faculty of Science, Physics Department, The Maharaja Sayajirao University of Baroda, Vadodara 390001, Gujarat, India
e-mail: sheshapatel30@gmail.com

J. P. Singh
e-mail: janardanmsu@yahoo.com

© Springer Nature Singapore Pte Ltd. 2021
P. K. Behera et al. (eds.), *XXIII DAE High Energy Physics Symposium*,
Springer Proceedings in Physics 261,
https://doi.org/10.1007/978-981-33-4408-2_153

present work, we calculate the coupling constants of both physical η and η' mesons with the nucleon using quark-flavor basis. We include characteristic contribution coming from the pseudoscalar gluonic operator.

153.2   The Sum Rule

Correlator of the nucleon current between vacuum and one $\eta^{(\prime)}$-state [11, 14–17]:

$$\Pi(q, p) = i \int d^4x e^{iqx} \langle 0 \mid T\{J_N(x), \overline{J_N}(0)\} \mid \eta^{(\prime)}(p)\rangle. \tag{153.1}$$

$\eta^{(\prime)}NN$ coupling constant $g_{\eta^{(\prime)}NN}$ is defined through the coefficient of the pole as

$$\bar{u}(qr)(\hat{q} - M_n)\Pi(q, p)(\hat{q} - \hat{p} - M_n)u(ks)\mid_{q^2 = M_n^2, (q-p)^2 = M_n^2} = i\lambda^2 g_{\eta^{(\prime)}NN}\bar{u}(qr)\gamma_5 u(ks), \tag{153.2}$$

where λ is the coupling constant of the proton current with one-proton state.

$$\langle 0 \mid J_N(0) \mid q\rangle = \lambda u(q). \tag{153.3}$$

Following [16] we define the projected correlation function

$$\Pi_+(q, p) = \bar{u}(qr)\gamma_0 \Pi(q, p)\gamma_0 u(ks). \tag{153.4}$$

Π_+ can be regarded as a function of q_0 in the reference frame in which $\vec{q} = 0$. The even and odd parts of the Π_+ satisfy dispersion relations as

$$\Pi_+^E(q_0^2) = -\frac{1}{\pi} \int dq_0' \frac{q_0'}{q_0^2 - q_0'^2} Im\,\Pi_+(q_0'), \tag{153.5}$$

$$\Pi_+^O(q_0^2) = -\frac{1}{\pi} \int dq_0' \frac{1}{q_0^2 - q_0'^2} Im\,\Pi_+(q_0').$$

On taking Borel transform with respect to q_0^2:

$$\hat{B}[\Pi_+^E(q_0^2)] = \frac{1}{\pi} \int dq_0' q_0' e^{\frac{-q_0^2}{M^2}} Im\,\Pi_+(q_0'), \tag{153.6}$$

$$\hat{B}[\Pi_+^O(q_0^2)] = \frac{1}{\pi} \int dq_0' e^{\frac{-q_0^2}{M^2}} Im\,\Pi_+(q_0'),$$

where M is the Borel mass parameter. The R.H.S of (153.6) is expanded in terms of the observed spectral function. The absorptive part of the projected correlation function can be written as

$$\text{Im}\Pi_+(q, p) = -\bar{u}(qr)i\gamma_5 u(ks)\pi\lambda^2 g(q_0, \vec{p}^2)\left[\frac{\delta(q_0 - M_n)}{q_0 - E_k - \omega_p} + \frac{\delta(q_0 - E_k - \omega_p)}{q_0 - M_n}\right]$$
$$+ \left[\theta(q_0 - s_{\eta^{(\prime)}}) + \theta(-q_0 - s_{\eta^{(\prime)}})\right]\text{Im}\Pi_+^{OPE}(q, p), \quad (153.7)$$

where $s_{\eta^{(\prime)}}$ is the effective continuum threshold. To include characteristic contribution coming from the pseudoscalar gluonic operator, we consider the light-cone expansion of the quark propagator [18]:

$$\langle 0 \mid T\{q(x)\bar{q}(0)\} \mid 0 \rangle \rightarrow \frac{\Gamma(-\epsilon)\alpha_s}{72 \times 4\pi(-x^2)^{-\epsilon}}G\tilde{G}\not{x}\gamma_5 + \cdots, \quad (153.8)$$

for dimension $d = 4 - 2\epsilon$, $G\tilde{G}$ is evaluated at the origin in the spirit of short distance expansion and ellipsis stand for other structures. Here $G\tilde{G} = \frac{1}{2}\epsilon^{\mu\nu\rho\sigma}G_{\mu\nu}^a G_{\rho\sigma}^a$, $\epsilon^{0123} = +1$.

Define matrix element [19]:

$$\langle 0 \mid T\{q(x)\bar{q}(0)\} \mid \eta^{(\prime)} \rangle = \frac{\Gamma(-\epsilon)}{72}(-x^2)^{\epsilon}a_{\eta^{(\prime)}}\not{x}\gamma_5 + \cdots \quad (153.9)$$

For expression of $a_{\eta^{(\prime)}}$, see [19]. The removal of divergence requires renormalization and we do this in MS scheme. In $\Pi(q, p)$, the operator $G\tilde{G}$ originates from only d-quark line and there is no contribution where gluons originate from two different quark lines. From (153.6) and (153.7), we get

$$\frac{1}{\pi}\int dq_0' e^{(-\frac{q_0'^2}{M^2})}\hat{B}\text{Im}\Pi_+(q', p) = \bar{u}(q)i\gamma_5 u(q - p)\frac{\lambda^2}{E_k + \omega_p - M_n}\left\{e^{-\frac{M_n^2}{M^2}}g(M_n, \mathbf{p}^2)\right.$$
$$\left. - e^{-\frac{(E_k + \omega_p)^2}{M^2}}g(E_k + \omega_p, \mathbf{p}^2) + \text{cont.}\right\},$$

$$\frac{1}{\pi}\int dq_0' q_0' e^{(-\frac{q_0'^2}{M^2})}\hat{B}\text{Im}\Pi_+(q', p) = \bar{u}(q)i\gamma_5 u(q - p)\frac{\lambda^2}{E_k + \omega_p - M_n}\left\{e^{-\frac{M_n^2}{M^2}}g(M_n, \mathbf{p}^2)M_n - \right.$$
$$\left. e^{-\frac{(E_k + \omega_p)^2}{M^2}}(E_k + \omega_p)g(E_k + \omega_p, \mathbf{p}^2) + \text{cont.}\right\}, \quad (153.10)$$

where last terms stand for the continuum contributions.

153.3 Calculations and Results

In OPE evaluation of the correlator $\Pi(q, p)$, operators up to dimension five have been included. Three types of nonlocal bilinear quark operators contribute to the vacuum-to-meson matrix elements: $\bar{q}(0)i\gamma_5 q(x)$, $\bar{q}(0)\gamma_\mu\gamma_5 q(x)$, and $\bar{q}(0)\gamma_5\sigma^{\mu\nu}q(x)$. Borel transform of coefficients of $\bar{u}(q)i\gamma_5 u(q - p)$ from OPE side of the correlator has been equated to the same from the phenomenological side in a Borel window to get sum rules. Details of the calculation can be found in [20].

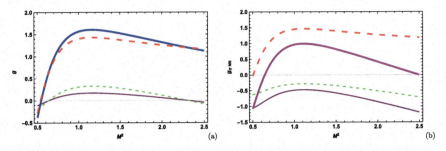

Fig. 153.1 Plots of our results for $g_{\eta NN}$ and $g_{\eta' NN}$ as a function of M^2 (thick solid line). Also plotted our results on $g_{\eta NN}$ and $g_{\eta' NN}$ without contribution from gluonic operator $\alpha_s G \tilde{G}$ (long-dashed line), only due to $\alpha_s G \tilde{G}$ (thin solid line) and due to part with f_s in $a_{\eta^{(\prime)}}$ (short-dashed line). **a** is for η when $\mathbf{p}^2$ is $-\frac{3m_\eta^2}{4}$. **b** is for η' when $\mathbf{p}^2$ is $-\frac{3m_{\eta'}^2}{4}$

Table 153.1 Comparison of our results on $g_{\eta NN}$ and $g_{\eta' NN}$ with results for the same from recent literature

References	$g_{\eta NN}$	$g_{\eta' NN}$	Comment
Present work	$(0.64 - 1.26)$	$(0.44 - 1.27)$	QCD sum rule
[25]	$(4.95 - 5.45)$	$(5.6 - 10.9)$	GT relation + Dispersion relation
[26]	(3.4 ± 0.5)	(1.4 ± 1.1)	Theory (GT relation)
[9]	2.241	–	Photoproduction
[4]	(3.78 ± 0.34)	$(1 - 2)$	Theory
[8]	$(0.39, 0.92)$	–	Photoproduction + (Isobar model, Dispersion relation)
[10]	0.89	0.87	Fitting photoproduction data
[12]	4.2 ± 1.05	–	QCDSR at unphysical point
[27]	4.399 ± 0.365	2.166 ± 0.312	Chiral quark-soliton model
[28]	6.852	8.66	Potential model

Working with Π_{+odd}, we have chosen the range of the Borel mass as $0.8\,\text{GeV}^2 < M^2 < 1.8\,\text{GeV}^2$ for η' and $1.5\,\text{GeV}^2 < M^2 < 2.5\,\text{GeV}^2$ for η. We have also used Π^m, the correlator used in the mass sum rule for the proton [21]. From the ratio of the Π_{+odd} sum rule to the nucleon mass sum rule, the coupling constant can be obtained as

$$\frac{g(M_n, \vec{p}^2)}{M_n(E_p + \omega_p - M_n)} = \frac{\hat{B}[\Pi_{+odd}^{OPE-cont.}]}{\hat{B}[\Pi^m]} - \frac{M^4}{(E_p + \omega_p)^2 - M_n^2} \frac{d}{dM^2}\left(\frac{\hat{B}[\Pi_{+odd}^{OPE-cont.}]}{\hat{B}[\Pi^m]}\right).$$

$$(153.11)$$

We have used the standard values of the parameters appearing in Π [11–13, 16, 21, 22] for making the estimate of the coupling constants $g_{\eta^{(\prime)} NN}$ (all quantities are in GeV unit): $\langle \overline{q}q \rangle = -1.65 \pm 0.15$, $\langle \frac{\alpha_s}{\pi} G^2 \rangle = 0.005 \pm 0.004$, $\delta^2 = 0.2 \pm 0.04$, $f_q = (1.07 \pm 0.02) f_\pi$, $f_s = (1.34 \pm 0.06) f_\pi$, $s_{\eta^{(\prime)}} = 2.57 \pm 0.03$, $s_0 = 2.5$, $m_0^2 = 0.8 \pm 0.1$, $f_{3\pi} = 0.0045$, $\phi = 40° \pm 1°$, $\frac{h_q}{m_q} = -4 \frac{f_q}{f_\pi} \langle \overline{q}q \rangle$, $\langle \overline{q} g \sigma \cdot G q \rangle = m_0^2 \langle \overline{q}q \rangle$. We have shown the plot for the coupling constant g obtained from (153.11) as a function of M^2 for $\overrightarrow{p}^2 = -3m_{\eta^{(\prime)}}^2 / 4$ in Fig. 153.1. Combining with a result for a similar plot for $\overrightarrow{p}^2 = -m_{\eta^{(\prime)}}^2 / 2$, we can extrapolate them to get the result at the physical point $\overrightarrow{p}^2 = -m_{\eta^{(\prime)}}^2 + \frac{m_{\eta^{(\prime)}}^4}{4M_{\tilde{n}}^2}$, which is not directly approachable due to its 0/0 form. We get

$$g_{\eta NN} = 0.96^{+0.16}_{-0.17} (M^2)^{+0.14}_{-0.15} (rest) \qquad (153.12)$$

$$g_{\eta' NN} = 0.76^{+0.27}_{-0.24} (M^2)^{+0.24}_{-0.24} (rest).$$

Errors are due to different phenomenological parameters and finite range of Borel mass. Substituting our results in the flavor-singlet Goldberger–Treiman relation for QCD [3], we get $g_A^{(0)} \simeq 0.22$ for the medium values of the coupling constants and $g_A^{(0)} \simeq 0.27$ for the highest values of the same. This may be compared with the results obtained by COMPASS ($Q^2 = 3 \, \text{GeV}^2$) [23]: $g_A^{(0)} = [0.26\text{–}0.36]$ and NNPDF poll. 1($Q^2 = 10 \, \text{GeV}^2$) [24]: $g_A^{(0)} = 0.25 \pm 0.10$ (Table 153.1).

References

1. E. Witten, Nucl. Phys. B **149**, 285 (1979); G. Veneziano, Nucl. Phys. B **159**, 213 (1979)
2. S.D. Bass, Phys. Lett. B **463**, 286 (1999); Phys. Scripta **T99**, 96 (2002). arXiv:hep-ph/0111180
3. G.M. Shore, G. Veneziano, Nucl. Phys. B **381**, 23 (1992)
4. G.M. Shore, Nucl. Phys. B **744**, 34 (2006); Lect. Notes Phys. **737**, 235 (2008). arXiv:hep-ph/0701171
5. R. Machleidt, Phys. Rev. C **63**, 024001 (2001)
6. ThA Rijken, Phys. Rev. C **73**, 044007 (2006)
7. K. Kawarabayashi, N. Ohta, Nucl. Phys. B **175**, 477 (1980)
8. I.G. Aznauryan, Phys. Rev. C **68**, 065204 (2003)
9. J. Nys et al., Phys. Rev. D **95**, 034014 (2017)
10. L. Tiator et al., Eur. Phys. J. A **54**, 210 (2018). arXiv:807.04525
11. H. Kim, T. Doi, M. Oka, S.H. Lee, Nucl. Phys. A **678**, 295 (2000)
12. J.P. Singh, F.X. Lee, L. Wang, Int. J. Mod. Phys. A **26**, 947 (2011)
13. J.P. Singh, Phys. Lett. B **749**, 63 (2015)
14. L.J. Reinders, H.R. Rubinstein, S. Yazaki, Phys. Rep. **127**, 1 (1985)
15. M.C. Birse, B. Krippa, Phys. Lett. B **373**, 9 (1996)
16. Y. Kondo, O. Morimastu, Nucl. Phys. A **717**, 55 (2003)
17. T. Doi, Y. Kondo, M. Oka, Phys. Rep. **398**, 253 (2004)
18. I.I. Balitsky, V.M. Braun, Nucl. Phys. B **311**, 541 (1989)
19. M. Beneke, M. Neubert, Nucl. Phys. B **651**, 225 (2003)
20. Janardan P. Singh, Shesha D. Patel, Phys. Lett. B **791**, 249 (2019)
21. B.L. Ioffe, V.S. Fadin, L.N. Lipatov, *Quantum Chromodynamics* (Cambridge Univ. Press, 2010)
22. P. Ball, J. High Energy Phys. **9901**, 010 (1999)
23. C. Adolph et al., Phys. Lett. B **753**, 18 (2016)

24. E.R. Nocera et al., Nucl. Phys. B **887**, 276 (2014). arXiv:1406.5539
25. N.F. Nasrallah, Phys. Lett. B **645**, 335 (2007)
26. T. Feldmann, Int. J. Mod. Phys. A **15**, 159 (2000)
27. G.-S. Yang, H.-C. Kim, Phys. Lett. B **785**, 434 (2018). arXiv:1807.09090
28. ThA Rijken, M.M. Nagels, Y. Yamamoto, Prog. Th. Phys. Suppl. **185**, 14 (2010)

Chapter 154
Effect of New Resonance States on Fluctuations and Correlations of Conserved Charges in a Hadron Resonance Gas Model

Subhasis Samanta, Susil Kumar Panda, Bedangadas Mohanty, and Rita Paikaray

Abstract We investigate the role of suspected resonance states that are yet to be confirmed experimentally on the fluctuation and correlation between conserved charges using ideal hadron resonance gas (I-HRG) model. We observe that the lattice QCD data of χ_S^2, χ_{BS}^{11} in the hadronic phase can be described well by I-HRG model with inclusion of additional resonances. We have also studied the beam energy ($\sqrt{s_{NN}}$) dependence of the fluctuation observables of net charge. Proper experimental acceptance cuts as applicable to data from the relativistic heavy-ion collider have been used in the model to compare with the data from experimental measurements. We find that experimental data do not show much deviation from the baseline calculation of I-HRG model. The effect of additional resonances on fluctuation observables at different $\sqrt{s_{NN}}$ has also been studied.

154.1 Introduction

Substantial efforts worldwide have been devoted to the investigation of strongly interacting matter under extreme conditions. At present, Relativistic Heavy-Ion Collider (RHIC) at Brookhaven National Laboratory (BNL), USA and Large Hadron Collider (LHC) at CERN, Switzerland, have been investigating the properties of Quantum Chromodynamic (QCD) matter at very high temperature and almost zero net baryon density. The temperature reached in such collisions at ultra-relativistic

S. Samanta (✉) · B. Mohanty
School of Physical Sciences, National Institute of Science Education and Research, HBNI, Jatni, Bhubaneswar 752050, India
e-mail: subhasis.samant@gmail.com

S. K. Panda · R. Paikaray
Department of Physics, Ravenshaw University, Cuttack 753003, India

B. Mohanty
Department of Experimental Physics, CERN, CH-1211, Geneva 23, Switzerland

© Springer Nature Singapore Pte Ltd. 2021
P. K. Behera et al. (eds.), *XXIII DAE High Energy Physics Symposium*,
Springer Proceedings in Physics 261,
https://doi.org/10.1007/978-981-33-4408-2_154

energy is very large (around 400–500 MeV). This super-hot matter is formed by quark and gluon degrees of freedom. This matter is called the quark-gluon plasma (QGP). The beam energy scan program of RHIC is investigating the QGP matter over a wide range of temperature and net baryon density [1]. After the creation of QGP, the system expands and cools and a transition from QGP phase to the hadronic phase occurs. The I-HRG model has been successful in describing the hadron yields created in heavy-ion collision experiments. The I-HRG model also gives a qualitative description of the lattice QCD data in the hadronic phase but discrepancies appear at a quantitative level. For example, I-HRG fails to describe LQCD result of baryon-strange correlation [2]. There could be many reasons for the discrepancy. To reduce this discrepancy, we have included additional hadrons in HRG model. These hadrons are predicted in various quark models but yet to be confirmed experimentally. In this work, we have studied temperature dependence of susceptibilities of conserved charges using I-HRG model. We have also studied the center of mass energy $\sqrt{s_{NN}}$ dependence of ratios of cumulants of conserved charge.

154.2 HRG Model

In the I-HRG model, the system of thermal fireball consists of hadrons and resonances given in the particle data book. Particles are point-like and non-interacting. The logarithm of the partition function of a hadron resonance gas in the grand canonical ensemble can be written as [3]

$$\ln Z = \sum_i \ln Z_i, \qquad (154.1)$$

where the sum is over all the hadrons. For particle species i,

$$\ln Z_i = \pm \frac{V g_i}{2\pi^2} \int_0^\infty p^2 \, dp \ln[1 \pm \exp(-(E_i - \mu_i)/T)], \qquad (154.2)$$

where V is the volume of the thermal system, g_i is the degeneracy, $E_i = \sqrt{p^2 + m_i^2}$ is the single particle energy, m_i is the mass of the particle, and $\mu_i = B_i \mu_B + S_i \mu_S + Q_i \mu_Q$ is the chemical potential. In the last expression, B_i, S_i, Q_i are, respectively, the baryon number, strangeness, and electric charge of the particle, μ's are the corresponding chemical potentials. "+" corresponds to fermions and "−" corresponds to bosons. In this model, a resonance behaves identically to that of a stable hadron. The pressure P of the system can be calculated using the standard definitions.

$$P = \sum_i \frac{T}{V} \ln Z_i = \sum_i (\pm) \frac{g_i T}{2\pi^2} \int_0^\infty p^2 \, dp \ln[1 \pm \exp(-(E_i - \mu_i)/T)]. \quad (154.3)$$

The susceptibilities of conserved charges can be calculated as [4]

$$\chi_{BSQ}^{xyz} = \frac{\partial^{x+y+z}(P/T^4)}{\partial(\mu_B/T)^x \partial(\mu_S/T)^y \partial(\mu_Q/T)^z},$$
(154.4)

where x, y, and z are the order of derivatives of the quantities B, S, and Q.

154.3   Results

We have done the calculations in I-HRG model with two different input hadronic spectra. For the first set, we have taken all the confirmed hadrons and resonances that consist of only up, down, and strange quarks listed in the PDG 2016 Review [7]. This list includes all the confirmed mesons listed in the Meson Summary Table [7] and all baryons in the Baryon Summary Table [7] with three- or four-star status. This is referred as PDG 2016 in this. The other set includes all the resonances from the previous set, i.e., PDG 2016 as well as the other unmarked mesons from the Meson Summary Table and baryons from the Baryon Summary Table with one- or two-star status which are not confirmed yet. We refer this set as PDG 2016+ [8]. Figure 154.1 shows temperature dependence of χ_S^2 and χ_{BS}^{11} at zero chemical potential [9]. The dotted and solid lines correspond to the calculations in HRG model with hadron spectra PDG 2016 and 2016+, respectively. We have compared our result with the continuum estimate of the lattice QCD result [2, 4, 5]. We observe improvement in results of HRG model for the hadronic spectrum PDG 2016+ where strange baryons are involved.

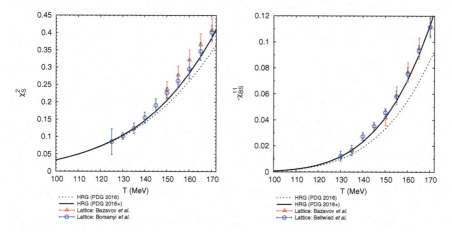

Fig. 154.1 Variation of χ_S^2 and χ_{BS}^{11} with temperature at zero chemical potential calculated in HRG model with hadron spectra from PDG 2016 and PDG 2016+. Results are compared to LQCD calculations [2, 4, 5]

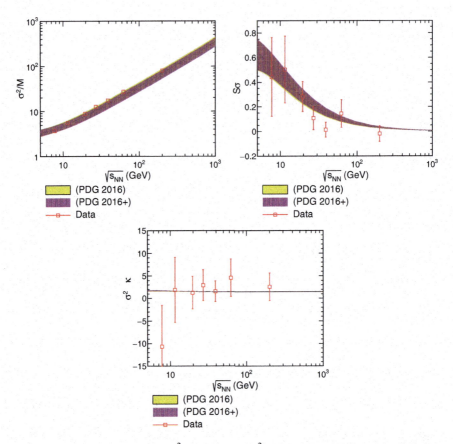

Fig. 154.2 Energy dependence of σ^2/M, $S\sigma$, and $\kappa\sigma^2$ of net charge. HRG model calculations are done with the hadron spectra PDG 2016 and PDG 2016+. Experimental data of fluctuations measured in Au+Au collisions in 0–5% centrality by STAR collaboration is taken from [6]

In Fig. 154.2, we have shown energy dependence of σ^2/M ($= \chi^2/\chi^1$), $S\sigma$ ($= \chi^3/\chi^2$), and $\kappa\sigma^2$ ($= \chi^4/\chi^2$) of net charge. We have compared our results with experimental data of net charge fluctuations for Au+Au collision in 0–5% centrality measured by the STAR collaboration [6]. The experimental data is measured within the pseudorapidity range $|\eta| < 0.5$ and within the transverse momentum range $0.2 < p_T < 2$ GeV (removing net protons of $p_T < 0.4$ GeV). The same acceptances have been used in the HRG model calculation [9]. It can be seen that experimental data do not show much deviation from the I-HRG model calculations. Further, two sets of hadronic spectra give almost similar result.

References

1. L. Adamczyk et al., STAR Collaboration. Phys. Rev. Lett. **112**, 032302 (2014)
2. S. Borsanyi, Z. Fodor, S.D. Katz, S. Krieg, C. Ratti, K. Szabo, JHEP **1201**, 138 (2012)
3. A. Andronic, P. Braun-Munzinger, J. Stachel, M. Winn, Phys. Lett. B **718**, 80 (2012)
4. R. Bellwied, S. Borsanyi, Z. Fodor, S.D. Katz, A. Pasztor, C. Ratti, K.K. Szabo, Phys. Rev. D **92**(11), 114505 (2015)
5. A. Bazavov et al., HotQCD Collaboration. Phys. Rev. D **86**, 034509 (2012)
6. L. Adamczyk et al., STAR Collaboration. Phys. Rev. Lett. **113**, 092301 (2014)
7. C. Patrignani et al. [Particle Data Group], Chin. Phys. C **40**(10), 100001 (2016)
8. S. Chatterjee, D. Mishra, B. Mohanty, S. Samanta, Phys. Rev. C **96**(5), 054907 (2017)
9. S. Samanta, S.K. Panda, B. Mohanty, Int. J. Mod. Phys. E **27**(10), 1850080 (2018)

Chapter 155
Inclusive Υ Production in p–Pb Collisions at $\sqrt{s_{NN}} = 8.16$ TeV with ALICE at the LHC

Wadut Shaikh

Abstract The Υ production in p–Pb collisions at $\sqrt{s_{NN}} = 8.16$ TeV is studied with the muon spectrometer of the ALICE detector at the CERN LHC. We report the inclusive Υ(1S) nuclear modification factors (R_{pPb}) as a function of rapidity, transverse momentum (p_T) and centrality of the collision and compare the results with those obtained at $\sqrt{s_{NN}} = 5.02$ TeV. Theoretical model predictions as a function of y_{cms} and p_T are also discussed. The results of Υ(2S) suppression integrated over y_{cms}, p_T and centrality are also reported and compared to the corresponding Υ(1S) measurement.

155.1 Introduction

Quarkonia are well-known probes to study the properties of the deconfined medium, called Quark–Gluon Plasma (QGP), created in ultrarelativistic heavy-ion collisions. The modification of quarkonium production in heavy-ion collisions with respect to the binary-scaled yield in pp collisions at the LHC is explained by suppression of quarkonia via color screening mechanism [1] and (re)generation of quarkonia. In the color screening mechanism, color charges present in the deconfined medium screen the binding potential between the q and $\bar{q}$ quarks, leading to a temperature-dependent melting of the quarkonium states according to their binding energies. For bottomonia, (re)generation effects are expected to be negligible due to the small number of produced b quarks. The cold nuclear matter effects (shadowing, parton energy loss, interaction with the hadronic medium) which are not related to the deconfined medium may also lead to a modification of quarkonium production. In order to disentangle the CNM effects from the hot nuclear matter effects, quarkonium production is studied in p–Pb collisions in which the QGP is not expected to be

Wadut Shaikh (for the ALICE collaboration).

W. Shaikh (✉)
Saha Institute of Nuclear Physics, HBNI, Kolkata 700064, India
e-mail: wadut.shaikh@cern.ch

© Springer Nature Singapore Pte Ltd. 2021
P. K. Behera et al. (eds.), *XXIII DAE High Energy Physics Symposium*,
Springer Proceedings in Physics 261,
https://doi.org/10.1007/978-981-33-4408-2_155

formed. The measurement has been performed reconstructing $\Upsilon(1S)$ and $\Upsilon(2S)$ mesons via their dimuon decay channel, in the forward (proton-going direction) and backward (lead-going direction) rapidity ranges down to zero transverse momentum with integrated luminosities of 8.4 ± 0.2 nb^{-1} and 12.8 ± 0.3 nb^{-1}, respectively.

155.2 Experimental Setup and Data Analysis

The detailed description of the ALICE Muon Spectrometer (MS) can be found in [2]. It consists of a set of absorbers, a dipole magnet, five tracking stations, and two trigger stations. The data sample, event selection criteria, analysis procedure are taken from the reference [3].

155.3 Results

The cold nuclear matter effects in p–Pb collisions can be quantified through the nuclear modification factor defined as

$$R_{\mathrm{pPb}} = \frac{N_{\Upsilon}}{\langle T_{\mathrm{pPb}} \rangle \times (A \times \varepsilon) \times N_{\mathrm{MB}} \times \mathrm{BR}_{\Upsilon \to \mu^+\mu^-} \times \sigma_{\Upsilon}^{\mathrm{pp}}} \tag{155.1}$$

where

- N_{Υ} is the number of Υ in a given y_{cms}, p_T or centrality bin.
- $\langle T_{\mathrm{pPb}} \rangle$ is the centrality-dependent average nuclear overlap function.
- $A \times \epsilon$ is the product of the detector acceptance and the reconstruction efficiency.
- N_{MB} is the number of collected minimum-bias events.
- $\mathrm{BR}_{\Upsilon \to \mu^+\mu^-}$ is the branching ratio of Υ in the dimuon decay channel ($\mathrm{BR}_{\Upsilon(1S) \to \mu^+\mu^-}$ $= 2.48 \pm 0.05\%$, $\mathrm{BR}_{\Upsilon(2S) \to \mu^+\mu^-} = 1.93 \pm 0.17\%$) [4].
- $\sigma_{\Upsilon}^{\mathrm{pp}}$ is the inclusive Υ production cross section for pp collisions at the same energy, y_{cms} and p_T interval as for p–Pb collisions. These values have been obtained by means of an interpolation procedure based on the LHCb pp Υ measurements [5].

The inclusive $\Upsilon(1S)$ nuclear modification factor at $\sqrt{s_{\mathrm{NN}}} = 8.16$ TeV for the two studied beam configurations is shown in Fig. 155.1 (left panel). In this figure, as well as all the other figures, the vertical error bars represent the statistical uncertainties and the open boxes represent the uncorrelated systematic uncertainties. The full boxes around $R_{\mathrm{pPb}} = 1$ show the size of the correlated systematic uncertainties. The measured R_{pPb} values indicate $\Upsilon(1S)$ suppression both at forward and backward rapidity. The R_{pPb} values are also compared with previous ALICE measurements at $\sqrt{s_{\mathrm{NN}}} = 5.02$ TeV [2] and results are compatible within the uncertainties.

The rapidity dependence of the R_{pPb} is shown in Fig. 155.1 (right panel). The suppression already observed in the integrated case is confirmed, the size of the

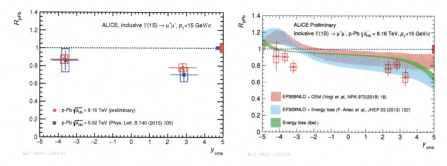

Fig. 155.1 Integrated R_{pPb} of ϒ(1S) in p–Pb collisions at $\sqrt{s_{NN}}$ = 8.16 TeV compared to 5.02 TeV (left). ϒ(1S) R_{pPb} as a function of y_{cms} at $\sqrt{s_{NN}}$ = 8.16 TeV and the values are compared to theoretical calculations based on different CNM effects (right)

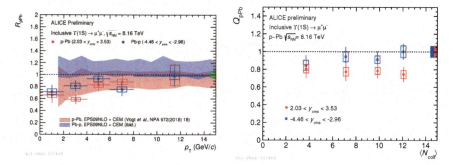

Fig. 155.2 ϒ(1S) R_{pPb} as a function of p_T at $\sqrt{s_{NN}}$ = 8.16 TeV compared to theoretical calculations based on different CNM effects (left). ϒ(1S) Q_{pPb} as a function of average number of binary nucleon–nucleon collisions ($\langle N_{coll} \rangle$) at $\sqrt{s_{NN}}$ = 8.16 TeV (right)

uncertainties does not allow us to draw more detailed conclusions on the rapidity dependence.

The p_T dependence of the ϒ(1S) R_{pPb} is shown in Fig. 155.2 (left panel). A suppression of low-p_T ϒ(1S) is observed both at forward and backward rapidities.

The rapidity and p_T dependencies of the R_{pPb} are compared to a next-to-leading order (NLO) CEM calculation using the EPS09 parameterization of the nuclear modification of the gluon PDF [6, 7] and to a parton energy loss calculation [8] with and without EPS09 gluon shadowing at NLO. The shadowing and energy loss calculations describe the p_T and rapidity dependent results at forward rapidity within uncertainties while they overestimate the data at backward rapidity.

The ϒ(1S) nuclear modification factor has also been studied as a function of the collision centrality. Q_{pPb} is used instead of R_{pPb} due to a possible bias in the centrality determination [9, 10] and it is defined the same way as R_{pPb} in 155.1. The centrality dependence of ϒ(1S) Q_{pPb} is shown in Fig. 155.2 (right panel). The Q_{pPb} is found to be independent of the centrality within uncertainties [3].

Fig. 155.3 Integrated R_{pPb} of $\Upsilon(1S)$ and $\Upsilon(2S)$ in p–Pb collisions at $\sqrt{s_{NN}} = 8.16$ TeV

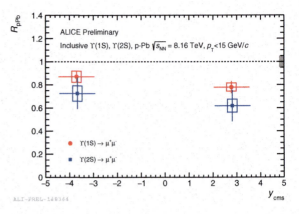

The lower $\Upsilon(2S)$ yield does not allow differential studies, hence only results integrated over y, p_T and centrality are presented in Fig. 155.3. The two resonances show a similar suppression, slightly larger for $\Upsilon(2S)$ [3]. The difference in the R_{pPb} of the $\Upsilon(2S)$ and $\Upsilon(1S)$ amounts to 1σ at forward-y and 0.9σ at backward-y. The CMS [11], ATLAS [12] and LHCb [13] collaborations also observed a larger suppression for the $\Upsilon(2S)$ than for the $\Upsilon(1S)$.

155.4   Conclusions

We have presented new results on the $\Upsilon(1S)$ and $\Upsilon(2S)$ nuclear modification factor in p–Pb collisions at $\sqrt{s_{NN}} = 8.16$ TeV, measured by ALICE. The R_{pPb} of $\Upsilon(1S)$ is similar at forward and backward rapidities with a hint for a stronger suppression at low p_T. In both rapidity intervals there is no evidence for a centrality dependence of the $\Upsilon(1S)$ Q_{pPb}. Models based on nuclear shadowing and coherent parton energy loss fairly describe the data at forward rapidity, while they tend to overestimate the R_{pPb} at backward rapidity. The $\Upsilon(2S)$ and $\Upsilon(1S)$ suppressions are compatible within 1σ.

References

1. T. Matsui, H. Satz, Phys. Lett. B **178**, 416 (1986)
2. ALICE Collaboration, Phys. Lett. B **740**, 105 (2015)
3. ALICE Collaboration, CERN-ALICE-PUBLIC-2018-008
4. Particle Data Group Collaboration, Chin. Phys. C **40**, 100001 (2016)
5. LHCb Collaboration, EPJC **74**, 2835 (2014), JHEP **11**, 103 (2015)
6. R. Vogt, Phys. Rev. C **92**, 034909 (2015)
7. J.L. Albacete et al., Nucl. Phys. A **972**, 18 (2018)
8. F. Arleo et al., JHEP **03**, 122 (2013)
9. ALICE Collaboration, JHEP **11**, 127 (2015)

10. ALICE Collaboration, CERN-ALICE-PUBLIC-2017-007
11. CMS Collaboration, JHEP **04**, 103 (2014)
12. ATLAS Collaboration, EPJC **78**, 1717 (2018)
13. LHCb Collaboration, JHEP **11**, 194 (2018)

Chapter 156
Event-by-Event Charge Separation in Au+Au Collisions at $\sqrt{s_{NN}}$ = 200 GeV Using AMPT

Anjali Attri, Jagbir Singh, and Madan M. Aggarwal

Abstract The strong magnetic field created in non-central heavy-ion collisions induces an electric field parallel/antiparallel to it. This induced electric field leads to the separation of positively and negatively charged particles along the axis of the magnetic field, the phenomenon is known as Chiral Magnetic effect (CME) [1–3]. We are reporting the measurement of event-by-event charge separation in Au+Au collisions at $\sqrt{s_{NN}}$ = 200 GeV using AMPT model. Event-by-event charge separation is obtained using Sliding Dumbbell Method (SDM). The three-particle correlator for the AMPT generated events is compared with those obtained by reshuffling the charges of particles in an event.

156.1 Introduction

High-energy heavy-ion collisions provide us the opportunity to study the particle interactions at the extreme conditions of temperature and energy density. In non-central heavy-ion collisions, fast moving ions create a strong magnetic field ($B \sim 10^{15}$ T) and a deconfined state of quarks and gluons. The interplay of the magnetic field and the deconfined state created in these collisions results in the phenomenon of CME. The observation of positive and negative charge separation along the axis of the magnetic field or perpendicular to the reaction plane would be an experimental evidence of this effect. To detect the charge separation a multi-particle correlator $\langle \cos(\phi_\alpha + \phi_\beta - 2\Psi_{RP}) \rangle$ has been proposed, where ϕ_α, ϕ_β denote the azimuthal angles of the particles α, β and Ψ_{RP} is the reaction plane angle [4]. The multi-particle correlator has been evaluated by the STAR experiment at RHIC and the ALICE experiment at LHC [5].

A. Attri (✉) · J. Singh · M. M. Aggarwal
Panjab University, Chandigarh 160014, India
e-mail: anjali@rcf.rhic.bnl.gov

© Springer Nature Singapore Pte Ltd. 2021
P. K. Behera et al. (eds.), *XXIII DAE High Energy Physics Symposium*,
Springer Proceedings in Physics 261,
https://doi.org/10.1007/978-981-33-4408-2_156

156.2 Analysis Technique

We are using the Sliding Dumbbell Method (SDM) to investigate the event-by-event charge separation, for which we calculate the observable $Db_\pm$, that can be defined as

$$Db_\pm = \frac{N_+^{forw}}{(N_+^{forw} + N_-^{forw})} + \frac{N_-^{back}}{(N_+^{back} + N_-^{back})} . \quad (156.1)$$

N_+^{forw} and N_-^{forw} are the numbers of positively and negatively charged particles on the forward side of the dumbbell, respectively. N_+^{back} and N_-^{back} are the numbers of positively and negatively charged particles on the backward side of the dumbbell, respectively. By sliding the $\Delta\phi = 60°$ dumbbell in the steps of $\delta\phi = 1°$, we scan the whole azimuthal plane and calculate the quantity $Db_\pm$ for each $\Delta\phi$ region to aquire the maximum value of $Db_\pm$. In each event we extract the maximum value of $Db_\pm$ ($Db_\pm^{max}$) along with the condition that asymmetry (asy) <0.25, where asy can be defined as follows:

$$asy = \frac{Pos_{ex} - Neg_{ex}}{Pos_{ex} + Neg_{ex}} \quad (156.2)$$

Here, $Pos_{ex} = N_+^{forw} - N_-^{forw}$ represents the excess of positive charges on the forward side of the dumbbell and $Neg_{ex} = N_-^{back} - N_+^{back}$ denotes the excess of negative charges on the backward side of the dumbbell. $Db_\pm^{max}$ distributions corresponding to different centrality intervals have been acquired and are further sliced into different groups depending upon the highest (0–10%) and lowest (90–100%) $Db_\pm^{max}$ values. The Q-cumulants method is used to get three-particle correlators. These three-particle correlators are obtained for both same-sign and opposite-sign charged pairs as a function of centrality and $Db_\pm^{max}$ binning. For the background estimation, the charges of the particles are reshuffled keeping θ and ϕ same and the results obtained by charge reshuffle are compared with those obtained from the simulated events.

156.3 Analysis Details

The AMPT (String melting On) ~2 million events for Au+Au collisions at the center of mass energy $\sqrt{s_{NN}} = 200$ GeV are generated [6]. The tracks in the pseudorapidity region $|\eta| < 1.0$ and transverse momentum range $0.15 < p_T < 2.0$ GeV/c in each event are used for the analysis.

156.4   Results and Discussions

The $Db_{\pm}^{max}$ distributions for the 40–50 and 50–60% central events are shown in Fig. 156.1a, b. The solid curve is for the AMPT simulated data and the dotted curve is for the charge reshuffle. The distributions are seen to be agreeing within the statistical errors. It has also been viewed that the $Db_{\pm}^{max}$ distributions shift toward higher values of $Db_{\pm}^{max}$ as we go from the central to peripheral collisions which indicates more charge separation in the peripheral collisions.

In Fig. 156.2, the three-particle correlator ($\gamma = < \cos(\phi_\alpha + \phi_\beta - 2\Psi_{RP}) >$) for same-sign and opposite-sign charged pairs as a function of centrality for both AMPT simulated data and charge reshuffle have been shown. The closed symbols are for the AMPT simulated data and open symbols are for the charge reshuffle. It is seen that the correlation for the same and opposite-sign charged pairs is negative. This has also been noticed that the correlations for the oppositely charged pairs are relatively smaller than those for the same-sign charged pairs and the correlation increases as we go from the central to the peripheral collisions. However, the correlations for charge reshuffle appear to be the same for both same and opposite charged pairs.

$\Delta\gamma$ ($\gamma_{opp} - \gamma_{same}$) correlator for each centrality interval, which is further divided into ten different groups depending upon the $Db_{\pm}^{max}$ values is shown in Fig. 156.3. The solid blue symbols are for the AMPT simulated data and open red symbols are for the charge reshuffle. It has been seen that the particles are strongly correlated for the top $Db_{\pm}^{max}$ bins. Also, the AMPT simulated data and the charge reshuffle are exhibiting a similar trend as there is no signal of CME expected in the AMPT model.

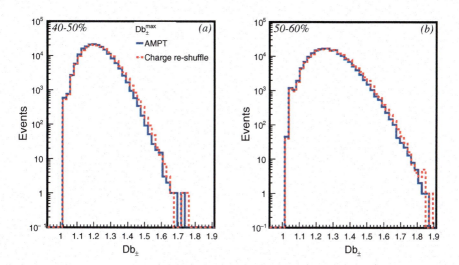

Fig. 156.1 Distributions of $Db_{\pm}^{max}$ for 40–50% (**a**) and 50–60% (**b**) central events for AMPT simulated data (solid curve) and charge reshuffle (dotted curve)

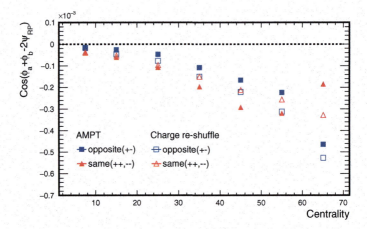

Fig. 156.2 Three-particle correlator for same and opposite-sign charged pairs of both AMPT simulated data (closed symbols) and charge reshuffle (open symbols) as a function of centrality

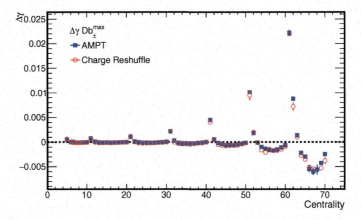

Fig. 156.3 The centrality dependence of $\Delta\gamma$ corresponding to different $Db_\pm^{max}$ values

Acknowledgements The financial assistances from the Department of Science and Technology and University Grants Commission of the Government of India, are gratefully acknowledged.

References

1. D.E. Kharzeev et al., Prog. Part. Nucl. Phys. **88**, 1–28 (2016)
2. I. Selyuzhenkov, (2009). arXiv:0910.0464
3. A. Bzdak, V. Koch, J. Liao, Phys. Rev. C **83**, 014905 (2011)
4. S.A. Voloshin, Phys. Rev. C **70**, 057901 (2004)
5. B.I. Abelev et al., STAR Collaboration. Phys. Rev. C **81** 054908 (2010); B. Abelev et al., ALICE Collaboration. Phys. Rev. Lett. **110**, 012301 (2013)
6. Z. Lin et al., Phys. Rev. C **72**, 064901 (2005)

Chapter 157
Production of Electrons from Heavy-Flavour Hadron Decays in Different Collision Systems with ALICE at LHC

Sudhir Pandurang Rode

Abstract Heavy-flavour quarks, due to their large masses, are produced in the early stages of the relativistic heavy-ion collisions via initial hard scatterings. Therefore, as they experience the full system evolution, heavy quarks are effective probes of the hot and dense medium created in such collisions. In pp collisions, the measurement of heavy-flavour hadron production cross-sections can be used to test our understanding of the Quantum ChromoDynamics (QCD) in the perturbative regime. Also, pp collisions provide a crucial reference for the corresponding measurements in larger systems. In Pb–Pb (Xe–Xe) collisions, the measurement of the nuclear modification factor of heavy-flavour hadrons provides information on the modification of the invariant yield with respect to pp collisions due to the produced cold and hot QCD matter. The possible mass dependence of the parton energy loss can be studied by comparing the R_{AA} of pions, charm and beauty hadrons. In this contribution, recent results from ALICE at the LHC are reported with a focus on the different measurements of the heavy-flavour electrons in pp collisions at 2.76, 5.02, 7 and 13 TeV and in Pb–Pb (Xe–Xe) collisions at 5.02 (5.44) TeV. The results include the differential production cross-sections and nuclear modification factors of heavy-flavour electrons at mid-rapidity. The comparison of experimental data with model predictions is discussed.

157.1 Introduction

A Large Ion Collider Experiment (ALICE) is one of the four major experiments carried out at the LHC at CERN. It is a dedicated heavy-ion experiment aiming at the study of the strongly interacting matter consisting of thermally equilibrated deconfined quarks and gluons, also called Quark–Gluon Plasma. Heavy-flavour quarks are

Sudhir Pandurang Rode (for ALICE collaboration).

S. Pandurang Rode (✉)
Indian Institute of Technology Indore, Indore 453552, Madhya Pradesh, India
e-mail: sudhir.pandurang.rode@cern.ch

© Springer Nature Singapore Pte Ltd. 2021
P. K. Behera et al. (eds.), *XXIII DAE High Energy Physics Symposium*,
Springer Proceedings in Physics 261,
https://doi.org/10.1007/978-981-33-4408-2_157

among the most important probes to understand the nature of QGP. Due to their large masses ($m_Q \gg \Lambda_{QCD}$), they are produced at the early stage of heavy-ion collisions via hard scattering making them crucial probes since they witness the full evolution of the Quantum ChromoDynamics (QCD) medium. Their production mechanism can be described theoretically by perturbative QCD in the full transverse momentum range [1]. The heavy-flavour hadrons can be measured via electrons originating from their semi-leptonic decay channel.

157.2 Heavy-Flavour Electron Measurement with ALICE

Measurements of electrons from heavy-flavour hadron decays require excellent particle identification capabilities and reconstruction efficiencies, which are provided by the ALICE detector. Charged particles are identified using the Inner Tracking System (ITS), Time Projection Chamber (TPC) and Time of Flight (TOF) detectors using specific energy loss and time of flight information of the particles. A more detailed description of the ALICE detector can be found here [2]. The contribution of heavy-flavour electrons to the inclusive electron distribution is estimated after eliminating the primary source of background, i.e. electrons from Dalitz decays and photon conversions using electron–positron pair selections [3].

In the case of electrons from beauty hadron decays, the distribution of the distance of closest approach (DCA) of the electrons to the primary vertex is used. A larger DCA distribution of the signal compared to the background allows their separation. DCA templates of electrons from different sources are obtained from Monte Carlo (MC) simulations. The DCA distribution of the electrons from the various sources are fitted to the DCA distribution of inclusive electrons in the data using a maximum likelihood fit method.

157.3 Results and Discussions

In this section, recent results on the production cross-section of electrons coming from heavy-flavour hadron decays and their nuclear modification factors (R_{AA}) in Pb–Pb and Xe–Xe collisions at various center-of-mass energies are shown.

157.3.1 Invariant Cross-Sections in Proton–Proton Collisions

The p_T-differential production cross-sections of heavy-flavour electrons are measured using the so-called data-driven method, i.e. photonic-electron tagging method at $\sqrt{s} = 2.76$, 5.02, 7 and 13 TeV. At 7 TeV, the data-driven method allows a reduction by a factor 3 of the systematic uncertainties in comparison to previous publications [4]. As shown in Fig. 157.1 (upper), all the measured cross-sections

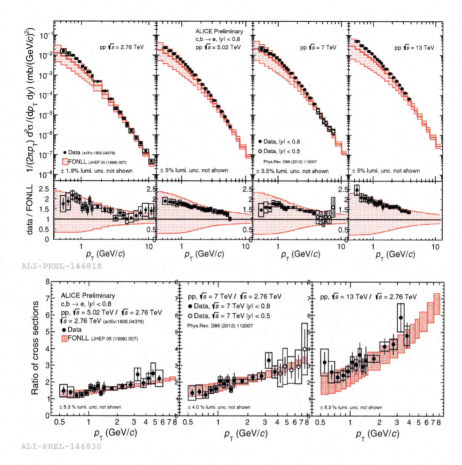

ALI-PREL-146818

ALI-PREL-146830

Fig. 157.1 p_T-differential cross-sections of heavy-flavour electrons in pp collisions at different colliding energies

are in agreement with the FONLL prediction [5]. The theoretical uncertainties are reduced when forming the ratio between the cross-sections at different energies, hence the ratios of the measured cross-sections are shown in the lower panel of Fig. 157.1 and are found to be in agreement with FONLL predictions.

157.3.2 R_{AA} of Electrons from Heavy-Flavour Decays in Pb–Pb and Xe–Xe Collisions

The R_{AA} of heavy-flavour electrons in Pb–Pb and Xe–Xe collisions at $\sqrt{s_{NN}} = 5.02$ and 5.44 TeV, respectively, are measured to study the energy loss of heavy quarks

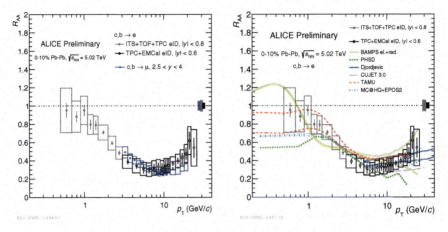

Fig. 157.2 Comparison of R_{AA} of heavy-flavour electrons in Pb–Pb collisions at $\sqrt{s} = 5.02$ TeV with R_{AA} of heavy-flavour muons (left) and with theoretical models (right)

inside the QGP medium. Similar to the analysis in pp collisions, the contribution from the heavy-flavour electrons is estimated using the photonic-electron tagging method.

The R_{AA} of heavy-flavour electrons is shown in Fig. 157.2 and is found to be compatible with the R_{AA} of heavy-flavour decay muons measured in central (0–10%) Pb–Pb collisions at $\sqrt{s_{NN}} = 5.02$ TeV at forward rapidity ($2.5 < y < 4$). In the very low p_T region, the measured R_{AA} is compatible with the TAMU [6] prediction. At higher p_T, the measurement agrees with others models (Djordjevic [7], CUJET 3.0 [8])

In Xe–Xe, the pp reference for R_{AA} estimation is obtained by interpolating the pp references measured at $\sqrt{s} = 5.02$ and 7 TeV [9]. The R_{AA} of heavy-flavour electrons is compared with the R_{AA} of heavy-flavour muons. Similar suppression is observed for the muons in both centrality classes as shown in Fig. 157.3.

157.3.3 R_{AA} of Electrons from Beauty Quark Decays in Pb–Pb Collisions

The R_{AA} of electrons from beauty hadron decays in 0–10% central Pb–Pb collisions at $\sqrt{s_{NN}} = 5.02$ TeV is shown in Fig. 157.4. The yield of electrons from beauty hadron decays is estimated using the DCA template fit method.

As shown in Fig. 157.4 (right), a smaller suppression of electrons from beauty quark decays with respect to electrons from heavy-flavour decays hints at a mass dependence of the quark energy loss in the QGP medium. This measurement is consistent with models that consider mass-dependent radiative and collisional energy losses [10, 11]. For R_{AA} calculation, the pp reference is obtained by an energy

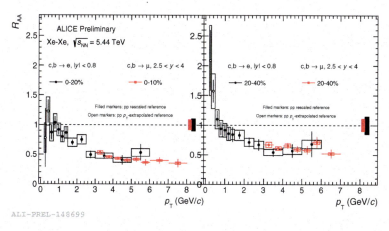

ALI-PREL-148699

Fig. 157.3 Comparison of R_{AA} of electrons from heavy-flavour hadron decays in Xe–Xe collisions with R_{AA} of muons from heavy-flavour hadron decays

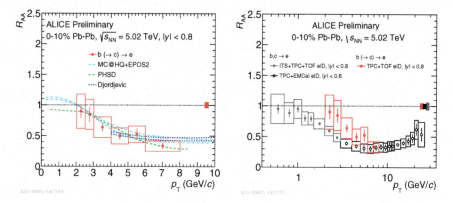

Fig. 157.4 Comparison of R_{AA} of electrons coming from beauty hadron decays in Pb–Pb collisions at $\sqrt{s} = 5.02$ TeV

interpolation from 7 to 5 TeV using FONLL. Data analysis of new pp reference at $\sqrt{s}$ = 5 TeV is ongoing, which would reduce the systematic and statistical uncertainties in the R_{AA} measurement and will give more precise results.

157.4 Summary

In summary, we have presented an overview of the measurements of electrons from the decay of heavy-flavour hadrons in pp, Pb–Pb (Xe–Xe) collisions at different energies. The p_T-differential production cross-sections of heavy-flavour electrons in pp collisions at 2.76, 5.02, 7 and 13 TeV are measured and are in agreement with

FONLL predictions. The nuclear modification factor of the heavy-flavour electrons is measured in Pb–Pb and Xe–Xe collisions at 5.02 and 5.44 TeV, respectively, and is found to be consistent with the R_{AA} of heavy-flavour muons at forward rapidity ($2.5 < y < 4$). The R_{AA} of beauty decay electrons, while compatible with one of the heavy-flavour electrons, hints at separation at low p_T that would points towards the mass-dependent energy loss of the quarks inside the medium. It is also consistent with theoretical predictions, which include collisional and radiative energy losses.

With the ongoing detector upgrades [12], the precision of the measurement will considerably increase. The improved impact-parameter resolution, together with the improved luminosity of the LHC accelerator complex, will improve the significance of the upcoming measurements.

References

1. R. Averbeck, Prog. Part. Nucl. Phys. **70**, 159 (2013)
2. K. Aamodt et al., ALICE Collaboration. JINST **3**, S08002 (2008)
3. S. Acharya et al., ALICE Collaboration. JHEP **1810**, 061 (2018)
4. B. Abelev et al., ALICE Collaboration. Phys. Rev. D **86**, 112007 (2012)
5. M. Cacciari, M. Greco, P. Nason, JHEP **9805**, 007 (1998)
6. M. He, R.J. Fries, R. Rapp, Phys. Lett. B **735**, 445 (2014)
7. M. Djordjevic, M. Djordjevic, Phys. Rev. C **92**(2), 024918 (2015)
8. J. Xu, J. Liao, M. Gyulassy, JHEP **1602**, 169 (2016)
9. S. Acharya et al., ALICE Collaboration. Phys. Lett. B **788**, 166 (2019)
10. M. Nahrgang, J. Aichelin, P.B. Gossiaux, K. Werner, Phys. Rev. C **89**(1), 014905 (2014)
11. E.L. Bratkovskaya, W. Cassing, V.P. Konchakovski, O. Linnyk, Nucl. Phys. A **856**, 162 (2011)
12. B. Abelev et al., ALICE Collaboration. J. Phys. G **41**, 087001 (2014)

Chapter 158
A 3D Kinetic Distribution that Yields Observed Plasma Density in the Inner Van Allen Belt

Snehanshu Maiti and Harishankar Ramachandran

Abstract A steady-state distribution is obtained that approximately yields the observed plasma density profile of the inner Van Allen radiation belt. The model assumes a collisionless, magnetized plasma with zero electric field present. The inner Van Allen belt consists of a plasma comprising high-energy protons and relativistic electrons. The particle trajectories are obtained from the collisionless Lorentz Force equation for different initial distributions. The resulting steady-state distributions obtained after particles lost to the loss cone are eliminated and are used to generate the density profile. The distribution's dependence on energy E and magnetic moment μ is adjusted to make the density profile agree with observations. For a distribution that is a function of energy times a function of magnetic moment, the calculation leads to the desired type of density profile. The kinetic distribution and the type of density profile obtained are presented.

158.1 Introduction

The inner Van Allen radiation belt exists approximately from an altitude of 1000–6000 km, (0.2–2) R_E above the Earth's surface and contains electrons in the range of hundreds of KeV and energetic protons exceeding 100 MeV, trapped by the strong geomagnetic field (relative to the outer belts) in the region [1]. The plasma is collisionless in nature and experiences Lorentz force within the magnetosphere. The particles are confined in a magnetic mirror and undergo gyro-motion, bounce motion and drift motion around the earth.

The plasma density observed is given by a commonly used form of an exponential (oxygen) plus a power law (hydrogen) [2].

$$n(r) = n_O e^{-(r-R_I)/h} + n_H r^{-1} \tag{158.1}$$

S. Maiti (✉) · H. Ramachandran
Indian Institute of Technology Madras, Chennai, India
e-mail: snehanshu.maiti@gmail.com

© Springer Nature Singapore Pte Ltd. 2021
P. K. Behera et al. (eds.), *XXIII DAE High Energy Physics Symposium*,
Springer Proceedings in Physics 261,
https://doi.org/10.1007/978-981-33-4408-2_158

Fig. 158.1 Observed density profile of the radiation belt with altitude at the equator

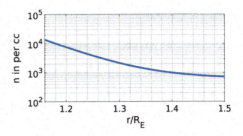

where r is the geocentric radius, $n_O = 10^5 \text{cm}^{-3}$ is the density of oxygen, $n_H = 10^3 \text{cm}^{-3}$ is the density of hydrogen, h = 400 km is the scale height and $R_I = 1.0314 R_E$, where R_E is radius of the earth.

The current aim is to find a steady-state kinetic distribution of the particles which yields a density distribution closely resembling the observed density function at the equator as in Fig. 158.1. We predict an initial f(E, μ) distribution analytically and use the same to do numerical test particle simulations by considering different combinations of perpendicular and parallel energy to arrive at and closely mimic the density profile of the radiation belt and compare the two methods.

A numerical model of the radiation belt established can help study the effect of disturbances in the belt due to whistlers, ULF waves, solar activities, seismo-electric activities, etc. and resulting particle precipitations from the belt to make predictions for IITM nano-satellite mission [3].

158.2 Analytical Model

The density distribution can be found out by integrating the phase space distribution function of the particles over the velocity space. This is written in a polar coordinate as follows and integrated over the entire ϕ direction from 0 to 2π.

$$n = \int\int\int f(v)v^2 \sin\theta \, dv \, d\theta \, d\phi, \qquad n = 2\pi \int\int f(v)v^2 \sin\theta \, dv \, d\theta \quad (158.2)$$

The above equation can be next converted to a function of E and μ by replacing (158.2) with (158.3) to obtain (158.4–158.6).

$$E = \frac{1}{2}mv^2, dE = mvdv, \quad v_\perp = v\sin\theta, dv_\perp = v\cos\theta d\theta, \quad \mu = \frac{mv_\perp^2}{2B}, d\mu = \frac{mv_\perp dv_\perp}{B}$$
$$(158.3)$$

$$n = \frac{2\pi}{m} \int \int f(E, \mu) dE \, v \sin\theta d\theta \tag{158.4}$$

$$n = \frac{2\pi}{m} \int \int f(E, \mu) dE \frac{v_\perp}{v \cos\theta} dv_\perp, \quad n = \frac{2\pi}{m} \int \int f(E, \mu) dE \frac{v_\perp}{v_\parallel} dv_\perp \tag{158.5}$$

$$n(s) = \frac{2\pi B(s)}{m^2} \int_0^\infty \int_0^{\frac{E}{B}} \frac{f(E, \mu) dE d\mu}{\sqrt{E - \mu B(s)}} \tag{158.6}$$

The plasma in the Van Allen belts is collisionless and hence chosen to have a near Maxwellian distribution of thermal energy as follows. A simple distribution function f(E, μ) is chosen such that

$$f(E) = E e^{\frac{-E}{KT}}, \quad f(\mu) = \mu e^{-\mu} \tag{158.7}$$

The density profile from the above-predicted distribution is presented in Fig. 158.6b as a comparison with observed density and numerical simulations.

158.3   Numerical Model

The inner radiation belt being located between L shell 1.5–2.5, the geomagnetic field line at L shell 1.5 is considered for this simulation. A dipole model of the earth's magnetic field is considered here which is a first-order approximation of the rather complex true earth's magnetic field and holds good for lower L shells [1]. In a dipole model, the geocentric radius r, the geomagnetic latitude θ considered northwards from the equator and the arc length s along L shell are related as

$$r = L \cos^2\theta, \quad ds^2 = dr^2 + (rd\theta)^2, \quad d\theta = \frac{ds}{L\sqrt{\sin^2 2\theta + \cos^4\theta}} \tag{158.8}$$

A polar plot of L shell 1.5 is presented in Fig. 158.2a using (refeq8). The 's' coordinate system is considered along the L shell with the origin at the equator. In the s coordinates, s_{max} (towards the poles) represents the value corresponding to a radial distance r = h + R_E or an altitude of h = 1000 kms above the surface of the earth (radius = RE) and where the magnetosphere ends. The value of s_{max} is 5044 kms in the s coordinates. The dipole model magnetic field strength in polar coordinates is given below and used in Lorentz force equation to simulate particle trajectories.

$$B(r) = B_0 (\frac{R_E}{r})^3 \sqrt{(1 + 3\sin^2\theta)} \tag{158.9}$$

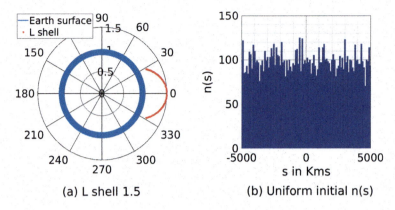

(a) L shell 1.5   (b) Uniform initial n(s)

Fig. 158.2 **a** Shows a polar plot of the lower magnetosphere L shell 1.5 and **b** shows particles distributed uniformly along the L shell

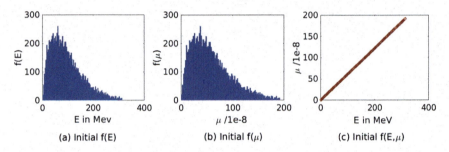

(a) Initial f(E)   (b) Initial f(μ)   (c) Initial f(E,μ)

Fig. 158.3 Initial energy and magnetic moment distribution of 10,000 protons

Particles are distributed uniformly in a position along the 1D s coordinate initially as obtained in Fig. 158.2b using the random number generator algorithm.

This 1D position distribution of the particles is transformed into 2D polar (r, θ) coordinates using (158.8) and is further converted into 3D Cartesian coordinates x, y, z as $x = r \cos \theta$, $y = 0$ and $z = r \sin \theta$.

Next an initial f(E, μ) is chosen. $f(E)$ is chosen as the analytical model. μ is $f(E)/(\alpha B_{max})$ where $E_\perp = \mu B$ is varied as $\alpha = \frac{E}{\mu B}$ for different cases to obtain different density functions. This initial f(E, μ) is presented in Fig. 158.3 for $\alpha = 20$ and is converted into 3D velocity space below.

The initial parallel and perpendicular velocity distribution can be resolved into v_x, v_y and v_z as

$$v_\perp = \frac{2\mu B}{m}, \quad v_x = v_\perp \cos \phi, \, v_y = v_\perp \sin \phi, \, v_z = \sqrt{v^2 - v_x^2 - v_y^2} \quad (158.10)$$

Test particle simulations are run with 10,000 protons for 2 s with the above initial distributions and the trajectory of the particles obey Lorentz force. The Runge–Kutta

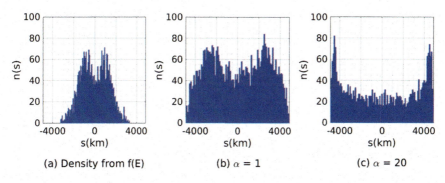

Fig. 158.4 Density distribution for different values of α shows best fit for $\alpha = 20$

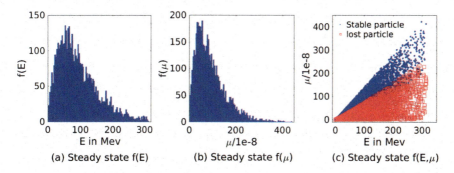

Fig. 158.5 Energy and magnetic moment distribution of 10,000 protons at steady state

4 method is used to solve the ODE. $T_{gyro} = \frac{2\pi m}{qB} = 2000\,\mu s$. So, the timestep of $10\,\mu s$ is chosen which satisfies Nyquist criteria and captures the particle trajectory accurately. The particles have an average bounce period of 0.2 s [1] and hence attain the steady state in Fig. 158.6a in a very short time as seen in n(t). Many particles are lost due to a smaller initial pitch angle and the rest of the particles attain steady state.

The density profile, n(s), is presented in Fig. 158.4 for different values of $\alpha = 1$ and 20 and also if density is obtained only from f(E) without varying f(E, μ). This shows a best fit to observed density for $\alpha = 20$.

The E-μ distribution at the steady state is presented in Fig. 158.5 for $\alpha = 20$. The lost particles (in loss cone) clearly separate out from the trapped particles in an orbit in Fig. 158.5c when μ is observed at their bounce points.

158.4   Conclusion

Figure 158.6b compares the observed, analytically predicted and numerically obtained densities.

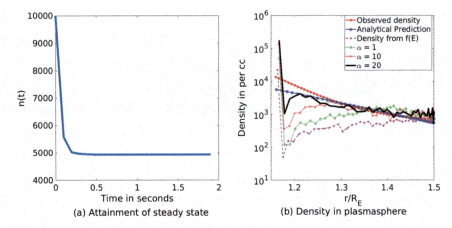

Fig. 158.6 **a** Shows the density profiles obtained for different values of α and **b** shows the attainment of steady state of particles through n(t)

An $\alpha = 20$ represents the best observed density. The analytical method gives approximate particle guiding centre trajectory whereas the numerical simulation takes care of true gyromotion. Hence, the loss cone could be properly studied using the current numerical simulations.

References

1. W. Hess, Energetic particles in the inner Van Allen belt. Space Sci. Rev. **1**(2), 278–312 (1962)
2. R.L. Lysak, Magnetosphere-ionosphere coupling by Alfven waves at midlatitudes. J. Geophys. Res. **109**, A07201 (2004)
3. N. Sivadas et al., A nano-satellite mission to study charge particle precipitation from the Van Allen radiation belts caused due to seismo-electromagnetic emissions (2014)

Chapter 159
Implications of Flux-Tube Formulation of Dual QCD for Thermalization, Quark Pair Creation and QGP

Deependra Singh Rawat, H. C. Chandola, Dinesh Yadav, and H. C. Pandey

Abstract Keeping in view of our earlier study, the multiflux-tube structure of topologically excited magnetically condensed dual QCD vacuum has been extended in the thermal domain to analyze the possible mechanism behind the QCD phase transition. The stability and annihilation of color flux-tubes in the intermediate energy regime has been investigated for QGP formation by discussing the flux-tube interaction process under varying temperature conditions at different hadronic boundaries by computing critical parameters of phase transition and condensed monopole density. Their implications in quark pair creation relevant to elaborate the phase structure of QCD and gain insight into the mechanism of the parton-hadron phase transition has also been discussed.

The vacuum dynamics of QCD especially the color confinement and its implications to explore the underlying phase structure of QCD still remains the subject of crucial investigation in theoretical high energy physics. To resolve the problem of color confining nature of QCD vacuum, people like Nambu, 't Hooft and Mandelstem [1] have argued that color electric flux confinement in gauge theory can be analogous to magnetic flux confinement in a superconductor which is realized as a result of Cooper pair condensation. In this scenario of the dual version of QCD, the monopoles are supposed to be condensed as a result of the dual Meissner effect that squeezed the color flux in the form of thin flux-tubes that paves the clear signal of color confinement in QCD. However, the appearance of the magnetically charged object in the QCD

D. S. Rawat (✉) · H. C. Chandola · D. Yadav
Department of Physics (UGC-Centre of Advanced Study), Kumaun University, Nainital, India
e-mail: dsrawatphysics@kunainital.ac.in

H. C. Chandola
e-mail: chandolahc@kunainital.ac.in

D. Yadav
e-mail: dinoyadav02@rediffmail.com

H. C. Pandey
Department of Physics, Birla Institute of Applied Sciences, Bhimtal, India
e-mail: hempandey@birlainstitute.co.in

© Springer Nature Singapore Pte Ltd. 2021
P. K. Behera et al. (eds.), *XXIII DAE High Energy Physics Symposium*,
Springer Proceedings in Physics 261,
https://doi.org/10.1007/978-981-33-4408-2_159

is a great huddle in its further development. There are several attempts made in this direction among which magnetic symmetry-based dual QCD [2–4] has been used here as an effective conjecture to explain the confining structure of colored quarks. In this dual formulation, the tube-like structures arise (which are capped by colored quarks at their ends) as a result of monopole (confining agents) condensation with the appearance of the mass mode of vector and scalar glueballs. This novel picture of confinement is expected to provide a further understanding of the dynamical structure of QCD vacuum and an ordered way to explore the complex phase structure of QCD, dynamics of phase transition with the possibility of QGP [5, 6]-like intermediate state under the extreme conditions of temperature and density. The purpose of this article is to explore the flux-tube structure by using the quartic potential which is an effective potential reliable in relatively weak coupling and is an appropriate choice to deal with the dynamics of phase transition. The numerical estimation of the critical parameters of phase transition, i.e. the critical radius, critical density and condensed monopole density along with their analytic behavior, is discussed for the optimal value ($\alpha_s = 0.12$) of the coupling constant. Its implications on thermalization, quark pair creation and QGP are also discussed.

In the (4+n)-dimensional unified space, the imposition of magnetic symmetry [7] as an additional internal isometry provides a gauge invariant investigation of topological character that develops the confining structure in the theory by bringing duality at the level of potential and field strength. The resulting dual Lagrangian in quenched approximation may then be expressed as [2–4],

$$\mathcal{L}_m^{(d)} = -\frac{1}{4} B_{\mu\nu}^2 + |[\partial_\mu + i\frac{4\pi}{g} B_\mu^{(d)}]\phi|^2 - 3\lambda\alpha_s^{-2}(\phi^*\phi - \phi_0^2)^2. \tag{159.1}$$

The effective potential introduced here generates the dynamical breaking of magnetic symmetry which leads to the magnetic condensation by ensuing the dual Meissner effect and establishes the color confinement in QCD. The field equation associated with the above Lagrangian may then be derived in the following form:

$$\mathcal{D}^\mu D_\mu\phi + 6\lambda\alpha_s^{-2}(\phi^*\phi - \phi_0^2)\phi = 0, \text{ and } \partial^\nu B_{\mu\nu} - i\frac{4\pi}{g}(\phi^* \overset{\leftrightarrow}{\partial}_\mu \phi) - 8\pi\alpha_s^{-1} B_\mu^{(d)}\phi\phi^* = 0 \tag{159.2}$$

which closely resembles with the Nielsen and Olesen [8] vortex-like solutions, and indicates a clear signal of the flux-tube configuration inside the QCD vacuum. We focus on the single flux-tube solution using the cylindrical symmetry and the longitudinal orientation of flux-tube, for which the dual gauge field and the monopole field are $\mathbf{B}_\mu^{(d)} = g^{-1}\cos\alpha(\partial_\mu\beta)\hat{m}$ and $\phi(x) = \exp(in\varphi)\chi(\rho)$, $(n = 0, \pm1, ...)$, respectively. With the uniqueness of the function $\phi(x)$ and using the Nelson–Olesen ansatz [8] for the dual gauge field and monopole field in static limit we have, $\mathbf{B}^{(d)}(\rho) = -\hat{\phi}B(\rho)$ and $B_t^{(d)} = B_\rho^{(d)} = B_z^{(d)} = 0$. This leads to the color electric field in the cylindrically symmetric configuration and is given by [2]

$$E_m(\rho) = -\frac{1}{\rho}\frac{d}{d\rho}(\rho B(\rho)). \tag{159.3}$$

The associated field equations governing the flux-tube structure may then be expressed in the following form:

$$\frac{1}{\rho}\frac{d}{d\rho}\left(\rho\frac{d\chi}{d\rho}\right) - \left[\left(\frac{n}{\rho} + (4\pi\alpha_s^{-1})^{\frac{1}{2}}B(\rho)\right)^2 - 6\lambda\alpha_s^{-2}(\chi^2 - \phi_0^2)\right]\chi(\rho) = 0,$$

(159.4)

$$\frac{d}{d\rho}\left[\rho^{-1}\frac{d}{d\rho}(\rho B(\rho))\right] + 8\pi g^{-1}\left(\frac{n}{\rho} - 4\pi g^{-1}B(\rho)\right)\chi^2(\rho) = 0.$$

(159.5)

In the asymptotic limit $\phi \to \phi_0$ as $\rho \to \infty$, the appropriate solutions for the dual gauge potential that ensures the formation of chromoelectric flux-tubes are as $B(\rho) = -ng(4\pi\rho)^{-1}[1 + F(\rho)]$ and $F(\rho) \xrightarrow{\rho \to \infty} C\rho^{\frac{1}{2}}\exp(-m_B\rho)$ where C is a constant and $m_B (= 4\pi g^{-1}\sqrt{2}\phi_0)$ is the vector glueball mass that determines the magnitude of the dual Meissner effect. The color flux-tube formation and the confinement of color isocharges can be visualized more effectively on the energetic grounds by investigating the energy per unit length of the flux-tube (string tension) [2–4] and introducing a new variable R as $\rho = R\sin\theta$ with energy minimization condition ($\varepsilon_C = \varepsilon_D$) that leads to the critical radius and critical density [3] of phase transition as $R_c = \left(\frac{8}{3}\pi n^2\alpha_s\right)^{\frac{1}{4}}m_B^{-1}$ and $d_c = \frac{1}{2\pi R_c^2} = \left(\frac{32}{3}\pi^3 n^2\alpha_s\right)^{-\frac{1}{2}}m_B^2$, respectively. For the optimal value of coupling constant (α_s) in the near infrared sector of QCD as $\alpha_s = 0.12$ with the vector glueball mass $m_B = 2.102$ GeV lead to, $R_c = 0.094$ fm and $d_c = 18.003$ fm^{-2}. In this case, below 0.094 fm, the system shifted toward a medium of unbound color charges and an entirely different mechanism is expected to take over. The flux-tube density in this sector increases rapidly and with a sufficiently dense flux-tube system, the flux-tube annihilation as a result of their interaction may take place which then leads to the generation of dynamical pairs of quarks and gluons. The gluon self-interactions are then expected to contribute a major role in the thermal domain of QCD matter and create a color conducting plasma, the so-called quark-gluon plasma in the intermediate energy regime. As a result of such flux-tube melting in the high momentum transfer sector of the QCD vacuum, the system is expected to evolve with an intermediate QGP phase. The variation of the associated flux-tube energy for $\alpha_s = 0.12$ is depicted in Fig. 159.1 and exhibit a close agreement with that obtained analytically. Under thermalization conditions using partition function approach formalism along with the mean-field treatment [4] where the associated thermodynamical potential transform the vector mass mode of the magnetically condensed QCD vacuum as $m_B^{(T)} = m_B\sqrt{[1 - (T/T_c)^2]}$, where $T_c = 2\phi_0\sqrt{3/(4\pi\alpha_s + 1)}$. It further modify the associated critical parameters of phase transition as $R_c(T) = R_c/\sqrt{[1 - (T/T_c)^2]}$ and $d_c(T) = d_c[1 - (T/T_c)^2]$. The QCD monopole of phase transition density participating in the vacuum condensation may be derived using the field equation for potential $B_\mu^{(d)}$ as $n_m(\phi) = (\alpha_s/8\pi)m_B^2[1 - (T/T_c)^2]$. The variation of condensed monopole density for the optimal coupling $\alpha_s = 0.12$ with temperature is given by Fig. 159.1 that clearly demonstrate a gradual decrease in monopole

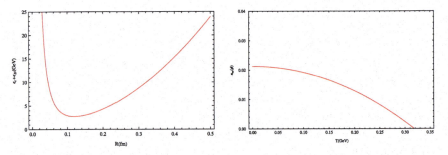

Fig. 159.1 (Color online) Variation of flux-tube energy ($\varepsilon_C + \varepsilon_D$) with R (left) and variation of condensed monopole density ($n_m(\phi)$) with T for $\alpha_s = 0.12$

density, and at a transition point it vanishes out. Such behavior of $n_m(\phi)$ is also supported by recent lattice results [6].

Thus, the thermal evolution of stable flux-tube structure has shown a smooth phase transition of hadronic matter to a full deconfined state of partons due to the melting/annihilation of flux-tubes with the dynamical quark pair creation in the intermediate energy region that may correspond to quark-gluon plasma of point-like colored quarks and gluons as its constituents. The study of critical radius and critical density have also shown the QCD system will acquire a strongly coupled confining phase at its low energy domain and appear in the deconfined state at a high energy regime. The thermal variation of monopole density leads to the gradual evaporation of condensed monopole that will be transformed into thermal monopoles during the QCD phase transition. These thermal monopoles are supposed to be responsible for the possible formation of QGP before acquiring its complete deconfined state. The dynamical quark pair creation in the intermediate state has its intimate connection with the confining structure of QCD vacuum as the creation of $q\bar{q}$-pair screens the confinement potential at a large distance scale that will be dealt in our forthcoming communications [9].

Acknowledgements One of the authors (DSR) is thankful to University Grants Commission, New Delhi, India for the financial assistance under its BSR (RFSMS) scheme. The authors (DSR and HCC) are also thankful to the organizers of XXIII DAE-BRNS (High Energy Physics) Symposium 2018, I.I.T. Madras for their hospitality during the course of the symposium.

References

1. Y. Nambu, Phys. Rev. D **10**, 4262 (1974); S. Mandelstam, Phys. Rep. **23C**, 145 (1976); G. 't Hooft, Proc. Eur. Phys. Soc. 1225 (1975)
2. H.C. Pandey, H.C. Chandola, Phys. Lett. B **476**, 193 (2000)
3. H.C. Chandola, D. Yadav, Nucl. Phys. A **829**, 151 (2009)

4. D.S. Rawat, H.C. Chandola, H.C. Pandey, D. Yadav, Springer Proc. Phys. **203**, 625 (2018); H.C. Chandola, D.S. Rawat, H.C. Pandey, D. Yadav, H. Dehnen. Adv. High Energy Phys. 4240512 (2020)
5. E.V. Shuryak, Nucl. Phys. A **750**, 64 (2005)
6. J. Liao, E.V. Shuryak, Phys. Rev. C **75**, 054907 (2007)
7. Y.M. Cho, Phys. Rev. **D21**, 1080 (1980); ibid 2415 (1981); Y.M. Cho, F.H. Cho, J.H. Yoon, Phys. Rev. **D87**, 085025 (2013)
8. H. Neilsen, B. Olesen, Nucl. Phys. B **61**, 45 (1973)
9. D.S. Rawat, H.C. Chandola (to be published)

Chapter 160
CP Phase Analysis Using Quark-Lepton Complementarity Model in 3 + 1 Scenario

Gazal Sharma, B. C. Chauhan, and Surender Verma

Abstract The existence of sterile neutrino is revolutionizing Physics from the smallest to the largest scale. After the recent reports from the MiniBooNE experiment at Fermi lab that observed far more ν_e appearance data than expected, suggesting the possible existence of the fourth generation of another generation of neutrinos. These results would provide challenges for the Standard Model of Particle Physics if it is confirmed in future experiments and will have imperative implications on cosmology and astroparticle physics. Also, this will require new neutrino mass models to accommodate these degrees of freedom. With respect to that, the current work is just the extension of our recent work toward the CP phase analysis of Quark-Lepton Complementarity (QLC) model in 3 + 1 scenario. The parametrization of non-trivial correlation between CKM_4 and $PMNS_4$ using Monte Carlo Simulation is used to estimate the texture of V_{m_4}. We have predicted the numerical ranges for sterile neutrino parameters and also investigated the values for Dirac CP violation phases and the CP re-phasing invariants using the model in 3 + 1 scenario. The consequences of the model are the predictions for CP-Violating phase invariant J, Dirac phase ϕ, and the sterile neutrino parameters. The results obtained numerically and analytically this paper stood in good agreement with the experimental data. The results of this work would be very important in view of future sterile neutrino experiments.

G. Sharma (✉) · B. C. Chauhan · S. Verma
Department of Physics and Astronomical Science, School of Physical and Material Sciences,
Central University of Himachal Pradesh (CUHP), Dharamshala, Kangra 176215, HP, India
e-mail: gazzal.sharma555@gmail.com

B. C. Chauhan
e-mail: chauhan@associates.iucaa.in

S. Verma
e-mail: s_7verma@yahoo.co.in

© Springer Nature Singapore Pte Ltd. 2021
P. K. Behera et al. (eds.), *XXIII DAE High Energy Physics Symposium*,
Springer Proceedings in Physics 261,
https://doi.org/10.1007/978-981-33-4408-2_160

160.1 Introduction

The results of neutrino oscillation experiments in the past several years have provided us a very strong sign about the massive nature of neutrinos as well as the mixing properties of various lepton flavors and their oscillations. Many ongoing experiments; LSND, MiniBooNE, MINOS, Daya Bay, IceCube, etc., are working on the investigation of the existence of sterile neutrinos at different mass scales. Quarks and Leptons being the fundamental particles of matter and Standard Model, might result in some symmetry at high-energy scale resulting in the complementarity (Quark-Lepton Complementarity) between the two of them.

The Quark-Lepton Complementarity (QLC) and its possible consequences have been widely studied and investigated in the literature. The simple correspondence between the U_{PMNS} and U_{CKM} matrices has been proposed and used by several authors and analyzed in terms of a correlation matrix V_c in particular. According to our analysis a very clear non-trivial structure of V_c and the strong indication of gauge coupling unification at a high scale triggered us to put constraints on Dirac CP violation phase and CP violation re-phasing invariant J.

The major motivation behind this work is testing our model in the fourth-generation scenario. After the successful results obtained in our previous papers, we have tried to extend our model and did our analysis in $3 + 1$ scenario. Also, one another factor that pushed us toward the extension of our model in $3 + 1$ scenario is that the results obtained in our previous works [1–6] (*and references therein*) are quite consistent with the recent results from NovA and IceCube which gives us a new ray of hopes in favour of our model and its stability. According to our investigations, there is a possibility of the existence and role of sterile neutrinos in the Quark lepton complementarity, which helped us to give some constrained results for two sterile neutrino mixing angles, i.e., θ_{24} and θ_{34}. Along with that also predicted the values for Dirac CP violation phase and the CP re-phasing invariants using the model in $3 + 1$ scenario.

In Sect. (160.2), we discuss in brief the theory of the QLC model along with that we describe in brief the theoretical framework of the QLC model in $3 + 1$ scenario. For the generation of 4×4 V_{m_4} matrix standard parametrizations were taken for U_{CKM_4} and U_{PMNS_4} which is included in Sect. (160.3) According to the model procedure, after using the most credible texture of the correlation matrix we derive the constraints on the θ_{24}^{PMNS} and θ_{34}^{PMNS} mixing angle along with the Φ (Dirac CP Violation phase) and J (Jarlskog Invariant) in the Sect. (160.4). Finally, in Sect. (160.5) we conclude and summarize our results.

160.2 Phenomenology of QLC Model in 3 + 1 Scenario

When we observe a pattern of mixing angles of quarks and leptons and combine them with the pursuit for unification, i.e., symmetry at some high energy leads the concept of quark lepton complementarity, i.e., QLC. Possible consequences of QLC have been widely investigated in the literature and in particular, a simple correspondence between the PMNS and CKM matrices has been proposed and analyzed in terms of a correlation matrix V_m. The generalized relation between them is

$$V_c = U_{CKM} \cdot U_{PMNS} \quad \text{to} \quad V_{m_4} = U_{CKM_4} \cdot \psi_4 \cdot U_{PMNS_4}$$

where V_c is the correlation matrix defined as a product of U_{PMNS} and U_{CKM}.

When sterile neutrinos are introduced in the $3 + N_s$ schemes, where the N_s is the number of new mass eigenstates along with ψ_4 in QLC the model equation the above equation takes the form

$$V_c = U_{CKM_4} \cdot \psi_4 \cdot U_{PMNS_4}$$

where the quantity ψ_4 is a diagonal matrix $\psi_4 = diag(e^{(\psi_i)})$ and the four phases ψ_i are free parameters as they are not restricted by present experimental evidence.

160.3 Structure of CKM_4 and $PMNS_4$

In order to calculate the structure of V_{m_4} we have used the U_{CKM_4} and U_{PMNS_4} taking reference from several works [7–10]. The CKM matrix in SM is a 3×3 unitary matrix while in the SM_4 (this is just a simple extension of the SM, and none of its essential features are disturbed. It obeys all the SM symmetries and does not introduce or include any new ones), the U_{CKM_4} matrix is 4×4, matrix can be written as

$$U_{CKM_4} = \begin{bmatrix} \tilde{V}_{ud} & \tilde{V}_{us} & \tilde{V}_{ub} & \tilde{V}_{ub'} \\ \tilde{V}_{cd} & \tilde{V}_{cs} & \tilde{V}_{cb} & \tilde{V}_{cb'} \\ \tilde{V}_{td} & \tilde{V}_{ts} & \tilde{V}_{tb} & \tilde{V}_{tb'} \\ \tilde{V}_{t'd} & \tilde{V}_{t's} & \tilde{V}_{t'b} & \tilde{V}_{t'b'} \end{bmatrix},$$

where all the elements of the matrix have their usual meanings except for b' and t', which we have already defined above.

The Dighe–Kim (DK) parametrization defines [7–10]

$\tilde{V}_{ud} = 1 - \frac{\lambda^2}{2}, \tilde{V}_{us} = \lambda, \tilde{V}_{ub} = A\lambda^3 C e^{i\delta_{ub}},$

$\tilde{V}_{ub'} = p\lambda^3 e^{-i\delta_{ub'}}, \tilde{V}_{cd} = -\lambda, \tilde{V}_{cs} = 1 - \frac{\lambda^2}{2},$

$\tilde{V}_{cb} = A\lambda, \tilde{V}_{cb'} = q\lambda^2 e^{-i\delta_{cb'}},$

$\tilde{V}_{td} = A\lambda^3 (1 - C e^{i\delta_{ub}}) + r\lambda^4 (q e^{-i\delta_{cb'}} - p e^{-i\delta_{ub'}}),$

$\tilde{V}_{ts} = -A\lambda^2 - qr\lambda^3 e^{-i\delta_{cb'}} + \frac{A}{2}\lambda^4 (1 + r^2 C e^{i\delta_{ub}}), \tilde{V}_{tb} = 1 - \frac{r^2\lambda^2}{2},$

$\tilde{V}_{tb'} = r\lambda, \tilde{V}_{t'd} = \lambda^3 (q e_{i\delta_{cb'}}) + Ar\lambda^4 (1 + C e^{i\delta_{ub}}),$

$\tilde{V}_{t's} = q\lambda^2 e^{-i\delta_{ub'}} + Ar\lambda^3 + \lambda^4 (-p e^{-i\delta_{ub'}} + \frac{q}{2} e_{i\delta_{cb'}} + \frac{qr^2}{2} e_{i\delta_{cb'}}),$

$\tilde{V}_{t'b} = -r\lambda \quad and \quad \tilde{V}_{t'b'} = 1 - \frac{r^2\lambda^2}{2},$

In this work, for the calculation of U_{PMNS_4} we consider the simplest $3 + 1$ scheme. In the presence of the sterile neutrino ν_s, the flavor ($\nu_\alpha, \alpha = e, \mu, \tau, s$) and the mass

Table 160.1 The limits obtained on sterile mixing angles

Parameters (GeV)	$\theta_{24}^{PMNS_4}$	$\theta_{34}^{PMNS_4}$
For $m_{t'} = 400$	$6.57° - 23.36°$	$1.53° - 31.59°$
For $m_{t'} = 600$	$6.87° - 23.15°$	$3.78° - 32.40°$
Parameters (GeV)	$\mid U_{\mu 4} \mid^2$	$\mid U_{\tau 4} \mid^2$
For $m_{t'} = 400$	$0.0003 - 0.0300$	$0.00 - 0.2031$
For $m_{t'} = 600$	$0.0001 - 0.0236$	$0.00 - 0.1432$

eigenstates (ν_i, $i = 1, 2, 3, 4$) are connected through a 4×4 unitary mixing matrix U, which depends on six complex parameters [11]. A particularly convenient choice of the parametrization of the mixing matrix is

$$U = \tilde{R}_{34} R_{24} \tilde{R}_{14} R_{23} \tilde{R}_{13} R_{12},$$

160.4 Results

In this work, a non-trivial correlation between CKM_4 and $PMNS_4$ mixing matrices is obtained by taking into account the phase mismatch between quark and lepton sectors as ψ_4, a diagonal matrix $\psi_4 = diag(e^{(\psi i)})$. The U_{CKM_4} matrix can be described using Dighe-Kim (DK) parametrization, where all the elements of U_{CKM} are unitary up to $\mathcal{O}(\lambda^4)$.

- The value of CP violation phases ϕ have been kept open varying freely between $(0 - 2\pi)$ and the reference values for θ_{24}, θ_{34}, $\mid U_{\mu 4} \mid^2$ and $\mid U_{\tau 4} \mid^2$ are assume to vary freely between $(0 - \pi/2)$.

  For the unknown phases ϕ and the four ψ_i, as they are not constrained by any experimental data, we vary their values between the interval $[0, 2\pi]$ in a flat distribution.

- After performing the Monte Carlo simulations we estimated the texture of the correlation matrix (V_{m_4}) for two different values of $m_{t'} = 400$ GeV & 600 GeV (where $m_{t'}$ is the mass of t') and implemented the same matrix in our inverse equation and obtained the constrained results for the sterile neutrino parameters. We obtained predictions for CP Violating re-phasing invariants and θ_{24} and θ_{34} and then compared our results with the current experimental bounds given by NoνA, MINOS, SuperK and IceCube-DeepCore [12–16] experiments.

  As we have divided our results in two parts, i.e., for $m_{t'} = 400$ GeV and $m_{t'} = 600$ GeV. The table below shows the comparison of upper limits obtained above via model with the four different experimental results (Tables 160.1 and 160.2).

Table 160.2 The upper limits of sterile mixing parameters obtained from model (QLC) and from NOνA, MINOS, Super-Kamiokande and IceCube- DeepCore

Experiment	θ_{24}^{PMNS4}	θ_{34}^{PMNS4}	$\mid U_{\mu4}\mid^2$	$\mid U_{\tau4}\mid^2$
NoνA	20.8	31.2	0.126	0.268
MINOS	7.3	26.6	0.016	0.20
SuperK	11.7	25.1	0.041	0.18
IceCube-DeepCore	19.4	22.8	0.11	0.15
QLC model	θ_{24}^{PMNS4}	θ_{34}^{PMNS4}	$\mid U_{\mu4}\mid^2$	$\mid U_{\tau4}\mid^2$
QLC (400 GeV)	23.36	31.59	0.030	0.203
QLC (600 GeV)	23.15	32.40	0.024	0.143

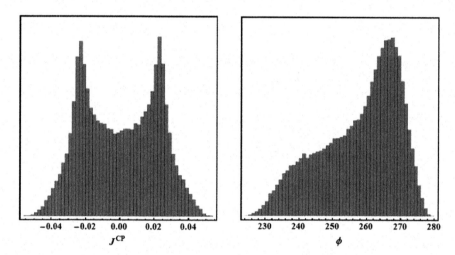

Fig. 160.1 Histograms of Jarlskog re-phasing invariant (J) and Dirac CP phase (ϕ)

160.5   Conclusion

Monte Carlo technique was used to conclude that the absolute values of the CP-violating invariants J and CP violation phase (Dirac Phase) is obtained as under (Fig. 160.1)

$$\mid J \mid = 0.001 - 0.055$$

$$\mid J \mid \text{ (Best fit) } = 0.021$$

$$\phi = 224.64° - 279.85°$$

$$\phi \text{ (Best fit) } = 267.62°$$

In the above Fig. (160.1), we have shown two different histograms for a varying range of Jarlskog invariant J and Dirac CP phase ϕ. From the plot on the left panel, one can clearly depict that the maximum number of J values lies near the values ~ -0.022 and 0.024, which is precisely our best fit values and is comparable to the recent particle data group value range [17]. While the plot on the right panel implies that the maximum number of values are gathered around $\phi = 267.62°$ which is our best value and lies very close to the bounds given by global data analysis [17].

Acknowledgements B. C. Chauhan acknowledges the financial support provided by the University Grants Commission (UGC), Government of India vide Grant No. UGC MRP-MAJOR-PHYS-2013-12281. We thank IUCAA for providing research facilities during the completion of this work.

References

1. M. Picariello, B.C. Chauhan, J. Pulido, E. Torrente-Lujan, Predictions from non trivial Quark-Lepton complementarity. Mod. Phys. Lett. A **22**, 5860 (2008)
2. B.C. Chauhan, M. Picariello, J. Pulido, E. Torrente Lujan Quark-lepton complementarity, neutrino and standard model data predict $\theta_{13} = 9^{+1}_{-2}°$. Euro. Phys. J. C **50**, 573 (2007)
3. G. Sharma, B.C. Chauhan, Quark-lepton complementarity predictions for θ_{23}^{pmns} and CP violation. JHEP **1607**, 075 (2016). [references contained therein]
4. G. Sharma et al., Quark-lepton complementarity model based predictions for θ_{23}^{PMNS} with neutrino mass hierarchy. Springer Proc. Phys. **203**, 251–256 (2018)
5. G. Sharma, B.C. Chauhan, Investigating the sterile neutrino parameters with QLC in 3 + 1 scenario. arXiv:1807.05785 [hep-ph]
6. G. Sharma, B.C. Chauhan, CP-violation phase analysis via non-trivial correlation of quarks and leptons in 3+1 scenario. arXiv:1904.02040 [hep-ph]
7. C.S. Kim, A.S. Dighe, Tree FCNC and non-unitarity of CKM matrix. Int. J. Mod. Phys. E **16**, 1445 (2007)
8. A.K. Alok, A. Dighe, S. Ray, CP asymmetry in the decays $B \rightarrow (X_s, X_d)\mu + \mu-$ with four generations. Phys. Rev. D **79**, 034017 (2009)
9. A.K. Alok et al., Phys. Rev. D **83**, 073008 (2011)
10. N. Klop, A. Palazzo, Imprints of CP violation induced by sterile neutrinos in T2K data. Phys. Rev. D **91**(7), 073017 (2015)
11. J. Schechter, J.W.F. Valle, Neutrino masses in $SU(2) \times U(1)$ theories. Phys. Rev. D **22**, 2227 (1980)
12. P. Adamson et al., [NoνA Collaboration], Search for active-sterile neutrino mixing using neutral-current interactions in NoνA. Phys. Rev. D **96**(7), 072006 (2017). FERMILAB-PUB-17-198-ND
13. E. Smith (for the collaboration), [NoνA Collaboration], Results from the NoνA experiment. Int. J. Mod. Phys. Conf. Ser. **46**, 1860038 (2018)
14. P. Adamson et al., [MINOS Collaboration], Search for sterile neutrinos mixing with muon neutrinos in MINOS. Phys. Rev. Lett. **117**, 151803 (2016)
15. K. Abe et al., [Super-Kamiokande Collaboration], Limits on sterile neutrino mixing using atmospheric neutrinos in Super-Kamiokande. Phys. Rev. D **91**, 052019 (2015)
16. M.G. Aartsen et al., [IceCube Collaboration], Search for sterile neutrino mixing using three years of IceCube DeepCore data. Phys. Rev. D **95**(11), 112002 (2017)
17. M. Tanabashi et al., Review of particle physics—particle data group. Phys. Rev. D **98**(3), 030001 (2018)

Chapter 161
Measurement of Higher Moments of Net-particle Distributions in STAR

Debasish Mallick

Abstract Studying fluctuations of conserved quantities, such as baryon number (B), strangeness (S) and electric charge (Q), provides insights into the bulk properties of matter created in high-energy nuclear collisions. The higher moments of multiplicity distributions of net proton, net kaon and net charge are expected to show large fluctuations near the QCD critical point. We present results on energy and centrality dependence of higher moments of net proton, net kaon and net charge multiplicity distribution over all BES-I energies from the STAR experiment. The net proton moments product ($S\sigma$ and $\kappa\sigma^2$) shows deviations from Poisson baseline (unity) in the lower beam energy region (below $\sqrt{s_{NN}} = 27$ GeV). In the most central (0–5%) collisions, the $\kappa\sigma^2$ of net proton distribution as a function of collision energy exhibit a non-monotonic behaviour and show deviation in the lower energy region from corresponding predictions from Poisson baseline, a transport model (UrQMD) and a thermal model, all of which do not include any physics of criticality.

161.1 Introduction

High-energy heavy-ion collision experiments primarily aim to study the properties of nuclear matter subjected to extreme conditions such as temperature and/or pressure and understand the nature of the phase transitions. The phase structure of this matter can be illustrated by the Quantum Chromodynamics (QCD) phase diagram characterized by temperature (T) and baryonic chemical potential (μ_B). Lattice QCD calculations show that quark-hadron phase transition at zero μ_B is a crossover [1, 2], while QCD-based models predict a first-order phase transition [3, 4] and the existence of the QCD critical point (QCP) at a large value of μ_B. By varying the centre of mass energy ($\sqrt{s_{NN}}$) of the colliding nuclei, various parts of the QCD phase diagram can be accessed in the experiments [5, 6]. Fluctuations of conserved charges such as net charge (Q), net strangeness (S) and net baryon (B) can serve as probes

D. Mallick (✉)
National Institute of Science Education and Research, HBNI, Jatni, India
e-mail: debasish.mallick@niser.ac.in

© Springer Nature Singapore Pte Ltd. 2021   1093
P. K. Behera et al. (eds.), *XXIII DAE High Energy Physics Symposium*,
Springer Proceedings in Physics 261,
https://doi.org/10.1007/978-981-33-4408-2_161

of the QCD phase transition and the critical point [7]. Moments of the conserved charge distributions are related to the correlation lengths of the system, ($\langle(\delta N)^2\rangle \sim \xi^2$, $\langle(\delta N)^3\rangle \sim \xi^{4.5}$ and $\langle(\delta N)^4\rangle \sim \xi^7$), which are expected to diverge near the critical point [8]. Moments are also related to susceptibilities of conserved charges calculated in lattice QCD and thermal models [9, 10]. In heavy-ion collisions, finite time and finite system size effects restrict the growth of the correlation length resulting in the finite enhancement of higher moments [11].

In this article, we report the results from phase I of the Beam Energy Scan (BES) program at RHIC, on the ratios of cumulants of net charge ($0.2 < p_T < 2$ GeV/c), net kaon ($0.2 < p_T < 1.6$ GeV/c) and net proton ($0.4 < p_T < 2$ GeV/c) distributions as a function of collision energy ($\sqrt{s_{NN}} = 7.7$–200 GeV), measured at midrapidity in Au+Au collisions in the STAR experiment [12–15]. Event-by-event net-particle distribution is found by taking the algebraic sum of the quantum numbers of positively charged particles and corresponding antiparticles. Cumulants up to the fourth order of the distribution are calculated using the relations: $C_1 = \langle N \rangle$, $C_2 = \langle(\delta N)^2\rangle$, $C_3 = \langle(\delta N)^3\rangle$ and $C_4 = \langle(\delta N)^4\rangle - 3C_2^2$, where $\delta N = N - \langle N \rangle$ and $\langle N \rangle$ is the mean of the distribution. To avoid the autocorrelation effect, the particles used in cumulant analysis are excluded from the centrality definition. To suppress the background effect such as initial volume fluctuations, the cumulants are calculated in multiplicity bins of unit width and then corrected using a method known as Centrality Bin-Width Correction (CBWC) [16]. Cumulants are corrected for the finite detector efficiency using a Binomial model [17]. Statistical errors are obtained using the Delta theorem [18] and Bootstrap [19] methods.

161.2   Results and Discussion

Figure 161.1 shows the raw (uncorrected for CBW and efficiency) multiplicity distributions for $\sqrt{s_{NN}} = 14.5$ GeV [14]. The mean values of the net-particle show an increasing trend from peripheral to central collisions, indicating the production of a higher number of positively charged particles than negatively charged particles in central collisions. Similarly, the standard deviation (σ) of the distribution also increases from peripheral to central collisions. At a given collision energy and centrality, net charge distribution has a larger value of σ compared to net kaon and net proton distributions which results in larger statistical uncertainties on net charge cumulants.

Figure 161.2 shows the energy dependence of cumulant ratios ($C_2/C_1 = \sigma^2/M$, $C_3/C_2 = S\sigma$, $C_4/C_2 = \kappa\sigma^2$) of net charge, net kaon and net proton distributions at midrapidity in Au+Au collisions measured by the STAR experiment. The $S\sigma$ for net proton and net charge are normalized with the corresponding Skellam expectations [20]. The cumulants used to construct moment product are corrected for CBW effect and detector efficiency. The σ^2/M for all the charges shows a monotonically increasing trend as a function of collision energy, $\sqrt{s_{NN}}$. However, the $\kappa\sigma^2$ of net charge and net kaon show a weak collision energy dependence. Similarly, the $S\sigma$ of

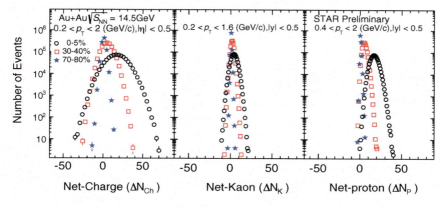

Fig. 161.1 Event-by-event uncorrected multiplicity distributions of net charge (left), net kaon (middle) and net proton (right) for Au+Au collisions at $\sqrt{s_{NN}} = 14.5$ GeV for 0–5% top-central (black circles), 30–40% mid-central (red squares) and 70–80% peripheral collisions (blue stars)

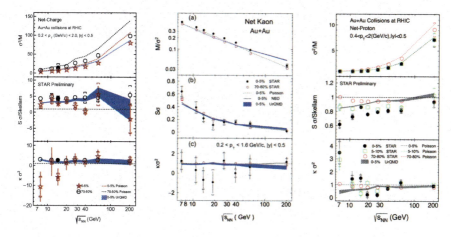

Fig. 161.2 Energy dependence of cumulant ratios of net charge, net kaon and net proton multiplicity distributions for 0–5%, 5–10% (green squares), and 70–80% centralities. The Poisson expectations are denoted as dotted lines and UrQMD calculations are shown as bands. The statistical and systematical error are shown in bars and brackets, respectively [12–14]

net kaon and $S\sigma$/Skellam of net charge weakly vary with energy. Within the large uncertainties, both $S\sigma$/Skellam and $\kappa\sigma^2$ of net kaon and net charge agree with the non-critical baselines such as Poisson expectations and UrQMD model calculations. In 0–5% centrality, the $\kappa\sigma^2$ of net proton shows a non-monotonic behaviour as a function of collision energy. Net proton $\kappa\sigma^2$ measured in 0–5% central collisions is close to the Poisson expectation, unity, for $\sqrt{s_{NN}} \geq 39$ GeV, it shows a dip at 27 and 19.6 GeV and then at 7.7 GeV it shows a large (16 times the value of that measured at 27 GeV) rise above the baselines. The trend of net proton $\kappa\sigma^2$ in 0–5% central collisions is similar to the predictions from QCD-based model calculations

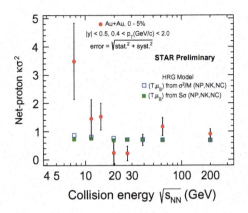

Fig. 161.3 Net proton $\kappa\sigma^2$ in 0–5% central Au+Au collisions measured by STAR experiment shown in red solid markers. The HRG model predictions with freeze-out conditions from σ^2/M (blue markers) and from $S\sigma$ (green markers) are also shown. Uncertainties on experimental data points are statistical and systematic errors added in quadrature

if the system traverses in the vicinity of the critical point [21]. Also, $S\sigma$/Skellam of net proton for central collisions shows a non-monotonic behaviour with collision energy and deviates from non-critical baselines in the low-energy region. The results for peripheral collisions at 70–80% centrality agree with the baselines. Figure 161.3 shows the net proton $\kappa\sigma^2$ in 0–5% centrality compared to the Hadron Resonance Gas (HRG) model calculations. With proper kinematic cuts as used in the experiments, the HRG model [22] calculations are performed using two sets of variables (T and μ_B) obtained from freeze-out of σ^2/M and $S\sigma$. Experimental data points on net proton $\kappa\sigma^2$ show deviations from the HRG model predictions in the low-energy region starting from 27 GeV. The HRG model, which successfully explains the various particle yields, fails to explain the trend for $\kappa\sigma^2$ of net proton which could indicate possible non-thermal contributions. In phase II of the BES program, the STAR experiment with detector upgrades is currently taking data with high statistics which would enable us to make a more precise measurement of higher order cumulants.

161.3 Summary

The $\kappa\sigma^2$ for central net proton distribution in Au+Au collisions at midrapidity shows a non-monotonic variation with collision energy. The $\kappa\sigma^2$ values deviate from the Poisson baseline and UrQMD model calculations. The HRG model, with the assumption of thermal equilibrium, cannot explain the fourth-order moments of net proton. Within the large statistical and systematic uncertainties, both $S\sigma$/Skellam and $\kappa\sigma^2$ of net kaon and net charge agree with Poisson baseline and UrQMD model [23] results.

References

1. Y. Aoki, G. Endrodi, Z. Fodor, S.D. Katz, K.K. Szabo, Nature **443**, 675 (2006)
2. A. Bazavov et al., Phys. Rev. D **85**, 054503 (2012)
3. S. Ejiri, Phys. Rev. D **78**, 074507 (2008)
4. E.S. Bowman, J.I. Kapusta, Phys. Rev. C **79**, 015202 (2009)
5. P. Braun-Munzinger, J. Stachel, Nature **448**, 302 (2007)
6. M.M. Aggarwal et al. [STAR Collaboration], arXiv:1007.2613 [nucl-ex]
7. M.A. Stephanov, K. Rajagopal, E. Shuryak, Phys. Rev. D **60**, 114028 (1999)
8. M.A. Stephanov, Phys. Rev. Lett. **102**, 032301 (2009)
9. S. Gupta, X. Luo, B. Mohanty, H.G. Ritter, N. Xu, Science **332**, 1525 (2011)
10. R.V. Gavai, S. Gupta, Phys. Lett. B **696**, 459 (2011)
11. B. Berdnikov, K. Rajagopal, Phys. Rev. D **61**, 105017 (2000)
12. L. Adamczyk et al. [STAR Collaboration], Phys. Rev. Lett. **113**, 092301 (2014)
13. L. Adamczyk et al. [STAR Collaboration], Phys. Lett. B **785**, 551 (2018)
14. X. Luo, N. Xu, Nucl. Sci. Tech. **28**(8), 112 (2017)
15. X. Luo, Nucl. Phys. A **956**, 75 (2016)
16. X. Luo, J. Xu, B. Mohanty, N. Xu, J. Phys. G **40**, 105104 (2013)
17. A. Bzdak, V. Koch, Phys. Rev. C **91**(2), 027901 (2015)
18. X. Luo, J. Phys. G **39**, 025008 (2012)
19. B. Efron, Ann. Stat. **7**, p1–26 (1979)
20. P. Braun-Munzinger, B. Friman, F. Karsch, K. Redlich, V. Skokov, Phys. Rev. C **84**, 064911 (2011)
21. M.A. Stephanov, J. Phys. G **38**, 124147 (2011)
22. D.K. Mishra, P. Garg, P.K. Netrakanti, A.K. Mohanty, Phys. Rev. C **94**, 014905 (2016)
23. M. Bleicher et al., J. Phys. G **25**, 1859 (1999)

Chapter 162
Inverse Magnetic Catalysis (IMC) in Vacuum to Nuclear Matter Phase Transition

Arghya Mukherjee and Snigdha Ghosh

Abstract We have studied the vacuum to nuclear matter phase transition in the presence of constant external background magnetic field with the mean field approximation in the Walecka model. The anomalous nucleon magnetic moment has been taken into account using the modified 'weak' field expansion of the fermion propagator. The critical temperature corresponding to the vacuum to nuclear medium phase transition is observed to decrease with the external magnetic field which can be identified as the inverse magnetic catalysis in the Walecka model whereas the opposite behaviour is obtained in case of the vanishing magnetic moment indicating magnetic catalysis.

162.1 Introduction

Extremely high magnetic field is expected to be produced in non-central heavy-ion collision. Magnetars with ultra-strong magnetic field can possess magnitude as high as $\sim 10^{18}$ Gauss. QCD being a confining theory at low energies, effective theories are employed to describe the low energy behaviour of the strong interaction. In such a theory, the quark condensate is described as the non-zero expectation value of the sigma field which is basically a composite operator of two quark fields. If the condensate is already present without any background field, the effect of its enhancement in the presence of the external magnetic field is described as *magnetic catalysis* (MC). Although most of the model calculations are in support of MC, some lattice results had shown inverse magnetic catalysis (IMC) where critical temperature follows the opposite trend. In the context of nuclear physics, the MC effect was discussed by Haber et al. [1]. There, the effect of the background magnetic field on

A. Mukherjee (✉)
Saha Institute of Nuclear Physics, 1/AF, Bidhannagar, Kolkata 700064, India
e-mail: arbp.phy@gmail.com

S. Ghosh
IIT Gandhinagar, Gandhinagar 382355, Gujarat, India
e-mail: snigdha.physics@gmail.com

© Springer Nature Singapore Pte Ltd. 2021   1099
P. K. Behera et al. (eds.), *XXIII DAE High Energy Physics Symposium*,
Springer Proceedings in Physics 261,
https://doi.org/10.1007/978-981-33-4408-2_162

the transition between vacuum to the nuclear matter at zero temperature was studied for the Walecka model. However, the *anomalous magnetic moment* (AMM) of the nucleons has not been taken into account. In this work, we incorporate the AMM considering the weak field regime of the external magnetic field and use the Walecka model within the mean field approximation to study the effective mass variation of the nucleons at finite temperature and density.

162.2   Results

In the Walecka model [2, 3], the nucleons interact with the scalar meson σ and vector meson ω. The effective mass variation of the nucleons can be obtained from the one loop self-energy (Σ_s and Σ_v) of nucleons. The corresponding Feynman diagrams are shown in Fig. 162.1. The propagators shown in bold lines are the medium as well as weak magnetic field modified propagators. The self-consistent equation governing the effective mass of the nucleon (m_N^*) is given by

$$m_N^* = m_N + \mathrm{Re}\Sigma_s(m_N^*) .$$

We have obtained the numerical solution of the self-consistent equation at different conditions. At zero temperature and zero density, the incorporation of anomalous magnetic moment is shown to favour the effective mass enhancement with the external magnetic field. It is observed that in the case of vanishing temperature within a dense nuclear medium, the effective mass decreases with the background magnetic field and this trend is shown to survive in case of non-zero temperature as well. There exists a particular temperature and chemical potential for which the effective nucleon mass suffers a sudden decrease corresponding to the vacuum to nuclear medium phase transition. It has been shown that this critical temperature decreases with the increase of B which can be identified as inverse magnetic catalysis in the Walecka model whereas the opposite behaviour is obtained in case of the vanishing magnetic moment (Figs. 162.2, 162.3 and 162.4) [4].

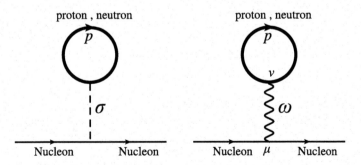

Fig. 162.1 Feynman diagram for scalar and vector meson interactions with nucleons

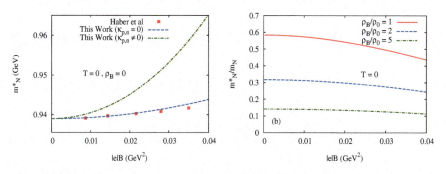

Fig. 162.2 Left panel shows the variation of the nucleon effective mass with eB at zero temperature and density. At $T = 0$, effective mass variation for different baryon densities are shown in the right panel

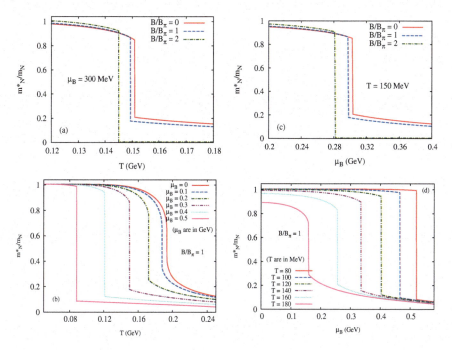

Fig. 162.3 Variation of effective mass of nucleon with temperature and baryonic chemical potential. Here $|e|B_\pi = m_\pi^2 = 0.0196\,\mathrm{GeV}^2$

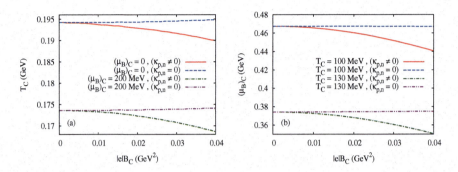

Fig. 162.4 a Variation of transition temperature with magnetic field at two different values of μ_B (0 and 200 MeV). **b** Variation of transition baryon chemical potential with magnetic field at two different values of T (100 and 130 MeV)

References

1. A. Haber, F. Preis, A. Schmitt, Phys. Rev. D **90**(12), 125036 (2014), arXiv:1409.0425
2. B.D. Serot, J.D. Walecka, in *Advances in Nuclear Physics*, vol. 16, ed. by J. Negele, E. Vogt (Plenum Press, New York, 1986), pp. 1–327
3. J. Alam, S. Sarkar, P. Roy, T. Hatsuda, B. Sinha, Ann. Phys. **286**, 159 [hep-ph/9909267] (2001)
4. A. Mukherjee, S. Ghosh, M. Mandal, S. Sarkar, P. Roy, Phys. Rev. D **98**, 056024 (2018), arXiv:1809.07028

Chapter 163
Thermoelectric Effect and Seebeck Coefficient of Hot and Dense Hadronic Matter in the Hadron Resonance Gas Model

Jitesh R. Bhatt, Arpan Das, and Hiranmaya Mishra

Abstract We study the thermoelectric effect of a baryon-rich plasma, with a temperature gradient, produced in heavy-ion collision experiments. We estimate the Seebeck coefficient for the hot and dense hadronic matter within the framework of relativistic kinetic theory using relaxation time approximation. For quantitative analysis, we use the hadron resonance gas model with hadrons and resonance states up to a cutoff in the mass as 2.25 GeV. The current produced due to temperature gradient can be a source of a magnetic field in heavy-ion collision experiments.

163.1 Introduction

Transport coefficients, e.g., shear viscosity (η), bulk viscosity (ζ), electrical conductivity (σ_{el}), etc., of strongly interacting matter produced in relativistic heavy-ion collision experiments (RHICE) have been investigated extensively [1–3]. In the present investigation, we study the thermoelectric behavior of the strongly interacting matter created in heavy-ion collisions. The strongly interacting matter created in heavy-ion collisions can exhibit thermoelectric effects due to a temperature gradient between the central and the peripheral regions of the medium. The phenomenon in which a temperature gradient in a conducting medium is converted to electrical current is known as the Seebeck effect. The Seebeck coefficient is defined as the electric field produced in a conducting medium due to a temperature gradient when the electrical current is set to zero [4]. In condensed matter systems, the thermoelectric effect requires only a temperature gradient as the ions in the lattice are stationary. However, in a pair plasma, e.g., in an electron-positron plasma, just having a temperature

J. R. Bhatt · A. Das (✉) · H. Mishra
Theory Division, Physical Research Laboratory, Navrangpura, Ahmedabad 380009, India
e-mail: arpan@prl.res.in

J. R. Bhatt
e-mail: jeet@prl.res.in

H. Mishra
e-mail: hm@prl.res.in

© Springer Nature Singapore Pte Ltd. 2021
P. K. Behera et al. (eds.), *XXIII DAE High Energy Physics Symposium*,
Springer Proceedings in Physics 261,
https://doi.org/10.1007/978-981-33-4408-2_163

gradient is not enough to lead to any net thermoelectric current due to the exact cancelation of the thermoelectric current of particles and their antiparticles. Hence, the Seebeck effect will not be manifested in a strongly interacting QCD medium with zero net baryon density. However, strongly interacting plasma with a finite baryon chemical potential will exhibit net thermoelectric current driven by the temperature gradient as the number of positive and negative charge carriers are not the same at finite baryon chemical potential. So one may expect a thermoelectric effect in a baryon-rich strongly interacting medium, to be produced at FAIR and in NICA [5]. For quantitative analysis, we calculate the Seebeck coefficient of the hot and dense hadronic medium produced in heavy-ion collision experiments, using the hadron resonance gas model [6].

163.2   Seebeck Coefficient for a Multicomponent System

Within the kinetic theory framework in the relaxation time approximation, the Seebeck coefficient ($S^{(i)}$) and electrical conductivity ($\sigma_{el}^{(i)}$) of the ith charged particle species with electric charge $e_{(i)}$, and the total Seebeck coefficient of the multicomponent system (S) can be shown to be [7]

$$S^{(i)} = \frac{\left(\mathcal{L}_{21}^{(i)} - \mu\mathcal{L}_{11}^{(i)}\right)}{e_{(i)}\mathcal{L}_{11}^{(i)}T}, \quad \sigma_{el}^{(i)} = e^2\mathcal{L}_{11}^{(i)}, \quad S = \frac{\sum_i S^{(i)}e_{(i)}^2\mathcal{L}_{11}^{(i)}}{\sum_i e_{(i)}^2\mathcal{L}_{11}^{(i)}}, \tag{163.1}$$

respectively. The integrals $\mathcal{L}_{11}$ and $\mathcal{L}_{21}$ for each species of charged particles as given in (163.1) in the Boltzmann approximation can be expressed as

$$\mathcal{L}_{11}^i = \frac{\tau_i g_i}{6\pi^2 T} \int_0^\infty \frac{k^4}{k^2 + m_i^2} \exp\left(-\frac{\left(\sqrt{k^2 + m_i^2} - \mu B^i\right)}{T}\right) dk, \tag{163.2}$$

and

$$\mathcal{L}_{21}^i = \frac{\tau_i g_i}{6\pi^2 T} \int_0^\infty \frac{k^4}{\sqrt{k^2 + m_i^2}} \exp\left(-\frac{\left(\sqrt{k^2 + m_i^2} - \mu B^i\right)}{T}\right) dk, \tag{163.3}$$

respectively, where m_i, g_i, τ_i and B^i is the mass, degeneracy, thermal averaged relaxation time and baryon number of the ith particle species, respectively. μ and T denote baryon chemical potential and temperature, respectively.

163.3 Results and Discussions

For quantitative estimates, we use the hadron resonance gas model of the hadronic medium. For the hadron resonance gas model, we include all the hadrons and resonances up to a mass cutoff $\Lambda = 2.25$ GeV [8]. To calculate the relaxation time, we use hard sphere scattering approximation between the hadrons and their resonances [9]. For hard sphere scattering, the important parameters which enter the calculation are the radii of the hadrons. We have taken a uniform radius of $r_h = 0.3$ fm for all the mesons and baryons. We have estimated the Seebeck coefficient using Eq. (163.1) for various values of baryon chemical potential $\mu = 60, 80, 100$ and 150 MeV. For each value of μ, we have considered temperature in a range 80–160 MeV. To get a feeling for $\mathcal{L}_{11}$, we have estimated the electrical conductivity $\sigma_{el} \equiv e^2 \mathcal{L}_{11}$ of pion. The left plot in Fig. 163.1 shows the variation of normalized electrical conductivity (σ_{el}/T) for π^+ with temperature (T). The behavior is similar to the previous result as obtained in [3].

In the right plot of Fig. 163.1, we show the variation of the total Seebeck coefficient (S) for the hadronic medium with normalized temperature (T/m_π) and baryon chemical potential (μ). From the right plot in Fig. 163.1, it is clear that the Seebeck coefficient of the hadronic medium increases with both temperature (T) and baryon chemical potential (μ). From (163.1), (163.2) and (163.3), it is clear that the individual Seebeck coefficients $(S^{(i)})$ are independent of the relaxation time (τ_i). However, the total Seebeck coefficient (S) of the system is dependent on τ_i, as it enters into the expression of S through the integral $\mathcal{L}_{11}^{(i)}$, which can be observed in (163.2). In the numerator of the total Seebeck coefficient (S) in (163.1), mesons do not contribute due to the fact that the Seebeck coefficient of the particle and the associated antiparticle is the same but opposite in sign. Hence, only the baryons contribute in the numerator of the total Seebeck coefficient (S) in (163.1). The mesons contribute only

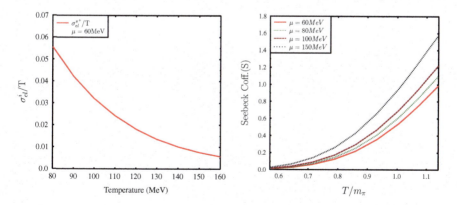

Fig. 163.1 Left plot: Variation of the electrical conductivity (σ_{el}/T) of π^+ with temperature (T) at $\mu = 60$ MeV. Right plot: Variation of Seebeck coefficient (S) of hadron resonance gas with temperature (T) and baryon chemical potential (μ) [7]

in the denominator of the total Seebeck coefficient (S) in (163.1). For the range of T and μ considered here, the dominant contribution to the total Seebeck coefficient (S) arises from protons. Proton Seebeck coefficient ($S^{(p)}$) decreases with increasing μ and T [7]. On the one hand, the quantity $\mathcal{L}_{11}^{(p)}$ for proton increases with μ and T [7]. This increasing behavior of $\mathcal{L}_{11}^{(p)}$ with increase in both μ and T is fast enough to make the product $S^{(p)}\mathcal{L}_{11}^{(p)}$ increasing with μ and T. Hence, the numerator for the total Seebeck coefficient (S) in (163.1) increases with both μ and T. On the other hand, in the denominator of the total Seebeck coefficient (S) in (163.1), the dominant contribution arises for the pions. $\mathcal{L}_{11}^{(pion)}$ decreases with increasing μ and T due to the decrease in relaxation time for pions with μ and T. Taken together, this explains the behavior of the total Seebeck coefficient with μ and T, as can be seen in Fig. 163.1 [7].

It is clear that a baryon-rich hadronic medium, to be produced in FAIR and NICA, can have a non-zero Seebeck coefficient. The net thermoelectric current ($j = \sigma_{el} S \nabla T$) can generate a transient magnetic field of the order of $\sim 10^{-3} m_\pi^2$ in the hadronic medium [7].

Acknowledgements A. Das would like to thank the organizer of the DAE-BRNS HEP symposium 2018 for giving the opportunity to present this work. The authors would like to thank Ajit M. Srivastava for suggesting the problem and subsequent extensive discussions. The authors would also like to thank the members of working group IV at WHEPP 2017, IISER Bhopal, for many discussions.

References

1. U.W. Heinz, R. Snellings, Annu. Rev. Nucl. Part. Sci. **63**, 123 (2013)
2. A. Wiranata, M. Prakash, Nucl. Phys. A **830**, 219C (2009)
3. M. Greif, C. Greiner, G.S. Denicol, Phys. Rev. D **93**, 096012 (2016)
4. T.J. Scheidemantel, C. Ambrosch-Draxi, T. Thonhauser, J.V. Badding, J.O. Sofo, Phys. Rev. B **68**, 125210 (2003)
5. See, https://www.gsi.de/en/researchaccelerators/fair.htm; http://nica.jinr.ru
6. P. Braun-Munzinger, K. Redlich, J. Stachel, nucl-th/0304013
7. J.R. Bhatt, A. Das, H. Mishra, Phys. Rev. D **99**, 014015 (2019)
8. C. Amsler et al. [Particle Data Group], Phys. Lett. B **667**, 1 (2008)
9. G. Kadam, H. Mishra, Phys. Rev. C **92**, 035203 (2015)

Chapter 164
Novel Wide Band Gap Semiconductor Devices for Ionizing Radiation Detection

Elizabeth George, Ravindra Singh, Pradeep Sarin, and Apurba Laha

Abstract We describe the fabrication and characterization of novel ionizing radiation detectors using wide band gap semiconductors like GaN and synthetic diamond. We have made interdigitated GaN metal-semiconductor-metal (MSM) detector with finger spacing of 8 μm with Schottky junctions on GaN epitaxial layers. Synthetic diamond grown by microwave plasma-assisted chemical vapor deposition (CVD) is a high-speed detector with applications in high radiation environment. In this paper, we present measurements of current-voltage (I-V) characteristics, charge transport with transient current technique (TCT) and charge collection efficiency measured on GaN and single crystal CVD diamond prototype detectors.

164.1 Introduction

Wide band gap semiconductors have gained wide interest in the field of charged particle detection due to their low intrinsic noise and high radiation tolerance. Silicon detectors traditionally used with a low band gap satisfy the necessary conditions for a tracking detector of high energy resolution. But the wide band gap semiconductor materials like diamond and GaN are important for applications where silicon performs poorly. The large band gap significantly suppresses thermal noise generated by intrinsic carriers, giving these devices a simple structure that can be fabricated in one or two lithographic steps compared to the multistep lithography required for doping of traditional silicon-based solid-state detectors. Furthermore, these materials have much higher radiation tolerance than silicon-based devices, which makes their application to charged particle radiation detection in particle physics especially attractive.

E. George (✉) · P. Sarin
Department of Physics, Indian Institute of Technology Bombay, Mumbai, India
e-mail: elizabeth14@iitb.ac.in

R. Singh · A. Laha
Department of Electrical Engineering, Indian Institute of Technology Bombay, Mumbai, India

© Springer Nature Singapore Pte Ltd. 2021
P. K. Behera et al. (eds.), *XXIII DAE High Energy Physics Symposium*,
Springer Proceedings in Physics 261,
https://doi.org/10.1007/978-981-33-4408-2_164

1107

We have chosen to investigate signal timing response and charge collection efficiency (CCE) in GaN (band gap: 3.39 eV) [1] and single crystal diamond (band gap: 5.5eV). With atomic displacement energies of 10–20eV (GaN) and 43 eV (diamond) these materials are expected to be very radiation hard, making their study important for future high luminosity high radiation environments.

164.2 Detector Prototype and Test Setup

High purity single crystal CVD diamond from IIa Technologies, Singapore is used as the test sample. The sample is 400 μm thick with a cross-section area of 5.6 mm × 5.3 mm. The sample is metallized with 50 nm Cr + 200 nm Au contacts on both sides after standard cleaning process [2].

GaN MSM devices are fabricated on a semi-insulating 3 μm thick GaN film grown epitaxially on a sapphire substrate by metal organic chemical vapor deposition (MOCVD). The inter-digital electrode (IDT) pattern is formed in the wafer by photolithographic technique. The dimensions of the fabricated IDT fingers are Finger width: 4 μm; Finger spacing: 8 μm; Finger length: 120 μm; Contact pad area: $150 \times 150 \ \mu m^2$. A Schottky junction is formed at the metal-GaN interface by depositing Ni/Pt/Au (10/30/80 nm) metal stack on the pattern. The electrode pattern in the device is such that GaN depletes sideways rather than depleting along the bulk. Figure 164.1a shows the detector prototypes of diamond and GaN used for the test. We made a direct measurement of the current pulse induced in the electrodes by the injection of stopping a particles from a Am-241 source in the diamond and GaN detectors with the transient current technique (TCT). Figure 164.1b shows the schematic of the setup used for TCT analysis of CVD diamond and GaN MSM detector. We also measured a energy spectrum of the GaN MSM device using a shaping amplifier for obtaining the charge collection efficiency of the detector.

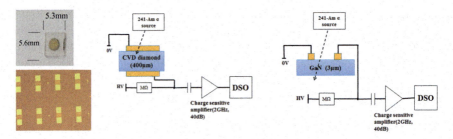

Fig. 164.1 **a** (Left top) sCVD diamond from IIa Technologies, (left bottom) Top view of fabricated GaN MSM devices. **b** (Right) Schematic of TCT setup used for the test

164.3   Results and Discussion

164.3.1   TCT Current Profile of IIa Diamond

Figure 164.2 shows the TCT current pulses for holes and electrons at room temperature for IIa CVD diamond. After the fast rise of the current signal from collection of carriers just under the electrode, the signal level is nearly flat during the drift of the charge carriers in bias field through the detector bulk. The slight slopes, positive for electrons and negative for holes imply a negative space charge effect present in the crystal. The falling edge shows the arrival of charge cloud at the opposite electrode with a longer tail for smaller bias voltages due to the increase in the diffusion width as drift time increases. From the TCT data, the charge carrier properties like mobility and drift velocity are calculated [3] and plotted in Fig. 164.3.

TCT signal response from GaN MSM device with different bias voltages at room temperature is plotted in Fig. 164.4. Signal response is Gaussian and the signal strength is comparable with that of diamond. With the increase in bias voltage the signal strength increases and saturates at a bias voltage of about −40 V.

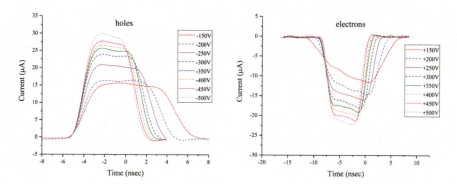

Fig. 164.2 Averaged current pulses for holes and electrons at room temperature for diamond detector

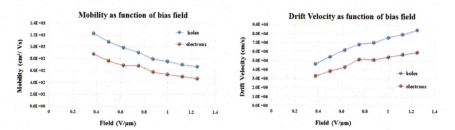

Fig. 164.3 Charge carrier mobility and drift velocity measured in diamond detector as a function of bias electric fields

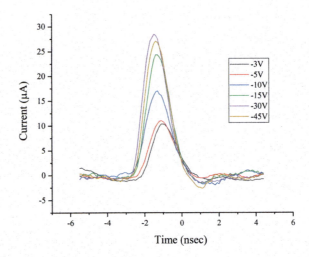

Fig. 164.4 TCT signal response from GaN

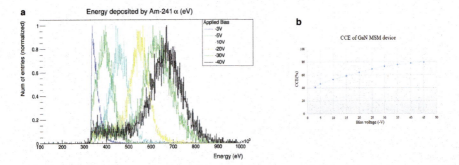

Fig. 164.5 a (Left) Am-241 α energy spectrum as function of bias voltage. **b** (right) Charge collection efficiency as a function of applied bias voltage

164.3.2 Signal Distribution and Charge Collection in GaN MSM Device

Using stopping and range of ions in matter (SRIM) simulation software, the energy loss in GaN of thickness 3 µm by α particles is evaluated to be 0.86 MeV [4]. Bragg peak in energy loss obtained from SRIM occurs at around 14 µm, hence α penetrates through the thin GaN layer. Figure 164.5a shows the α energy spectrum of recorded signals by a charge shaping amplifier as a function of applied bias voltages ranging from -3 to -40 V. Our measurement of charge collection efficiency of the GaN device as shown in Fig. 164.5b indicates an increase in CCE with applied bias voltage and reaching a maximum value close to 78% at -40 V bias.

164.4 Conclusion

We have measured the TCT signal profile in CVD diamond and GaN MSM device and studied the charge carrier properties in the samples. TCT signal response is excellent in thin GaN MSM device at low reverse bias voltages and is comparable with diamond signal strength. Charge collection efficiency of GaN MSM device saturates close to 78% at −40 V bias.

References

1. J. Wang et al., Review of using gallium nitride for ionizing radiation detection. Appl. Phys. Rev. **2**, 031102 (2015)
2. F. Bachmair, CVD diamond sensors in detectors for high energy physics. Ph.D. thesis, Section 2.3.4. ETH Zürich (2016)
3. S.M. Sze, K.K. Ng, *Physics of Semiconductor Devices*, 3rd ed. (Wiley, 2007)
4. SRIM—Stopping Range of Ions in Matter, http://www.srim.org, software by Jonathan Zeigler
5. M. Cerv et al., Diamond detector for beam profile monitoring in COMET experiment at J-PARC. JINST **10**, C06016 (2015)

Chapter 165
Simulation Study on Feasibility of RPC Operation in Low Gain Mode

Abhik Jash, Varchaswi K. S. Kashyap, and Bedangadas Mohanty

Abstract A resistive plate chamber can perform efficiently up to a maximum particle flux of $\sim$1 kHz/cm^2. One of the ways to increase its rate capability is to operate it in low gain or low charge production mode. We have simulated the produced charge in a standard RPC due to passage of muons and explored the conditions to operate it in low charge production mode. The effect of applied field and SF$_6$ content of the gas mixture on the produced charge has been studied. The conditions to produce charges as low as 100 fC have been established and the efficiency of the detector at this condition for different electronic thresholds has been calculated. Some preliminary experiments were performed to achieve the same goal.

165.1 Introduction

Resistive Plate Chambers (RPCs) [1] are widely used in high energy and nuclear physics experiments due to their very good efficiency, time resolution, and moderate position resolution. The easy and robust fabrication method has allowed it to be built in large sizes suitable for experiments requiring large coverage areas. The RPCs are generally deployed in low particle flux areas because of their low rate handling capability ($\sim$100 Hz cm^{-2}). Usage of low resistive electrodes and operating the detector in avalanche mode have made it possible to handle rates up to a few kHz cm^{-2} with more than 90% efficiency [2, 3]. RPCs with rate handling capabilities of 15 kHz cm^{-2} or more are needed for future accelerator-based high energy physics experiments, like CBM, ATLAS (HL-LHC), etc. Its rate capability can be increased [4] by reduc-

A. Jash (✉) · V. K. S. Kashyap · B. Mohanty
School of Physical Sciences, National Institute of Science Education and Research, HBNI, Jatni 752050, Odisha, India
e-mail: abhik.jash@niser.ac.in

V. K. S. Kashyap
e-mail: vkashyap@niser.ac.in

B. Mohanty
e-mail: bedanga@niser.ac.in

© Springer Nature Singapore Pte Ltd. 2021
P. K. Behera et al. (eds.), *XXIII DAE High Energy Physics Symposium*,
Springer Proceedings in Physics 261,
https://doi.org/10.1007/978-981-33-4408-2_165

ing the electrode resistivity further or by lowering its thickness; but they have adverse effects like increase in leakage current (faster aging) and reduced mechanical stability. The rate handling capability can be increased [4] by operating the detector in low gain mode which implies production of less charge within the detector. An order of magnitude reduction in the charge ($\sim$100 fC) would be necessary to achieve the required rate capability. In the present work we have simulated the processes leading to signal generation in a standard RPC geometry due to passage of muons and explored the conditions to operate it in low charge production mode. The efficiency of the detector has been calculated for different charge thresholds. Some preliminary experiments were performed on a glass RPC of same geometry to operate it within the desired limit.

165.2 Method

A RPC having 2 mm gas gap, 3 mm thick electrodes with readout strips of width 2.8 cm was modeled in Garfield++ [5]. Passage of muons through the RPC gas chamber produces primary electrons and ions which in presence of an electric field get multiplied depending on their kinetic energy. HEED++ [6] was used to find the properties of primary ionization created by muons of energy 2 GeV passing through the RPC gas chamber. The Townsend, attachment, and electron diffusion coefficients were calculated using Magboltz++ [7]; Owing to its parallel plate geometry, a constant electric field is expected to be present within the RPC gas chamber [8]. The field values within the gas chamber were calculated for different applied voltages using COMSOL Multiphysics [9] and supplied manually to Garfield++. Using all these inputs, Garfield++ calculated the induced current signal on a RPC readout strip. The flow chart of the simulation framework is shown in Fig. 165.1 (left). ROOT [10] was used to analyze the obtained signals. The current signal was integrated within 50 ns window to find the induced charge. A distribution of charges for 1000 muons is shown in Fig. 165.1 (right). The general method of selecting valid signals and not the detector noise is to pick only those signals which can produce a charge greater than a specific threshold. The efficiency of RPC in detecting muons has been calculated as the percentage of events that can produce a charge greater than a specific threshold, Q_{th}, compared to the total number of events.

165.3 Result

The variation of mean charge induced on a RPC readout strip due to passage of muons with the applied field within RPC gas chamber is shown in Fig. 165.2 (left), when the RPC is operated with gas mixtures containing different percentage of SF_6 along with $\sim$95% Freon 134a and 4.5% i-C_4H_{10}. The charges produced within the RPC chamber gain higher kinetic energy when the applied field is high and give rise

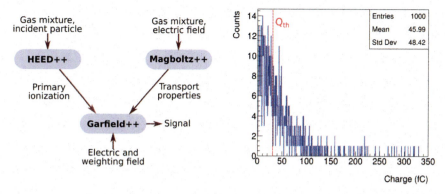

Fig. 165.1   Schematic of the used simulation framework, and a typical charge distribution spectrum obtained from simulation

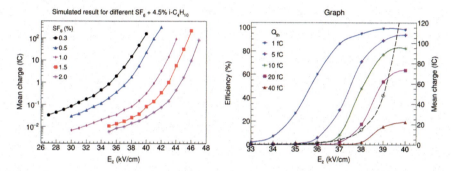

Fig. 165.2   (Left) Variation of mean charge with the applied field for different percentages of SF_6 in the gas mixture. (Right) Variation of RPC efficiency (solid lines) with the applied field for different charge thresholds, calculated for the gas mixture containing 0.3% SF_6. The black dashed line shows the variation of mean charge

to more secondary ionizations, thus producing higher amount of charge. SF_6 being a highly electro-negative gas absorbs the electrons and reduces the charge production. Figure 165.2 (right) shows the variation of RPC efficiency with the applied field for different charge thresholds, calculated for the gas mixture containing 0.3% SF_6. Variation of mean charge is also shown in the same figure using a black dashed line. It can be seen that to operate the detector with mean charge less than 100 fC the detector needs to be operated below the field of 39.6 kV/cm. The efficiency reduces with the choice of higher threshold. In an experiment, the threshold is decided depending on the level of noise and this is expected to decide the maximum achievable efficiency as well as the operating condition of the detector.

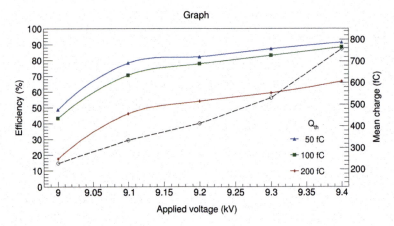

Fig. 165.3 Experimental result on variation of RPC efficiency (solid lines) with the applied voltage when RPC is operated with gas mixture containing 95.2% Freon 134a, 4.5% i-C_4H_{10}, 0.3% SF_6. The black dashed line shows the variation of mean charge

165.4 Experimental Study

A glass RPC of dimension 30 cm × 30 cm with 2 mm gas gap and 3 mm thick electrodes was operated with the gas mixture containing 95.2% Freon 134a, 4.5% i-C_4H_{10} and 0.3% SF_6. Signals from one readout strip of dimension 2.8 cm × 30 cm, arranged in telescopic coincidence with two scintillator paddles, were amplified using a NIM fast amplifier and recorded in an oscilloscope (1 GHz, 10 GS/s). For each signal, the values of voltage at each instant was divided by 50 Ω (characteristic impedance of the readout strip) to obtain the corresponding current values. The current signals were then analyzed to obtain the induced charge and detector efficiency following the same method as described in Sect. 165.2. Figure 165.3 shows the variation of mean induced charge and detector efficiency for different charge thresholds as a function of total applied voltage across the RPC.

165.5 Summary and Future Plan

Although simulations show a possibility of operating a RPC with charge production as low as 100 fC, experimentally, we find it difficult to go below 220 fC because of inherent detector and electronic noise. We have been able to achieve 50% efficiency with the choice of 50 fC threshold. In future, we plan to explore the sources of detector noise and minimize them, and also use electronics with lower noise and better amplification. We also plan to explore the possibility of using RPCs in multigap configuration to increase the efficiency, without increasing the noise level.

Acknowledgements   DAE and DST, Govt. of India, are acknowledged for financial support. BM also acknowledges the support from J C Bose National Fellowship of DST. We thank Mr. Subasha Rout for his assistance with the experimental setup. We thank the INO group at TIFR, India for providing some components to construct the RPC.

References

1. R. Santonico, R. Cardarelli, Nucl. Instrum. Methods A **187**, 377 (1981)
2. C. Bacci, C. Bencze, R. Cardarelli et al., Nucl. Instrum. Methods A **352**, 552 (1995)
3. R. Arnaldi, A. Baldit, V. Barret et al., Nucl. Instrum. Methods A **456**, 73 (2000)
4. G. Aielli, P. Camarri, R. Cardarelli et al., JINST **11**, 07014 (2016)
5. H. Schindler, R. Veenhof, https://garfieldpp.web.cern.ch/
6. I.B. Smirnov, Nucl. Instrum. Methods A **554**, 474 (2005)
7. S.F. Biagi, Nucl. Instrum. Methods A **273**, 533 (1988)
8. A. Jash, N. Majumdar, S. Mukhopadhyay, S. Chattopadhyay, JINST **10**, 11009 (2015)
9. COMSOL: a multiphysics simulation tool, https://www.comsol.co.in/
10. R. Brun, F. Rademakers, Nucl. Instrum. Methods A **389**, 81 (1997)

Chapter 166
Performance of a Low-Resistive Bakelite RPC Using PADI Electronics

Mitali Mondal, Jogender Saini, Zubayer Ahammed,
and Subhasis Chattopadhyay

Abstract An R&D programme has been going on at VECC, Kolkata with a low-resistive prototype bakelite RPC to be used in the CBM experiment at FAIR, Germany. It has been operated in an avalanche mode gas mixture to withstand the high flux environment of upcoming CBM experiment. A self-triggered front-end electronics PADI has been used to acquire the data. In this paper, the characterization of the detector such as IV characteristics, efficiency, and noise rate have been presented.

166.1   Introduction

Compressed Baryonic Matter (CBM) experiment at the future Facility of Anti-Proton and Ion Research (FAIR) is a fixed target experiment. It aims to explore the nuclear matter at high net-baryon densities and moderate temperatures with the beam energy range between 4 and 45 AGeV at a very high interaction rate of 10 MHz by measuring the bulk observables and rare diagnostic probes such as charmed particles and vector mesons decaying into dilepton pairs [1]. Thus the experiment requires very fast and radiation hard detectors, a novel data readout and analysis concept including free streaming front-end electronics, and a high performance computing cluster for online event selection [2]. Muon Chamber (MuCh) in CBM is an alternate layers of absorbers and triplet-detector systems to identify the muons from J/ψ and light vector mesons in a high particle density environment [3]. Gas Electron Multipliers (GEM) and Resistive Plate Chambers (RPC) are the active detectors for the 1st–2nd and the 3rd–4th MuCh stations, respectively. Excellent muon detection efficiency 95%, time resolution $\sim$1 ns, and spatial resolution ($\sim$mm) along with robustness, low cost, easy fabrication, ability to scale to large area makes RPC a good choice

M. Mondal (✉) · J. Saini · Z. Ahammed · S. Chattopadhyay
Variable Energy and Cyclotron Centre, 1/AF, Bidhannagar, Kolkata 700064, India
e-mail: mitalimondal.phy@gmail.com

© Springer Nature Singapore Pte Ltd. 2021
P. K. Behera et al. (eds.), *XXIII DAE High Energy Physics Symposium*,
Springer Proceedings in Physics 261,
https://doi.org/10.1007/978-981-33-4408-2_166

1119

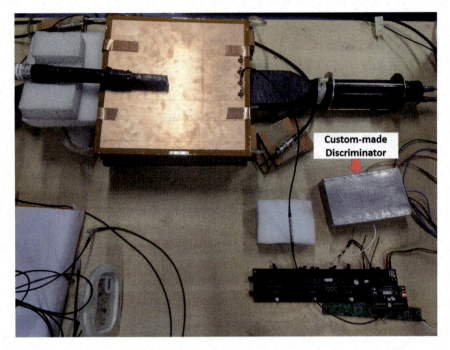

Fig. 166.1 Cosmic ray test setup of RPC with PADI electronics

for CBM. But, a very high particle rate in the third and fourth stations of MuCh (15 kHz/ cm^2 and 5.6 kHz/cm^2, respectively) became a big challenge for RPC detectors. The rate handling capability of RPC is limited due to the voltage drop across the electrodes in an RPC and equal to $\delta V = \rho \phi qd$, where ρ is the resistivity of the electrodes, d is the thickness of the electrodes, q is the charge in each avalanche and ϕ corresponds to particle flux. By reducing the bulk resistivity of electrodes as well as that of the gain of the detector and moving a part of the signal amplification to a sensitive front-end electronics and thereby reducing the avalanche charge, RPCs can be operated in a high flux environment.

166.2 Readout Chain: PADI

A PreAmplifier-DIscriminator (PADI) ASIC has been designed with 0.18 μm CMOS technology as the readout front-end board for timing RPCs in the CBM Experiment at FAIR. This ASIC has the following features: fully differential design, 50 Ω input impedance, preamplifier gain ~100, preamplifier bandwidth >300 MHz, peaking time <1 ns, noise related to input <25 μV RMS, comparator gain >100, low power consumption 17 mW/channel, common DC feedback loop for signal and threshold stabilization, threshold range related to input ± 500 mV [4]. The threshold of RPC signal can be adjusted by a potentiometer present on the FEE board. Due to the absence of pulse stretcher in PADI, digital outputs of low amplitude signals are not detected by conventional level translators LVDS to LVTTL, etc. GET-4 ASIC [5], designed by CBM-TOF group has been fabricated for integration with PADI FEE. Due to unavailability and complexity of the CBM-TOF readout chain, a custom-made discriminator has been built to process the output signal from PADI.

166.3   Description of the Prototype RPC and Experimental Setup

The prototype RPC has a single gas-gap (2 mm) and the electrodes are made of low resistive (4×10^{10} Ω cm) 2 mm thick bakelite plates. The inner surfaces of electrodes are coated by double layer linseed oil. The pick up panel (30 cm $\times$ 30 cm) of strip-with 2.3 cm has been made of 1.5 mm FR4 sheet sandwiched between two 35 μm copper layers. The detector has been operated in an avalanche mode gas mixture ratio of R134a:iC$_4$H$_{10}$:SF$_6$::94.2 : 4.7 : 1.1 with a flow rate of 3 sccm at 20 °C room temperature and 45% humidity. A cosmic test setup (Fig. 166.1) has been made using two paddle scintillators and a finger scintillator (7.5 cm $\times$ 1.5 cm). Scintillator pulses have been processed through a leading edge discriminator and a coincidence logic unit. Therefore, a 4-Fold signal has been made with the coincidence of scintillator pulses and the NIM logic signal of RPC from the custom-made discriminator. The efficiency of the detector has been calculated as

$$Efficiency = \frac{4 - Fold\ Counts\ *\ 100}{3 - Fold\ counts\ of\ Scintillators} \qquad (166.1)$$

166.4   Experimental Results

The I-V characteristics of the detector as shown in Fig. 166.2 has shown that the breakdown of gas happened at 10 kV. At different high voltages, the efficiency of the RPC has been measured. It has reached plateau and achieved an efficiency of 95% at 10.5 kV at a noise rate of 1.1 Hz/cm^2. The PADI on-board threshold was kept at 298 mV.

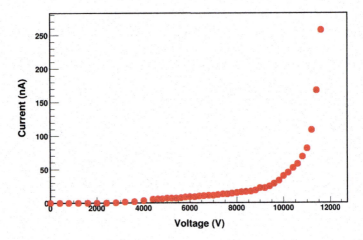

Fig. 166.2 IV characteristics of the RPC

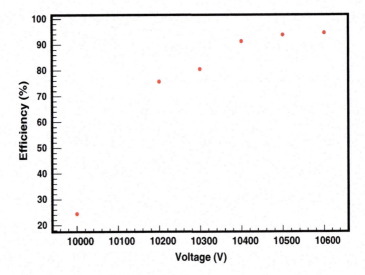

Fig. 166.3 Efficiency of the RPC as a function of high voltage

166.5   Conclusions

We have tested a low-resistive RPC module successfully with PADI electronics and achieved 95% cosmic muon detection efficiency at a dark count rate of 1.1 Hz/cm^2 shown in Figs. 166.3 and 166.4. Further tests of time resolutions are ongoing. Another self-triggered electronics STS/MuCh XYTER is also being tested with RPC.

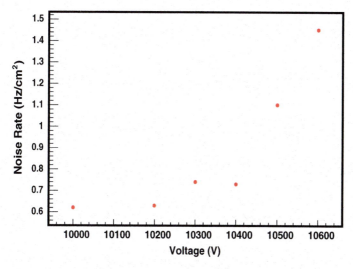

Fig. 166.4 Dark count rate of the RPC as a function of high voltage

References

1. C. Simon et al. [CBM Collaboration], Proc. Sci., https://pos.sissa.it/
2. https://www.gsi.de/work/forschung/cbmnqm/cbm.htm
3. B. Friman et al. (Eds.), *The CBM Physics Book*, Lecture notes in physics, vol. 814 (Springer, 2011)
4. M. Ciobanu et al., PADI-2, -3 and -4: The second iteration of the fast preamplifier—discriminator ASIC for time-of-flight measurements at CBM. IEEE NSS N13-44, 401–406 (2009)
5. H. Flemming, H. Deppe, *The GSI Event-Driven TDC with 4 Channels GET4* (IEEE NSS, Orlando, FL, 2009), pp. 1082–3654

Chapter 167
Study of Multi-gap Resistive Plate Chambers (MRPCs) as a Potential Candidate for Development of a PET Device

M. Nizam, B. Satyanarayana, R. R. Shinde, and Gobinda Majumder

Abstract The Multi-gap Resistive Plate Chambers (MRPCs) provide excellent timing as well as position resolutions at relatively low cost. We have designed and fabricated several six-gap glass MRPCs and extensively studied their performance. In this paper, we describe the detector, the electronics, and the data acquisition system of the setup. We present the data analysis procedure and initial results of our studies to measure the absolute position of a radioactive source (^{22}Na). We use Time Of Flight (TOF) as well as the hit coordinate information to demonstrate potential applications of MRPCs in medical imaging.

167.1 Introduction

The Multi-gap Resistive Plate Chamber (MRPC) is a modified version of RPC detector wherein the gas gap between the electrodes is further divided into multiple gaps by introducing electrically floating highly resistive plates. The MRPC was first conceptualized and developed in 1996 [1]. These detectors consist of many highly resistive plates (e.g., glass) and very thin gas gaps between them. The high voltage is applied only on the outermost electrodes and the inner electrodes are all electrically floating. The time resolution of these detectors improves with narrower gaps. Studies done by several groups have shown a time resolution lower than 100 ps for various MRPC configurations [2]. We have constructed several six-gap glass MRPCs of dimensions 305 mm × 305 mm × 7.5 mm. A schematic of the detector geometry with dimensions of various components is shown in Fig. 167.1c. The area of the internal glass

M. Nizam (✉)
Homi Bhabha National Institute, Mumbai 400094, India
e-mail: mohammad.nizam@tifr.res.in; nizamphys@gmail.com

M. Nizam · B. Satyanarayana · R. R. Shinde · G. Majumder
Tata Institute of Fundamental Research, Mumbai 400005, India

© Springer Nature Singapore Pte Ltd. 2021
P. K. Behera et al. (eds.), *XXIII DAE High Energy Physics Symposium*,
Springer Proceedings in Physics 261,
https://doi.org/10.1007/978-981-33-4408-2_167

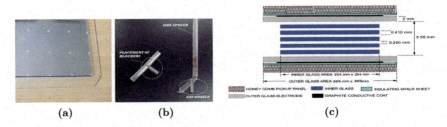

Fig. 167.1 **a** Placement of spacers, **b** blockers and side spacers and **c** the design of our detector

plates are of dimension 254 mm × 254 mm × 0.410 mm. Glass sheets of 2 mm thickness, coated with a conductive layer using graphite and paint of the NEROLAC brand, were used for the outer electrodes. The surface resistance of the conductive coat was in the range $(0.5–1)\,M\Omega/\square$. Two-sided non-conducting adhesive tapes were pasted on both sides of a mylar sheet to make small circular spacers of diameter 4 mm and thickness $\approx 250\,\mu$m. Twenty-five spacers were used to maintain each gas gap. Figure 167.1a shows the placement of the spacers. There is a space of around 2.7 cm between the edges of external and internal electrodes. The gas mixture (R134a (91.2%), C_4H_{10} (4.8%), SF_6 (4%)) can flow through this empty space of thickness 3.55 mm, instead of flowing through the 0.250 mm narrow gaps which would offer much higher resistance to the gas flow. In order to ensure a proper flow through the gaps, we introduced some blockers at appropriate places (one each near the gas inlets and two each near the gas outlets). This is illustrated in Fig. 167.1b. The pickup panels consist of plastic honeycomb laminated with eight copper strips of width 2.8 cm placed orthogonal to each other. A detailed description of our detector's performance can be found in [3].

167.2   The Experimental Setup

We have mounted two six-gap MRPCs horizontally separated by some distance which can be varied and a radioactive source (^{22}Na) is placed asymmetrically between the two detectors. ^{22}Na emits a positron which annihilates with an electron almost at rest and two gammas of 511 keV are produced with opposite momenta. The photons are detected by two detectors in coincidence with each other. Lines of Response (LOR) can be obtained by joining the hits coordinates. The time of flight information gives the exact position of the source on the line of response. The radioactive source is placed asymmetrically between the two detectors. A pair of scintillator paddles of dimensions 44 cm × 44 cm is placed above the top MRPC and below the bottom MRPC such that the MRPCs are well within the area of scintillator paddles. We read only three X-central and three Y-central strips of each MRPC to form the trigger. The X-strips and Y-strips are ORed separately and the ORed signals of X- and Y-planes are ANDed. The resultant AND output of each MRPC are finally ANDed to form the trigger. The AND signal of all four scintillator paddles is used as a veto to remove the cosmic muon background. We have used only the X-plane timing data in this study. The pickup strips are read by NINO ASIC [4]. NINO gives a differential (LVDS) output which is converted to ECL. The ECL signals are fanned out for the coincidence unit, scalers, and the TDC (Phillips Scientific 7186 H Time-to-Digital Converter). In the coincidence unit path, ECL signals are further converted to NIM. NIM signals are further discriminated and fanned out for individual scaler readings and Majority Logic Unit (MLU). As we mentioned in the previous section that we are reading only three central strips of each plane of both MRPCs, the X- and Y-strips of MRPCs are ORed separately. The ORed X and Y are ANDed. Finally the AND output of both MRPCs is given to the MLU (LeCroy 365AL Dual 4-Fold Majority Logic Unit) for coincidence. The AND of all four scintillator paddles is given as a veto to the MLU. The final output from MLU is given to the gate of TDC. The ECL output of each strip is delayed by adding appropriate lengths of cables such that the signal is well within the trigger window and then they are used as TDC stop.

167.3 Data Analysis

We operated our system at different high voltages to optimize signal/background ratio and time resolution. Table 167.1 summarizes the sig/bkg ratio with different high voltages and Fig. 167.2 shows the corresponding plots. We choose 15.0 kV as the operating voltage to get better time resolution as well as good sig/bkg ratio.

The X- and Y-coordinates of hits are recorded along with the time of arrival of the photon at the detector. The difference between timings of two opposite photons is calculated as $\Delta t = t_{MRPC1} - t_{MRPC2}$. To avoid mismatch in cable lengths and the TDC path of different channels, we have taken two reading for a fixed distance between the MRPCs. First, the source is kept at the bottom MRPC and we obtain $\Delta t_1 = t_{MRPC1} - t_{MRPC2}$. The same reading is repeated but with source just below the top MRPC and we obtain $\Delta t_2 = t_{MRPC1} - t_{MRPC2}$. Finally, we calculate the time of flight $TOF = (|\Delta t_1 - \Delta t_2|)/2$ so that the offset between TDC paths of Top and bottom MRPCs are canceled out. Our results of time of flight calculation have been summarized in Table 167.2. The Δt obtained from this study at 15.0 kV operating voltage is 1.141 ± 0.035 ns which include ~ 120 ps of electronic jitter.

Table 167.1 Signal to background ratio for different high voltages

H.V (kV)	Signal/h	sig/bkg	Δt (25 ps/count)
16.0	98.84	0.56	41.76 ± 0.10
15.0	108.73	0.82	44.25 ± 0.50
14.0	27.66	0.89	48.34 ± 2.83

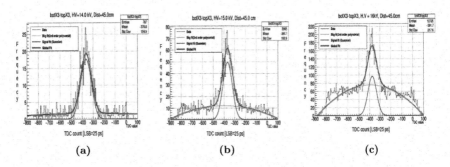

(a)   (b)   (c)

Fig. 167.2 Difference between the TDC data of the central strip ("X3") of the bottom MRPC and the TDC data of the central strip ("X3") of the top MRPC at different high voltages

Table 167.2 Time of flight for different distances between MRPCs

Distance (cm)	Δt_1	Δt_2	TOF	TOF (expected) (ns)
30	-335.10 ± 2.00	-249.40 ± 2.60	1.07 ± 0.08 ns	1.0
45	-369.50 ± 1.70	-256.80 ± 1.50	1.41 ± 0.06 ns	1.5
60	-386.70 ± 2.80	-224.10 ± 2.50	2.03 ± 0.09 ns	2.0
75	-405.30 ± 2.90	-205.10 ± 2.60	2.50 ± 0.10 ns	2.5

167.4 Summary and Future Plan

The time of flight obtained from this study is in good agreement with the actual calculated values. We obtained a Δt of 1.141 ± 0.035 ns at 15.0 kV operating voltage, which include $\sim$120 ps of electronic jitter. The time resolution improves with the increase in the operating high voltage of MRPC. We have seen a significant improvement in the time resolution after applying the time walk correction [3] and applying correction for the electronic jitter. We plan to get both analog and digital output from the front end preamplifiers to make time walk corrections using pulse height information.

References

1. E. Cerron Zeballos et al., Nucl. Instrum. Methods A **374**, 132 (1996)
2. P. Fonte et al. [ALICE Collaboration], Nucl. Instrum. Methods A **443**, 201 (2000); **478**, 183 (2002)
3. M.M. Devi, N.K. Mondal, B. Satyanarayana, R.R. Shinde, Eur. Phys. J. C **76**, 711 (2016). https://doi.org/10.1140/epjc/s10052-016-4570-2
4. F. Anghinolfi et al., Nucl. Instrum. Methods A **533**, 183–187 (2004)

Chapter 168
Measurement of Ion Backflow with GEM-Based Detectors

S. Swain, P. K. Sahu, S. K. Sahu, Surya Narayan Nayak, and A. Tripathy

Abstract A systematic study is performed to measure ion backflow fraction with GEM-based detectors. The ion current along with detector gain is measured with various voltage configurations and with different gas proportions. The observed ion backflow fraction is found to be very sensitive toward the applied drift field and the effective gain of the detector. The charge transparency in the GEM also depends on the gas mixtures. Here our main goal is to optimize the detector for the minimum ion backflow current at an optimized detector gain. So the gain and ion backflow values are calculated for a single and a quadruple GEM setups. It is observed that the ion backflow is very much dependent on the configuration of drift field and the GEM voltage. The induction field has a minor effect on ion backflow value. With an increase in gain, the ion backflow fraction decreases proportionally. From the scanning over gas ratios, we also observed that with increasing quencher proportion the gain decreases and ion backflow value is quite high.

168.1 Introduction

In recent years, there has been significant advances in the field of Micro Pattern Gas Detector (MPGD) due to their features of high rate capability, excellent spatial and time resolution, stability, radiation hardness, etc. [1, 2]. The Gas Electron Multiplier (GEM) is one of the micro-pattern detector, which is the first choice of a researcher whenever operation in high luminosity environment, stability over performance, and high radiation resistance are required. GEMs were first introduced by Fabio Sauli at CERN in 1997. The simplest GEM setup consists of a standard GEM foil placed in between a drift (acts as the cathode) and an induction (acts as the anode) plane enclosed within a chamber for gas flow. Generally, negative voltages are applied

S. Swain (✉) · P. K. Sahu · S. K. Sahu · A. Tripathy
Institute of Physics, HBNI, P.O.: Sainik School, Sachivalaya Marg, Bhubaneswar 751005, India
e-mail: sagarika.swain@cern.ch

S. Swain · S. N. Nayak
School of Physics, Sambalpur University, Jyoti Vihar, Burla, Sambalpur 768019, India

© Springer Nature Singapore Pte Ltd. 2021
P. K. Behera et al. (eds.), *XXIII DAE High Energy Physics Symposium*,
Springer Proceedings in Physics 261,
https://doi.org/10.1007/978-981-33-4408-2_168

across these electrodes with anode being grounded. A typical potential of a few hundred volts is applied across the GEM to create a strong electric field to initiate the avalanche processes. When an ionizing particle enters the detector, ionization occurs mainly in the drift volume and the primary electrons are drifted into the GEM holes. The holes act as amplification channels for the electrons inside the detector.

During the avalanche process, positive ions created inside the hole drift along the opposite field direction. While some ions get deposited in the top copper layer, some start moving toward the drift volume. In this process, the ions start accumulating near the cathode plane. This gradually creates a space charge effect, which distorts the original drift field and also the detector performance. Therefore, for a stable detector operation, it is very important to have minimum ions feedback. Ions backflow fraction is determined by measuring the ratio of the number of ions collected at the drift plane to the number of electrons accumulated at the anode readout (induction plane), keeping all other parameters fixed [3, 4]. In the beginning, a single GEM setup is used to measure its gain and ion backflow fraction by changing different parameters. This measurement is then extended for the quadruple GEM.

168.2 Experimental Setup

168.2.1 Single GEM

The single GEM setup consists of a standard double mask GEM foil having an active area of $10 \times 10 \, cm^2$, placed in between drift and an induction plane with 3.5 mm and 2 mm gaps, respectively. The schematic diagram of the setup is given in Fig. 168.1. The GEM foils used in the setup are obtained from CERN and assembled here locally. The drift plane is a 50 μm kapton foil with a copper layer of 5 μm on one side which faces the top of GEM, and the readout plane is a two-dimensional PCB with 120 readout pads each of area $9 \times 9 \, mm^2$. A 128 pin sum-up connector with a female lemo output is used to add the signals from all pads.

168.2.2 Quad GEM

The quad GEM setup used in this work consists of four layers of $10 \times 10 \, cm^2$ GEM foils obtained from CERN. The drift gap (between the drift plane and GEM1), three transfer gaps (between the GEMs) and induction gap (between GEM4 and the induction plane) are made 3, 2, 2, 2 and 2 mm, respectively. The schematic diagram of the setup is given in Fig. 168.2. For the quad GEM setup, the drift plane and electrodes of the GEM4 are biased with individual power supply HV1, HV3, and HV4 are also shown in this figure. For the distribution of correct voltages in rest of the electrodes, a voltage divider circuit is used. The resistance values are mentioned

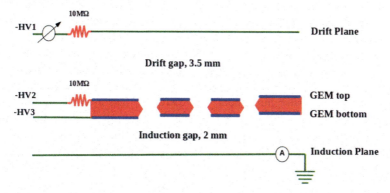

Fig. 168.1 Schematics diagram of the single GEM detector. Voltage scheme is designed with high voltage connections at the drift plane, top and bottom layer of GEM with induction plane is being grounded. Anode current reading is taken from a Kiethley Pico ammeter

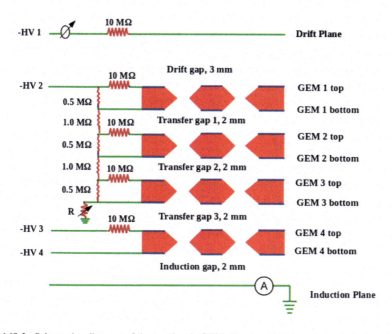

Fig. 168.2 Schematics diagram of the quadruple GEM detector. Voltage scheme is designed with four main high voltage connection at four electrodes i.e., at the drift, top layer of GEM1, top and bottom layers of GEM4. GEM3 bottom plane is grounded through a proper resistor and the induction plane is grounded and acts as anode. Anode current reading is taken from a Kiethley Pico ammeter

in the schematic diagram. The transfer fields between the GEMs are determined through the intermediate resistance values.

In the beginning, ultra-pure N_2 (99.999%) gas is passed through the detector, and gradually detector biasing is done to achieve conditioning at the maximum stable operating voltage. Then Ar/CO_2 gas mixture is passed through the chamber with an optimized flow rate value throughout the experiment. The measurements are done for both the setup with the different gas mixture ratio of Ar/CO_2 70:30, 80:20, and 90:20 to determine the influence of the gas proportions on the ion backflow fraction values. Throughout the experiment, the corresponding flow rates, as well as the ambient parameters such as temperature, pressure, and relative humidity are remained constant and recorded with a data logger [5].

168.3   Result

In the beginning, Fe^{55} source of 5.9 keV energy is used on the single GEM detector and the measurements are taken with different applied fields and Ar/CO_2 gas mixtures ratio with the various quencher proportions. Later the same source is used on the quad GEM detector for similar measurements.

168.3.1   Single GEM

The ion backflow fraction values are observed with changing drift field and induction field. To make a comparison of the performances of different gas mixtures the ion backflow fraction are displayed as a function of induction field E_d and E_i in Fig. 168.3a and b, respectively. An overall ion backflow scanning is done with detector gain as shown in Fig. 168.4.

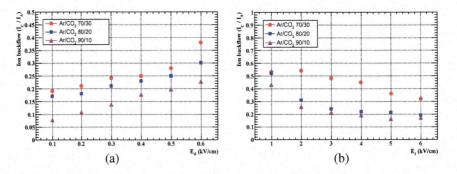

Fig. 168.3 Ion backflow fraction as a function: **a** E_d with $E_i = 4$ kV/cm. **b** E_i with $E_d = 0.4$ kV/cm, for different gas mixtures

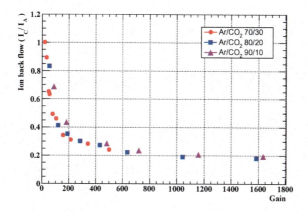

Fig. 168.4 Ion backflow fraction as a function of Gain for fixed $E_d = 0.4$ kV/cm and $E_i = 4$ kV/cm with different gas mixtures

168.3.2   Quad GEM

Next to keep the gain in the order of 10^4, we fix the GEM voltages ΔV 320 V, 300 V, and 270 V for 70:30, 80:20, and 90:10 Ar/CO_2 gas mixture ratios, respectively. The plots of Ion backflow with E_d and E_i are displayed in Fig. 168.5a, Fig. 168.5b, Fig. 168.6a for 70:30, 80:20, and 90:10 ratios, respectively. The gain values are given inside the figures. The low values of ion backflow fraction are observed from the low drift regions; however, the induction field which is responsible for high gain is not much affecting the ion backflow fraction for the quadruple GEM detector. The minimum ion backflow fraction values are observed in these figures and these are 3.1%, 3.0%, and 3.5% with drift field 0.1 kV in Ar/CO_2 in 70:30, 80:20, and 90:10 gas mixture ratios, respectively.

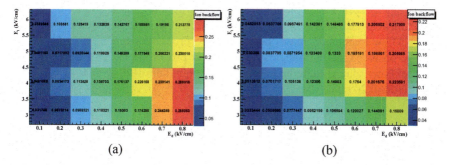

(a)   (b)

Fig. 168.5 Ion backflow fraction as a function: **a** E_d with $E_i = 4$ kV/cm. **b** E_i with $E_d = 0.4$ kV/cm, for different gas mixtures

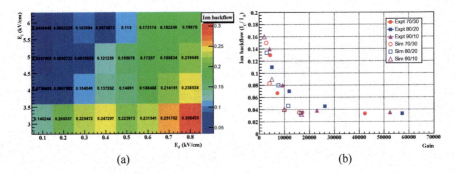

Fig. 168.6 Ion backflow fraction as a function: **a** E_d with $E_i = 4$ kV/cm. **b** E_i with $E_d = 0.4$ kV/cm, for different gas mixtures

In order to study the properties of the detector and compare them with the experimental results, a full and detailed simulation have been performed with Garfield++ simulation package [6]. ANSYS is used to create the geometry of the detector and the meshing needed for the field calculations [7]. Ion backflow study is done for quadruple GEM detector with variation of voltage configuration and gas ratio. Ion backflow of 3.5% is achieved with drift field 0.1 kV/cm.

168.4   Conclusion

The charge transparency in the GEM mainly depends on field configurations and gas mixtures. So a detailed measurement is done for the study of ion backflow with GEM-based detectors. The gain and ion backflow are calculated for a single and quadruple GEM detector. The dependencies of ion backflow fraction are carefully observed with E_d, E_i, ΔV and with the variation of gas ratio. It is found that the ion backflow is very much dependent on the configuration of drift field and the GEM voltage. The induction field has a minor effect on ion backflow value. With an increase in gain, the ion backflow fraction decreases proportionally. From the scanning over gas ratios, we also observed that with increasing quencher proportion the gain decreases and ion backflow value is quite high. We found a minimum IBF of 3.0% with E_d 0.1 kV/cm for 80:20 ratio.

References

1. F. Sauli, A new concept for electron amplification in gas detectors. Nucl. Instrum. Methods A **386**, 531–534 (1997)
2. F. Sauli, The gas electron multiplier (GEM): operating principles and applications. Nucl. Instrum. Methods A **805**, 2–24 (2016)

3. A. Bondar, Study of ion feedback in multi-GEM structures. Nucl. Instrum. Methods A **496**, 325–332 (2003)
4. H. Natal da Luz, Ion backflow studies with a triple-GEM stack with increasing hole pitch, JINST 13 P07025 (2018)
5. S. Sahu, Design and fabrication of data logger to measure the ambient parameters in gas detector R&D, JINST 12 C05006 (2017)
6. R. Veenhof, Garfield++. http://garfieldpp.web.cern.ch/garfieldpp/, 2010–2012
7. Ansys Software, Maxwell, http://www.ansys.com/products/electronics/ansys-maxwell

Chapter 169
Study of Boosted Hadronic Top Tagger in the Context of $t\bar{t}H$ Analysis

Saikat Karmakar

169.1 Introduction

The dominant decay mode of a top quark is $t \rightarrow bW$. Hadronic decay of the top quark represents the decay of the W boson into two quarks. A top quark is called boosted in the lab frame if the momentum of the top quark is high enough so that the decay products move in a direction close to that of the parent top quark. In the non-boosted case, the three decay products (b and the two quarks coming from W) are well separated and the top quark can be reconstructed easily from these three jets. For boosted top quarks, these three decay products merge together and cannot be separately identified. Reconstruction of the top quark is difficult in such a scenario. At the current 13 TeV centre-of-mass energy of LHC, there are many top quarks which have very high transverse momentum and are, therefore, boosted. Using a special algorithm (boosted hadronic top tagger), an attempt is made to reconstruct such top quarks. One such algorithm, HEPTopTaggerv2 [1], is presented here.

169.2 HEPTopTaggerV2 Algorithm

One of the most well-known algorithm for boosted hadronic top quark reconstruction is HEPTopTaggerv2. This algorithm works well when the p_T of the top quark is more than 200 GeV. Natural direction for finding boosted top is to look into the subjet analysis of the fat jet.

S. Karmakar (✉)
Tata Institute of Fundamental Research, Mumbai 400005, India
e-mail: saikat.karmakar@tifr.res.in

© Springer Nature Singapore Pte Ltd. 2021
P. K. Behera et al. (eds.), *XXIII DAE High Energy Physics Symposium*,
Springer Proceedings in Physics 261,
https://doi.org/10.1007/978-981-33-4408-2_169

1. First define a Cambridge-Aachen (C/A) [2] fat jet with radius parameter $R_{fat} = 1.5$.

2. Next Identify all subjets with mass drop criteria:

   - Undo the clustering of the fat jet j into two subjets j_1 and j_2 with $m_{j_1} < m_{j_2}$.
   - if $m_{j_2} < f_{drop} m_j$ keep both j_1 and j_2; otherwise keep only j_2. The value of f_{drop} is taken to be 0.8.
   - Continue decomposing until $m_{j_i} < 30$ GeV.

3. Find possible top candidates:

   - Iterate through all triplets of three hard subjets and filter with resolution $R_{filt} = min[0.3, R_{jk}/2]$, where R_{jk} is the distance between the jth and kth subjets.
   - Find the 5 hardest constituents.
   - Recluster these 5 into exactly three jets and assume they come from the decay of a top quark.
   - Reject triplets outside $m_{123} = [150, 200]$ GeV.

4. Acceptance of a top candidate:

   - Arrange the three jets j_1, j_2, and j_3 in decreasing order of p_T.
   - j_1, j_2, j_3 must satisfy one of the following criteria:

$$0.2 < arctan\frac{m_{13}}{m_{12}} < 1.3 \text{ and } R_{min} < \frac{m_{23}}{m_{123}} < R_{max}$$

$$R_{min}^2 \left(1 + \left(\frac{m_{13}}{m_{12}}\right)^2\right) < 1 - \left(\frac{m_{23}}{m_{123}}\right)^2 < R_{max}^2 \left(1 + \left(\frac{m_{13}}{m_{12}}\right)^2\right)$$

$$R_{min}^2 \left(1 + \left(\frac{m_{12}}{m_{13}}\right)^2\right) < 1 - \left(\frac{m_{23}}{m_{123}}\right)^2 < R_{max}^2 \left(1 + \left(\frac{m_{12}}{m_{13}}\right)^2\right)$$

   where $R_{min/max} = (1 \mp f_W) \frac{m_W}{m_t}$ and f_W is taken to be 0.15.

5. Accept the triplet for which m_{123} is closest to m_t.
6. Do a consistency check requiring $p_{T,t} > 200$ GeV.

169.3 Event Selection

For this study, $t\bar{t}$ sample, produced using Pythia8 [3] at 13 TeV centre-of-mass energy, has been used. Here one t decays leptonically and the other t decays hadronically. The fat jet was clustered with FastJet [4] using Cambridge-Aachen (C/A) clustering algorithm. The selected fat jets must have $p_T > 200$ GeV and $|\eta| < 2.5$. After decomposing the fat jet, the subjets are required to have $p_T > 20$ GeV and $|\eta| < 2.5$.

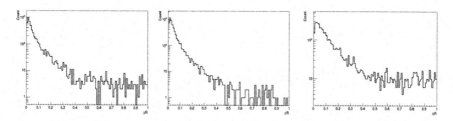

Fig. 169.1 ΔR matching of reco b jet (left), reco W_{lead} jet (middle) and reco $W_{sublead}$ jet (right) with the corresponding generator level quarks

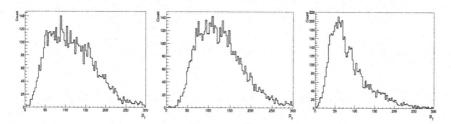

Fig. 169.2 p_T distribution of the reco b jet (left), reco W_{lead} jet (middle) and reco $W_{sublead}$ jet (right)

169.4   Results

To find the efficiency of top quark reconstruction we try to match the subjets of the top quark candidate fat jet with the decay products of the top quark at the generator level. The matching criterion is $\Delta R = \sqrt{(\Delta \eta)^2 + (\Delta \phi)^2} < 0.25$. The fat jet is identified as a top quark jet if all the three subjets match with any of the three generator level parton within $\Delta R < 0.25$.

Figure 169.1 shows the ΔR between reco b jet and generator level b quark, reco W_{lead} jet and generator level W_{lead} quark, and reco $W_{sublead}$ jet and generator level $W_{sublead}$ quark, respectively, from left to right. For b jet 78% of the reco jet matches with the generator level quark, for W_{lead} jet, 96% of reco jet matches with the generator level quark and for $W_{sublead}$ jet, 48% of reco jet matches with the generator level quark. But there are only 30% events where all three reco jets match with the corresponding generator level quark. So only 30% of the boosted tops are correctly reconstructed by the HEPTopTaggerv2 algorithm.

The main reason for the low efficiency of top quark reconstruction is the low p_T of the reco $W_{sublead}$ jet.

Figure 169.2 shows the p_T distribution of the all three reconstructed jets. For the b jet and W_{lead} jet the p_T is quite high compared to the $W_{sublead}$ jet. As a result, this jet is often going outside the fat jet of radius 1.5. So the HEPTopTaggerv2 algorithm can not identify this jet properly, causing the low efficiency of top quark reconstruction.

References

1. G. Kasieczka, T. Plehn, T. Schell, T. Strebler, G.P. Salam, Resonance searches with an updated top tagger, arXiv:1503.05921 [hep-ph]
2. R. Atkinn, Review of jet reconstruction algorithms. J. Phys.: Conf. Ser. **645**, 012008
3. P. Skand, T. Sjöstrand et al., An introduction to PYTHIA 8.2, arXiv:1410.3012 [hep-ph]
4. M. Cacciari, G.P. Salam, G. Soyez, FastJet user manual, arXiv:1111.6097 [hep-ph]

Chapter 170
Simulation Study for Signal Formation with Single GEM Detector

Sanskruti Smaranika Dani, Sagarika Swain, and Surya Narayan Nayak

Abstract Gas Electron Multiplier (GEM) (Sauli in Nucl Instrum Methods A 386:31, 1997 [1]) is widely chosen as a particle tracking devices in heavy ion physical experiments. Many experimental measurements and tests have been performed to investigate the characteristics and performances of GEM detectors, but for a better understanding of this kind of detector, the computer simulation is a very important tool. In this report, we will discuss the variation of GEM signal with different detector parameters by simulation.

170.1 Introduction

Gas Electron Multiplier (GEM) was invented by Fabio Sauli in 1997 at the gas detector development group at CERN [2] which belongs to the class of Microporous gas detector. GEM operation is mainly based on avalanche multiplication in a gas medium and effectively achieving amplification of originally ionized charges. Because of its increased beam luminosity, higher rate capability, fast timing ion suppression feature and good position resolution, GEM detectors are widely chosen as particle tracking devices in heavy ion physical experiments [3–5].

170.2 Single GEM Simulation

A typical GEM hardware configuration consists of a gas filled chamber inside the GEM foils. GEM foil consists of two thin layers of copper (5 μm) separated by a dielectric medium Kapton (50 μm) [6]. These holes have a typical diameter of 50 μm

S. S. Dani (✉) · S. Swain
Institute of Physics, PO: Sainik School, HBNI, Sachivalaya Marg, Bhubaneswar 751005, India
e-mail: sanskruti.dani28@gmail.com

S. N. Nayak
Jyoti Vihar, Burla, Sambalpur 768019, Odisha, India

© Springer Nature Singapore Pte Ltd. 2021   1141
P. K. Behera et al. (eds.), *XXIII DAE High Energy Physics Symposium*,
Springer Proceedings in Physics 261,
https://doi.org/10.1007/978-981-33-4408-2_170

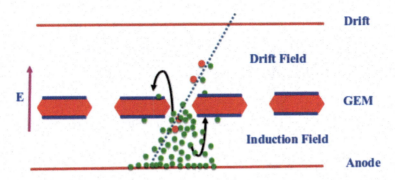

Fig. 170.1 Single GEM with avalanche mechanics

at the centre and 70 μm at the surface and pitch 140 μm based on needs, while each hole has a by-conical shape. Electrons produced in the avalanche leave the multiplication region and are collected in the readout. The gaps above and below the GEM foil are drift and induction gap, which have been taken as 3 cm and 1 cm, respectively. The detector is filled with a gas mixture of Ar/CO_2 in a 70/30 ratio that has been simulated and studied (Fig. 170.1).

170.3 Simulation Software

For the first instance, we have used a single GEM setup for the simulation model. So here we have used ANSYS [7] for building detector geometry and appropriate field configuration. ANSYS is an engineering software used to calculate the potential for a fixed set of boundary conditions. The resulting solution of ANSYS is stored in .lis files. The output .lis files of ANSYS are accessed through "GARFIELD++" programme [8], which is a computer simulating tool for 2-3D drift chamber developed in CERN in 1984. It includes different classes for calculation of primary ionization (Heed) [9], charge transport (MagBoltz) and signal. Gas mixture can be assigned through MagBoltz class [10] (Figs. 170.2 and 170.3).

170.4 Results

Signal will be induced on the induction plate during the time swarn of the electron created in the avalanche drift from the GEM hole to the anode. As electron and ion are created together, so both charges will contribute to the signal on an anode. This electron is known as the primary electron, drifting towards the GEM holes and producing electrons in the avalanche. That signal which we get at the anode is a raw

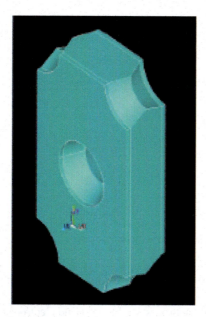

Fig. 170.2   Basic building block of GEM

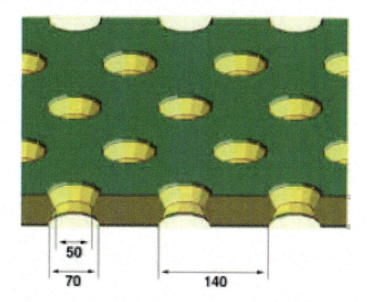

Fig. 170.3   GEM foil with dimension in micrometer

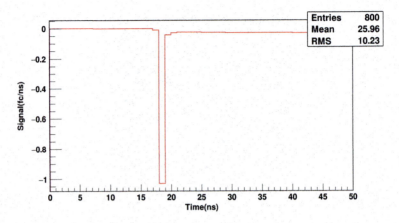

Entries	800
Mean	25.96
RMS	10.23

Fig. 170.4 Raw signal of current versus time at readout

signal. To analyze the signal, we start by simulating the induced current from a single electron in the drift region (Fig. 170.4).

In order to model signal processing by front-end electronics, the "raw signal" that is the induced current can be convoluted by using a "transfer function". The transfer function which we used in the simulation is given by

$$f(t) = \frac{t}{\tau} e^{1-t/\tau} \quad \text{where } \tau = 25$$

From Fig. 170.5, we can observe how by using a transverse function in the simulation, we can get a convoluted signal from the raw signal.

Figure 170.6 shows signal amplitude is almost negligible with a change in drift field strength. From this, we can conclude that the signal height is independent of drift field strength for a particular gas mixture.

By changing both induction field strength with gas mixture ratio, we observed from Fig. 170.8 that signal amplitude is directly proportional to induction field strength, as well as with Ar content in the gas mixture.

170.5 Conclusion

By using a transverse function in the simulation model, we got a convoluted signal from the raw signal. From this simulation, we concluded that signal amplitude is not dependent on the drift field, it remains constant with the variation of the drift field. We study the variation of signal height with induction field in different gas ratios. From this, we observed that current is directly proportional to the Ar content in the gas mixture and it is also directly proportional to the induction field.

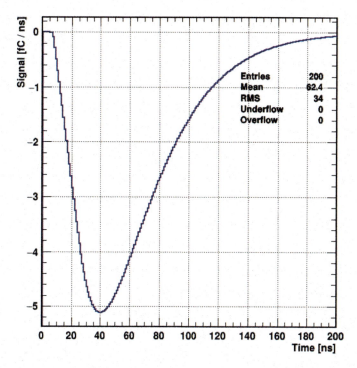

Fig. 170.5 Convolute signal of current vs time by using the transfer function

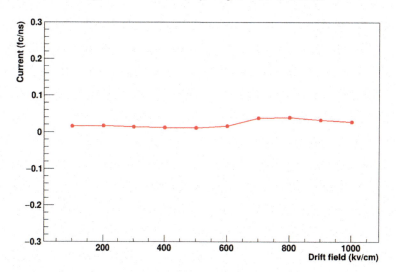

Fig. 170.6 Shows the variation of signal hight with drift field in Ar-CO_2 gas mixture of 70:30

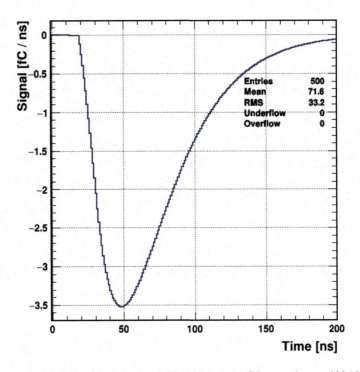

Fig. 170.7 Convolute signal for induction field 6 kV/cm in Ar-CO$_2$ gas mixture of 90:10

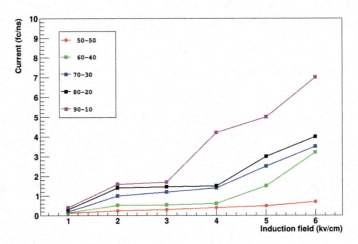

Fig. 170.8 Variation of current with Induction field in different gas ratios of Ar-CO$_2$ (i) 50:50, (ii) 60:40 (iii) 70:30 (iv) 80:20 (v) 90:10 which is mark in different colours

References

1. F. Sauli, Nucl. Instrum. Methods A **386**, 31 (1997)
2. F. Sauli, Nucl. Instrum. Methods A **805**, 2 (2016)
3. G. Bencivenni et al., Nucl. Instrum. Methods A **488**, 493 (2002)
4. M. Alfonsi et al., Nucl. Instrum. Methods A **525**, 17 (2004)
5. M. Ball et al., JINST **9**, C04025 (2014)
6. Gas Electron Multiplier. The Gas Detectors Development Group. [Online] CERN, https://gdd. web.cern.ch/GDD/
7. ANSYS Inc., ANSYS version 11 Electromagnetics, 2010
8. https://cern.ch/garfield
9. https://cern.ch/heed
10. https://cern.ch/magboltz

Chapter 171
Effect of Surface Resistivity of Electrode Coating on the Space Dispersion of Induced Charge in Resistive Plate Chambers (RPCs)

S. H. Thoker, B. Satyanarayana, and W. Bari

Abstract Resistive plate chamber (RPCs) is a parallel plate gaseous detector built using the electrodes of high volume resistivity. Operation and performance of the RPCs mostly depend upon the graphite conductive coating acting as the high voltage provider. Therefore, it is imperative that we study the effect of the surface resistivity of conductive coating of graphite on RPC characteristics and performance in detail. In this report, the effects of surface resistivity of conductive coating on the space dispersion of the induced charge of the three fabricated prototype RPCs of surface resistivity $1\,M\Omega/\square$, $100\,k\Omega/\square$, $40\,k\Omega/\square$ have been studied experimentally. All the three RPCs have been studied in avalanche mode of operation.

171.1 Introduction

The detailed structure of the RPC is shown in Fig.171.1 [1]. We studied RPC of glass (3 mm thick) of dimensions 30×30 cm^2 with a gas gap of 2 mm between them maintained by polycarbonate buttons. When the incoming charged particle ionizes the gaseous medium, current signal is induced on the external pickup strips due to the propagation of charges produced inside the gas gap. Uniform electric field is produced by two parallel plate electrodes of very high bulk resistivity [2]. The electric field inside the gas gap and charge induced on the external metallic strips depend also on the conductive coating on the glass electrodes.

S. H. Thoker (✉) · W. Bari
University of Kashmir, Srinagar, India
e-mail: shamsulthoker@gmail.com

W. Bari
e-mail: baritak@gmail.com

B. Satyanarayana
Tata Institute of Fundamental Research, Mumbai, India
e-mail: bsn@tifr.res.in

© Springer Nature Singapore Pte Ltd. 2021
P. K. Behera et al. (eds.), *XXIII DAE High Energy Physics Symposium*,
Springer Proceedings in Physics 261,
https://doi.org/10.1007/978-981-33-4408-2_171

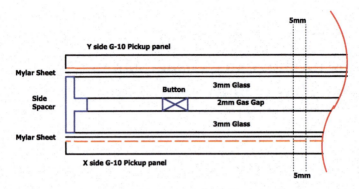

Fig. 171.1 Basic construction schematic of a resistive plate chamber

171.2 Details of the Experimental Setup

Fig.171.2 shows the schematic layout of the electronic circuit and Data acquisition system (DAQ) setup. The DAQ system houses (a) TDC for timing measurement (b) QDC (Model-2249A) for measurement of charge deposited (c) Scalars for strip hit data. The trigger criteria for this setup is

$$Trigger = (\bar{V}_1 + \bar{V}_1).(P_1.P_2.P_3) \tag{171.1}$$

Out of the five scintillators, two act as Veto paddles (V1, V2, both of dimensions 40 cm × 20 cm) placed at the top of the RPC with 5 mm gap in between them roughly at the center. P_1, P_2, and P_3 are finger paddles (with dimensions 3 cm × 40 cm). The three RPCs were tested with different high voltages ranging from ±5.0 kV to ±11.8 kV. Cosmic ray Muons were used as the charged particle source to test the performance of the RPCs. The induced current signals from the RPCs were picked up by a current-sensitive preamplifier for signal amplification. Out of the total number of 10 pickup strips (strip width 3cm), the study is done with only three strips (Left, Main and Right) of the RPC under investigation. Telescope window is mounted on the central or Main strip only.

171.3 Efficiency

The Efficiency of RPCs with different surface resistivities was studied under the same environmental conditions. The efficiency was calculated using the following equation

$$Efficiency = \frac{6F count}{5F count} \times 100\% \tag{171.2}$$

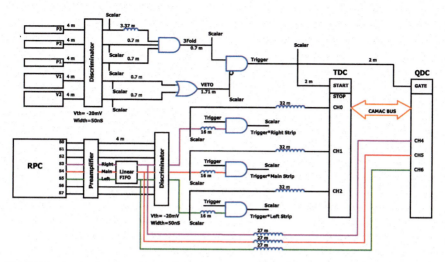

Fig. 171.2   Schematic of the experimental arrangement and electronic circuit for measuring RPC parameters

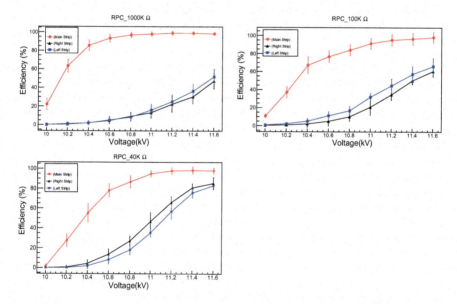

Fig. 171.3   Efficiency for main, right and left strips at different voltages for the RPCs of different surface resistivities (RPC1 $M\Omega/\square$, RPC100 $k\Omega/\square$, RPC40 $k\Omega/\square$) for the gaseous proportion, $C_2H_2F_4/C_4H_{10}/SF_6$: 95.2/4.5/0.3. The error bars are in multiples of 10

where 5 F is the trigger signal, which is the coincidence of the signals from three scintillator paddles (P1, P2, P3) and anti-coincidence of two veto (V1, V2) paddles, while as 6 F is the coincidence of the trigger signal and the RPC strip signal. Figure 171.3 shows the efficiency as a function of applied high voltage for the three RPCs of different surface resistivities. Efficiency increases with voltage and gets saturated at higher voltages. At applied voltage of (± 10.0 to ± 11.6 kV), the efficiency of 40 k$\Omega/\square$ RPC is higher than 100 k$\Omega/\square$, which in turn is higher than 1 M$\Omega/\square$ RPC for Left and Right strips. The asymptotic efficiency (e_{max}) for Main strip was found to be more than 95% for all the three RPCs. The knee voltage is the value for which 96% of the asymptotic efficiency is reached. The knee voltage is also found to be shifted to the higher voltage for the RPCs with lower surface resistivity.

171.4   Space Charge Dispersion with Different Surface Resistivity of RPCs

171.4.1   *Experimental Setup and Measurement with G10-Based PCB Pickup Strip*

The experimental setup for the measurement of the space dispersion of induced charge is same as shown in Fig. 171.2. To study the avalanche size and space dispersion of induced charge in millimeters, we replaced our old standard pickup panel (strip width 3 cm) with new G10-based PCB pickup panel (strip width 5 mm and inter-strip separation of 1 mm). To avoid reflections in G10-based PCB pickup panel, 68 Ω termination resistor is used. The G10-based pickup panel has a total number of the 50 channels. Out of the 50 channels, we instrumented only 25 channels (12 Left, 12 Right and Central or Main strip) for the study of charge dispersion measurements. For the experimental setup, the basic structure of our RPCs is same except that the graphite conductive coating of their carbon films are different in the range 1 M$\Omega/\square$, 100 k$\Omega/\square$, 40 k$\Omega/\square$. The trigger rate in this setup will drastically come down because the telescope window is now mounted on a central strip of narrow width.

171.4.2   *Experimental Results*

Figure 171.4 is the charge spectra of RPC with surface resistivity 40 k$\Omega/\square$. Similar charge distributions with less multiplicity are expected from RPCs with surface resistivity 100 k$\Omega/\square$ and 1 M$\Omega/\square$. Figure 171.5 is the charge distributions of the events with different incident positions from three RPCs with different surface resistivities. A double gaussian function has been used to fit the data of all the three RPCs [3]. One gaussian function has a narrower distribution (the sigma of the gaussian function is called σ_1) and the other function has a relatively wide distribution (the sigma is

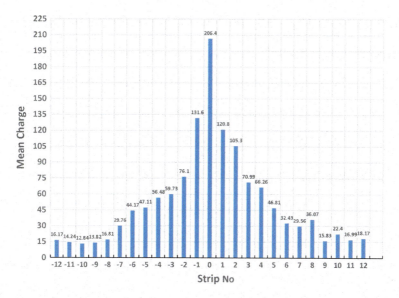

Fig. 171.4 The spectra of the charge distribution of 25 channels of RPC 40 kΩ/□ at 11.2 kV

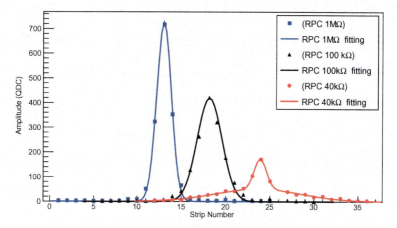

Fig. 171.5 The charge distributions for three events from three RPCs of different surface resistivity fitted with a double Gaussian. An offset has been added to look these charge distributions distinctly

called σ_2). Table 171.1 shows the summary of the values of σ_1 and σ_2 corresponding to the experimental data fitted with double Gaussian functions as shown in Fig. 171.5. From Table 171.1, it is clear that σ_1 does not change much while as σ_2 get broader. The broadening of σ_2 can be attributed to the variation of surface resistivity of conductive coating on the glass electrodes of the RPCs.

Table 171.1 Summary of the experimental values of σ_1 and σ_2. The strip number is multiplied by the pitch (6 mm) to get the total space dispersion of induced charge for RPCs of different surface resistivities

RPC	σ_1	σ_2
1 MΩ/$\square$	3.4 ± 0.07	15.16 ± 0.52
100 kΩ/$\square$	4.3 ± 0.09	28.13 ± 2.53
40 Ω/$\square$	4.5 ± 0.08	34.58 ± 5.21

171.5   Conclusion

From our experimental results, we concluded that the change in the surface resistivity has a remarkable influence on Efficiency and Space dispersion of induced charge. Space dispersion of induced charge has been studied in detail with G10-based PCB readout of 6mm pitch. RPC of lower surface resistivity shows more dispersion of induced charge as compared to the RPC of higher surface resistivity. Fitting to a function consisting of two Gaussians is a proper data analysis method. Statistical error of Gaussian function of narrower width can be interpreted as the position resolution of the RPC detector.

Acknowledgements Special thanks to Mr. Vishal, Mr. Chavan, and Mr. Pethuraj in fabricating RPCs and in the design of the experimental setup.

References

1. India-based Neutrino Observatory (INO), http://www.ino.tifr.res.in
2. W. Riegler, C. Lippmann, The physics of resistive plate chambers. Nucl. Instrum. Meth. A **518**, 86 (2004)
3. Y. Jin et al., Studies on RPC position resolution with different surface resistivity of high voltage provider. IEEE Nucl. Sci. Symp. Conf. Rec., 917 (2008)

Chapter 172
Study of Particle Multiplicity by 2 m × 2 m Resistive Plate Chamber Stack at IICHEP-Madurai

Suryanarayan Mondal, V. M. Datar, Gobinda Majumder, N. K. Mondal, S. Pethuraj, K. C. Ravindran, and B. Satyanarayana

Abstract An experimental setup consisting of 12 layers of glass Resistive Plate Chambers of size 2 m × 2 m has been built at IICHEP-Madurai to study the long-term performance and stability of RPCs produced on large scale in Industry. In this study, the data obtained by this setup was analysed to find out the events where more than one particle passed through the stack within a single trigger window. The results obtained from observation was then compared with simulation using CORSIKA and GEANT4.

172.1 Introduction

High energetic primary cosmic rays originating at out space continuously interacts with earth's atmosphere. This primary cosmic rays consist of mostly protons with a smaller fraction of higher Z-Nuclei elements. Upon interacting with earth's upper atmosphere, this primary cosmic rays result in showers of secondary particles which mostly consist of pions $\left(\pi^{\pm,0}\right)$ and Kaons $\left(K^{\pm}\right)$. These particles are short-lived and mostly do not reach the surface of the Earth. These particles further decay into $\mu^{\pm}$, e and γ. The γ, e do not reach the detector directly as they interact with the roof of the Lab and create showers.

Charged particle multiplicity in observational data obtained at IICHEP-Madurai was studied in this article. The main aim of this article is to test recent CORSIKA simulation using standard hadronic model and compare the result with observation.

S. Mondal (✉) · S. Pethuraj
NPD, Homi Bhaba National Institute, Anushakti Nagar, Mumbai, India
e-mail: suryamondal@gmail.com

S. Mondal · V. M. Datar · G. Majumder · S. Pethuraj · K. C. Ravindran · B. Satyanarayana
DHEP, Tata Institute of Fundamental Research, Homi Bhaba Road, Mumbai, India

N. K. Mondal
HENPPD, Saha Institute of Nuclear Physics, Bidhannagar, Kolkata, India

© Springer Nature Singapore Pte Ltd. 2021
P. K. Behera et al. (eds.), *XXIII DAE High Energy Physics Symposium*,
Springer Proceedings in Physics 261,
https://doi.org/10.1007/978-981-33-4408-2_172

172.2 Detector Construction

The Resistive Plate Chamber (RPCs) [1] stack operational at IICHEP, Madurai consisting of 12 RPCs stacked horizontally with a inter-layer gap of 16 cm. These RPC gap is made of two glass electrodes of thickness 3 mm with a gap of 2 mm between them. This gap is maintained using 2 mm thick poly-carbonate buttons. The RPCs are operated by applying a differential supply of ± 5 kV to achieve the desired electric field. The avalanche created by the ionization energy loss of charged particles in the RPCs induces signals in the two orthogonal pickup panels placed on both sides of the glass gaps labelled as X-side and Y-side. The pickup panels are made of parallel copper strips of width 28 mm with 2 mm gap between two consecutive strips. An event typically contains hit (one logic bit per strip indicating the signal in that strip is above the threshold value) for each strip and 16 time signal for each layer. One TDC channel records time signal coming from every alternating eighth strips on one side of the layer.

172.3 Monte-Carlo Simulation

In the CORSIKA [2] package, the several different hadronic interaction models are available. In this study, for simulating the behaviour of hadrons for higher energy range, the QGSJET (Quark Gluon String model with JETs) has been adopted and for the low energy range, the GHEISHA model has been used. The energy of the primary rays in CORSIKA is generated using the power-law spectrum, $E^{-2.7}$, within the energy range of 10–10^6 GeV for different primaries. The zenith and azimuth angle of primary particles are generated uniformly within the range of 0–85° and 0–360°, respectively. The magnetic rigidity cutoff has been implemented according to the location of the detector site. The particles generated by CORSIKA at the observation surface are provided as an input to the detector simulation. All detector parameters (inefficiency, noise, strip multiplicity, etc.) were calculated from observational data which are then used in GEANT4 simulation. A detailed study on this parameters can be found in literature [3]. The data from simulation and from observation were then analysed using the same algorithm to detect multiple tracks.

172.4 Event Reconstruction and Data Selection

For the event reconstruction, the strips hits are analysed separately, in the two-dimensional projections, namely, X–Z and Y–Z plane. In the present study, the clusters are formed with a maximum of four consecutive strips as the position resolution for higher multiplicities is found to be worse. In the first step of track reconstruction, the clusters associated with different tracks are grouped using the method of

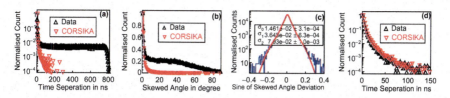

Fig. 172.1 **a** Time separation of two tracks for all events, **b** Skewed angle between two tracks originating outside of the detector, **c** Skewed angle difference between generated and reconstructed tracks fitted with triple-gaussian function and **d** Time separation of two tracks for events with only parallel tracks

Hough Transformation [4]. This method can detect all the tracks avoiding the noise hits. The projections from both X–Z and Y–Z planes are combined to produce final three-dimensional track(s). Any ghost tracks formed while combining are discarded by using the timing information. The events of interest for this analysis are the events with more than one reconstructed three-dimensional track.

The distribution of time separation between each pair of tracks for both simulation and data is shown in Fig. 172.1a. In the case of data, it can be observed that there is a large number of events where multiple particles are reaching the detector with large relative time delay. The particles originating in the different cosmic showers is the cause for these events. The distribution of the skewed angle between each pair of tracks for both simulation and data is shown in Fig. 172.1b.

Based on these observations shown in Fig. 172.1c, the tracks with a skewed angle less than 45 m rad are considered as parallel to each other. In the current study, to select the tracks generated from the particles originating from the same cosmic ray shower, only the parallel tracks are considered in the analysis. The time difference between a pair of tracks for both simulated and observed data after selecting only parallel tracks is shown in Fig. 172.1d. It can be observed that the events from the random coincidences disappear after rejecting the events with non-parallel tracks.

172.5   Results

The normalised fraction of events with two, three, and four tracks are calculated to be $6.35 \times 10^{-3}\%$, $5.8 \times 10^{-5}\%$ and $2 \times 10^{-6}\%$, respectively, from the cosmic ray data. The normalised fraction of the events with two, three, and four tracks are also calculated from the CORSIKA simulation for different types of cosmic ray primaries (H and He) and for different hadronic interaction models (QGSJET-II-04 and QGSJET01d). In order to compare the results between the data and simulation, the ratio of these normalised fractions of events between them is presented in Table 172.1 for all different simulation scenarios mentioned above.

Table 172.1 Ratio of multiple track fractions between simulation and observation for different primaries (H and He) and different physics packages (QGSJET-II-04 and QGSJET01d)

Tracks	QGSJET-II-04		QGSJET01d	
	H	He	H	He
2	0.34 ± 0.020	0.74 ± 0.030	0.34 ± 0.018	0.75 ± 0.021
3	0.17 ± 0.027	0.52 ± 0.056	0.16 ± 0.025	0.67 ± 0.069
4	0.08 ± 0.053	0.4 ± 0.26	0.04 ± 0.028	0.3 ± 0.18

172.6   Conclusion

The results of the current study reflect that the current physics models of interactions at the earth atmosphere are unable to reproduce the air showers accurately. The earlier measurements of muon multiplicity along with the present result can be used to improve the parameters of the hadronic model at high energies and/or cosmic ray spectral index.

References

1. R. Santonico, R. Cardarelli, Development of resistive plate counters. Nucl. Instrum. Methods Phys. Res. **187**, 377–380 (1981)
2. D. Heck et al., *1998 CORSIKA: A Monte Carlo Code to Simulate Extensive Air Showers* (Forschungszentrum Karlsruhe Report FZKA 6019)
3. S. Pethuraj et al., Measurement of cosmic muon angular distribution and vertical integrated flux by 2 m × 2 m RPC stack at IICHEP-Madurai. J. Cosmol. Astropart. Phys. **09**, 021–021 (2017)
4. N. Li-Bo, L. Yu-Lan, H. Meng, H. Bin, L. Yuan-Jing, Track reconstruction based on Hough-transform for nTPC. Chin. Phys. C, **38**(12) 126201

Chapter 173
Photo-Neutron Calibration of SuperCDMS Detectors

Vijay Iyer

173.1 SuperCDMS Detector Physics and WIMP Search

Super Cryogenic Dark Matter Search (SuperCDMS) [1] is a direct dark matter search experiment that has pioneered the use of low-temperature solid-state detectors to search for rare scattering of Weakly Interacting Massive Particles (WIMPs) [2] with atomic nuclei. The SuperCDMS Soudan experiment consisted 15 germanium detectors with a total mass of 9 kg.

WIMPs and neutrons elastically scatter off a nucleus causing a nuclear recoil (NR) and generate electron-hole pairs. Gammas and β particles scatter primarily off electrons causing electron recoils (ER). A bias voltage V applied to the detector makes the charges drift along the electric field lines. This mechanical work done on the electron-hole pairs by the field is released to the lattice as Neganov-Trofimov-Luke (NTL) phonons [3] with an energy $E_{NTL} = qVN$, where q is the charge of an electron and N is the number of electron-hole pairs created by the scatter event. All of the energy associated with a scatter event is eventually released in the form of phonons. E_{total} is the sum of recoil energy, E_R, and NTL phonon energy E_{NTL} that we measure in our detectors. The ionization yield Y is a relative measure of the number of electron-hole pairs created compared to the ER case. It is possible to write the total phonon energy as $E_{\text{total}} = E_R(1 + qVY/\epsilon)$, where ϵ is the average amount of recoil energy to create an electron-hole pair in the ER case (3.0 eV in Ge). NRs produce fewer electron-hole pairs than ERs and thus have an ionization yield less than 1. SuperCDMS detectors measure N and E_{total}.

Vijay Iyer on behalf of the SuperCDMS collaboration.

V. Iyer (✉)
School of Physical Sciences, National Institute of Science Education and Research, HBNI, Jatni 752050, India
e-mail: vijayiyercdms@gmail.com

© Springer Nature Singapore Pte Ltd. 2021
P. K. Behera et al. (eds.), *XXIII DAE High Energy Physics Symposium*,
Springer Proceedings in Physics 261,
https://doi.org/10.1007/978-981-33-4408-2_173

The SuperCDMS Soudan detectors were operated in two modes, the interleaved Z-sensitive Ionization and Phonon detectors (iZIP) operated at 4V and the CDMS low ionization threshold experiment (CDMSlite) detectors which were operated at higher voltages between 25 and 70 V. The E_{total} produced from an event can be amplified by the applied voltage. The higher the bias voltage, the more the contribution from E_{NTL} per electron-hole pair. This reduces the recoil energy threshold, but the ability to distinguish between ER and NR is sacrificed because N is too small to be measured independently the total phonon energy signal. Hence a direct measurement of Y is not possible for CDMSlite detectors. On the other hand, iZIPs possess the ability to discriminate between ERs and NRs [4].

173.2   Motivation and Photo-Neutron Calibration Concept

Dark matter search results are usually shown in the parameter space of dark matter-nucleon interaction cross section as a function of dark matter mass. If no dark matter-like events are found, an exclusion limit or curve is set based on the physical reach of an experiment. The motivation for a dedicated calibration of the SuperCDMS detectors comes from the fairly wide uncertainty in the exclusion curve [4] using CDMSlite data shown in Fig. 173.1a. The uncertainty comes from limited precision in determining the nuclear recoil energy scale. This stems from the inability of the CDMSlite detectors to measure the ionization yield as described in the previous section. An accurate understanding of the nuclear recoil energy scale is necessary for establishing the WIMP mass scale. CDMSlite makes use of the Lindhard model to predict the yield. The yield [5] can be written as follows:

$$Y = \frac{k \cdot g(\epsilon)}{1 + k \cdot g(\epsilon)} \qquad (173.1)$$

where $g(\epsilon) = 3\epsilon^{0.15} + 0.7\epsilon^{0.6} + \epsilon$, $k = 0.133Z^{2/3}A^{-1/2}$, $\epsilon = 11.5E_R Z^{-7/3}$.

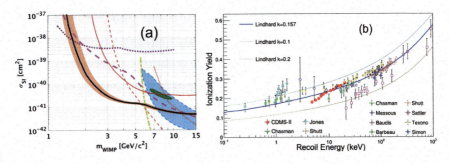

Fig. 173.1 **a** DMSlite R2 [4] exclusion limits black solid line with its associated uncertainty peach shaded band. **b** Various existing calibration points in the ionization yield vs nuclear recoil energy plane in Germanium [5]

Here Z is the atomic number and A is the atomic mass number of the recoiling nucleus, ε a reduced energy term, E_R is the recoil energy in keV and k describes the electronic energy loss. While the value of k is usually fixed at 0.157 for Ge using the original Lindhard theory, k is treated as a free parameter in this analysis. The model works reasonably well at recoil energies greater than 10 keV, as seen in Fig. 173.1b. More data points below 1 keV are needed to ascertain the reliability of this model at lower recoil energies. The photo-neutron calibration will provide data as low as ∼1 keV.

The calibration technique here involves using a high rate ^{88}Y or ^{124}Sb γ source placed next to a Be absorber. The γ source and Be wafer were placed above the detector. ^{124}Sb and ^{88}Y emit γs of 1.69 MeV and 1.84 MeV, respectively. Following a ^{9}Be(γ,n) reaction the ^{124}Sb+Be and ^{88}Y+Be sources produce neutrons between 22.8–24.1 keV and 151–159 keV, respectively. In case of ^{124}Sb, the maximum recoil of the Ge nucleus is around 1.3 keV and for ^{88}Y around 8.1 keV. By studying the recoils of Ge nuclei from these quasi-monoenergetic neutrons, the measured total phonon energy E_{total} of our detectors can be calibrated to the recoil energy E_R.

173.3 Analysis

The entire analysis can be divided in to three main parts: (i) Data selection, (ii) Simulation, and (iii) Yield extraction.

Data was taken in two modes:(i) with the γ + Be source (neutron on) and (ii) with just the γ source(neutron off). The first part of the analysis starts by subjecting the data to several quality checks to remove electronic glitches, low frequency noise, poorly reconstructed events, and bad periods of data. The neutron on and neutron off data for yttrium from iZIP detectors are shown in Fig 173.2a. To obtain the total phonon energy NR spectrum, neutron off data was subtracted from the neutron on data. Fig 173.2b shows the residual spectrum that is dominated by neutrons for yttrium beryllium data.

With the knowledge of the relationship between E_{total} and E_R as discussed in Sect. 173.2, one can then deduce the corresponding yield. The photo-neutron sources allow the determination of the ionization yield over the energy range covered by the resulting recoil spectrum. The second part of the analysis is simulating the expected E_R spectrum in the detector. The third part of the analysis is to compare the simulated neutron on E_{total} spectrum to the measured neutron on E_{total} spectrum. To determine the best fit, we use a negative log likelihood (NLL) function given by

$$- \ln L = - \sum_{i=1}^{N_D} \ln \{\rho_D (E_{\text{total}})_i\}, \tag{173.2}$$

where N_D is the number of events in a dataset and ρ_D is the energy-dependent PDF of the data spectrum. To obtain the simulated E_{total} spectrum from the simulated E_R spectrum, a yield function $Y(E_R)$ which takes the form shown in (173.1) is used. The

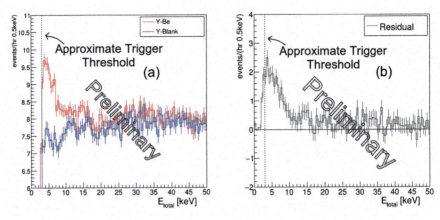

Fig. 173.2 **a** Comparison of neutron on and neutron off total phonon energy spectrum E_{total} for Yttrium beryllium iZIP data. **b** Residual spectrum after subtracting the neutron off from neutron on E_{total} spectrum

detector resolution and cut efficiencies are applied to the simulated E_{total} spectrum. The value of k in (173.1) is allowed to vary such that the negative log likelihood function is minimum. Statistical uncertainty on the simulated data is calculated by repeating the procedure with 500 simulated data samples of equal statistics and finding out the standard deviation of the best fit value of k in each case. Systematic uncertainty in the energy-dependent neutron-nucleus scattering cross section in germanium detector also needs to be accounted for. This can be done by obtaining cross section files for all the stable isotopes of Ge from the TENDL-2017 [6] nuclear database and repeating the NLL method to show the deviation in the best fit values of k.

173.4 Summary

The Lindhard model is known to have limited accuracy at low energies. Given the significant progress in lowering the recoil energy threshold for SuperCDMS, determination of yield at lower recoil energies has become increasingly important for SuperCDMS [1]. The results from the work described in this analysis will be helpful in putting more precise limits on SuperCDMS dark matter search results.

References

1. R. Agnese et al., SuperCDMS collaboration. Phys. Rev. D **95**, 082002 (2017)
2. G. Jungman, M. Kamionkowski, K. Griest, Phys. Rept. **267**, 195–373 (1996)

3. B. Neganov et al., Otkryt. Izobret **146**, 215 (1985)
4. R. Agnese et al., SuperCDMS collaboration. Phys. Rev. Lett. **116**, 071301 (2016)
5. P. Luke, J. Appl. Phys. **64**, 6858 (1988)
6. D. Barker, D.-M. Mei, Astropart. Phys. **38**, 1 (2012)

Chapter 174
Design and Development of Gas Mixing Unit for Gas Electron Multiplier (GEM) Chamber

Hemant Kumar, Asar Ahmed, Mohit Gola, Rizwan Ahmed, Ashok Kumar, and Md. Naimuddin

Abstract Gaseous detectors are the important components in each High Energy Physics (HEP) Experiment [1]. The operation of such detectors depends upon the mixture of various gases such as Ar, CO_2, CF_4, etc. However, the purity and the appropriate mixture of these gases is always the key component and has a direct impact on various properties of the detectors like gain, spatial, timing, and energy resolutions. The present work describes the design and construction of flexible and cost-effective Gas Mixing Unit (GMU) which is very useful for providing the appropriate mixture to various gaseous detectors like drift tubes (DTs), Gas Electron Multipliers (GEMs), Resistive Plate Chambers (RPCs), etc. We also present some preliminary results to demonstrate its stability with the changes in ambient conditions. The results were obtained by using this newly developed GMU with GEM 10 cm × 10 cm as a test detector.

174.1 Introduction

Traditional gas mixing unit (GMU) requires time to time gas calibration as it totally depends on the input gas pressure to gas mass flow controller (MFC) and the ambient conditions which makes them time-consuming to provide stable flow. Nowadays, fast, efficient, safe, and contamination free GMU is the need of all the research program which uses the variant gas mixture. A step towards the development of such a redundant system is taken by us to fulfill the demand. The key component of any GMU is the MFC. It must be fast and accurate to establish stable flow along with safety towards the back and overpressure across it. We used Alicat based MFC [2] which is pre-calibrated for over 80 gases and provides a flow range of 0–500 SCCM

H. Kumar (✉)
Department of Physics & Astrophysics, University of Delhi, Delhi, India
e-mail: hemant.nit.ec@gmail.com

A. Ahmed · M. Gola · R. Ahmed · A. Kumar · Md. Naimuddin
University of Delhi, Delhi, India

© Springer Nature Singapore Pte Ltd. 2021
P. K. Behera et al. (eds.), *XXIII DAE High Energy Physics Symposium*,
Springer Proceedings in Physics 261,
https://doi.org/10.1007/978-981-33-4408-2_174

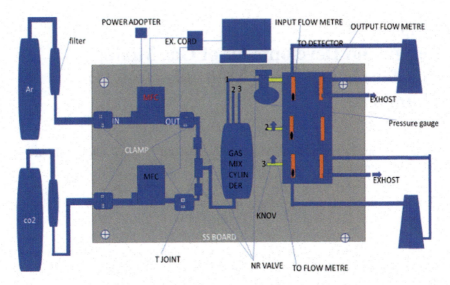

Fig. 174.1 Schematic diagram of the gas mixing unit

with 50–100 ms of response time. All the safety features are kept in mind to make GMU sustainable, which is explained in Sect. 174.2.

Figure 174.1 is the schematic view of the GMU assembled in the GEM lab at the University of Delhi.

174.2 Components

In this section, the features of various components are mentioned in designing and developing the GMU. Figure 174.2 depicts the components installed on a movable panel that makes it flexible to install at any place.

174.2.1 Mass Flow Controller (MFC)

MFC is used to adjust the precise flow of various input gases. The fraction of gases can be adjusted manually, as well as remotely through the software, which makes it very easy to operate without a traditional calibration procedure. In this setup, we used Alicat MFC which has a very wide range of flow with 0.5 SCCM resolution and is independent of the variation in input pressure up to the maximum recommended pressure of 145 psi.

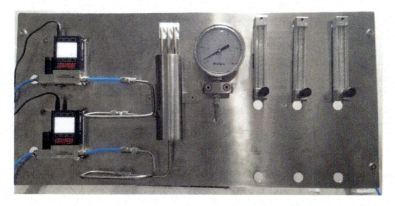

Fig. 174.2 Functional gas mixing unit

174.2.2   Gas Mixing Cylinder

The output of gases from various MFC is then fed to the mixing cylinder for the proper mixing of the input gases. We have used three output flow lines from it for the operation of three detectors at a time if required.

174.2.3   Flow Meter

Since different detectors need a different flow rate for the operation, to adjust the input flow rate and see the output flow rate of the gases from the detector, we used flow meters. It also helps in to see the leakage of the gases from the detector by measuring the difference in flow rate.

174.2.4   Pressure Gauge

To see the pressure at which gas is flushed in the detectors, a pressure gauge is installed for the real-time pressure measurement.

174.2.5   Dust Filter

Dust is like cancer for the micro-pattern gaseous detectors (GEMs). To prevent dust or any contamination produced in the input line, dust filter is used.

174.2.6 Non Returning Valve

There is always a risk of generation of overpressure inside the detector, to reduce the risk and working of the system under safe limit, pressure reducing valve (PRV) is used. If the pressure exceeds a certain limit (in this case 50 mbar), it bypasses the gas through it to maintain safe pressure and prevents any damage to the detector due to overpressure.

174.3 Results

To verify the sustained and reliable performance of the GMU, we fixed the output flow of MFC at 50 SCCM and varied the input pressure across the MFC from the gas cylinder. No change in output flow was observed for the input pressure range (60–100) psi shown in Fig. 174.3.

To see the gas leakage from the system, Ar and CO_2 were chosen. Gas from each cylinder was fed to MFC at 70 psi of the input pressure, passing through the gas filters. To make Ar and CO_2 ratio 70:30, MFCs were set at 58.3 SCCM and 25 SCCM, respectively, which makes the flow rate 5 l/h. The output of MFCs is fed to the mixing cylinder for proper mixing of gases. Through one of the output lines, the mixture was passed through the flow meter, where we observe the ~5l/h flow rate, that reflects the system gas leak tight.

For the precision in the mixing of the gases, we used the gas chromatography technique using Agilent 7890B GC [3]. Figure 174.4 shows the output of the gas chromatographer in the area (left) and the percentage of the gas in the mixture (right). We can see clearly the output percentage (69.9:29.3) of the gases which is very close to the set value (70:30) of the gases in the mixture.

This newly developed GMU was used for the study of the performance of $10\,cm \times 10\,cm$ triple GEM detector. For the tolerance of the detector towards high

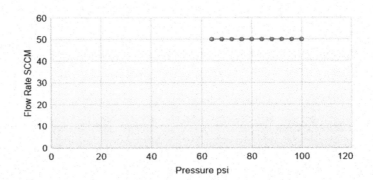

Fig. 174.3 Variation in flow rate with varying input gas pressure across MFC

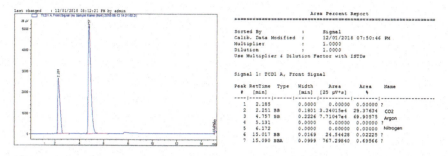

Fig. 174.4 Output of the gas chromatographer for Ar:CO_2 (70:30) mixture

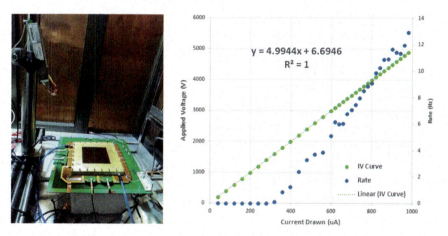

Fig. 174.5 GEM detector under study for high voltage test (left), I-V characteristic and spurious signal rate

voltage, the detector was flushed with pure CO_2 at 3l/h and measured the I-V characteristics and spurious signal rate as shown in Fig. 174.5. The stability of the gas flow rate was observed from the flow meter for several hours.

174.4   Conclusion

We have manufactured and assembled different components needed for the gas mixing unit, keeping in mind the precision, flexible optimization, cost-effectiveness, and safety factors. To see the stability of flow and precision of mixing, the output flow rate is measured at various input pressure and analyzed the output of GMU with a gas chromatograph. A few quality checks of $10\,cm \times 10\,cm$ triple GEM detector were done using this GMU and the results obtained were as per expectations. The setup is ready for providing the gas mixture in various gaseous detectors without calibration at a stable rate for a number of gases.

References

1. Maerschalk et al., Study of Triple-GEM detector for the upgrade of the CMS muon spectrometer at LHC (Université libre de Bruxelles, 2016)
2. https://www.alicat.com/product/gas-mass-flow-meters
3. https://www.agilent.com/en/products/gas-chromatography/gc-analyzers/gas-impurities-/bulk-gas-impurities-analyzers

Chapter 175
Construction of a Single GEM Detector Using Indigenous Anode Plate

A. Tripathy, S. Swain, P. K. Sahu, and S. Sahu

Abstract A prototype of single Gas Electron Multiplier (GEM) detector is fabricated in our laboratory using a single GEM foil of size $10 \times 10 \, cm^2$, a cathode plate and an anode plate (which is designed in our laboratory). The anode plate used is a single readout pad. The detector is operated using Argon and CO_2 gas mixtures in proportion of 70:30. High voltage connections are provided individually to the drift plane, GEM foil and induction plane. The anode plate used in this work is a single readout pad. Preliminary testing results show that this detector can withstand a voltage up to 460 V across the GEM foil and the anode plate without any spark. The tested results are presented in this article.

175.1 Introduction

The Gas Electron Multiplier (GEM) [1] is remarkable among other Micro-Pattern Gaseous Detectors (MPGD) [2] for its good energy resolution, position resolution, stable high gain and low ion feedback. GEMs are being adopted in many high energy physics experiments such as ALICE, CMS, CBM, etc.

In this article, we present the procedure of assembling a single GEM using an anode plate designed in our laboratory and its testing results using a ^{55}Fe radioactive source.

175.2 Gas Electron Multiplier

A GEM consists of a polymer foil with copper coating on both sides. The foil is perforated with a large number of identical holes, typically 50–100 per mm^2 as shown in Fig. 175.1. The holes are pierced on the foil using chemical etching and

A. Tripathy (✉) · S. Swain · P. K. Sahu · S. Sahu
Institute of Physics, Sachivalaya Marg, P.O. Sainik School, HBNI, Bhubaneswar-751005, India
e-mail: alekhika.t@gmail.com

© Springer Nature Singapore Pte Ltd. 2021
P. K. Behera et al. (eds.), *XXIII DAE High Energy Physics Symposium*,
Springer Proceedings in Physics 261,
https://doi.org/10.1007/978-981-33-4408-2_175

photolithography technique. In our experiment we are using a single mask GEM foil. The outer and inner hole diameters are respectively 70 μm and 50 μm. The hole pitch of the foil is 140 μm. A suitable potential difference is applied between the two copper sides of a GEM foil. When the radiation ionizes the gas, the primary electrons are collected, multiplied and then driven towards the anode plane.

175.3 Experimental Details

A single GEM detector prototype is assembled. The GEM foils and other necessary components were procured from CERN. First thermal stretching of the GEM foil and the drift plane is done. Then the GEM foil, the drift plane and the anode plate are assembled creating a drift gap and an induction gap of 3 mm and 2 mm respectively. A layout of the anode plate we have used is shown in Fig. 175.2b. A solid frame of G10 material is placed around the whole setup. A 100 μm thin Kapton window is placed on the top side of the G10 frame and finally all the metal screws are gently tightened. The experimental setup of the detector is shown in Fig. 175.2a.

In this study Ar/CO_2 gas in 70/30 volume ratio is used. Individual power supply is provided to the drift plane and top and bottom of the GEM foil. Here we have not used any sum-up board to measure the anode current. The anode plate is a single readout pad. The signal is taken directly from the anode plate and a single input is fed to a charge sensitive preamplifier. The signal from the preamplifier is used to measure the rate and to visualize the signal produced by the detector using an oscilloscope.

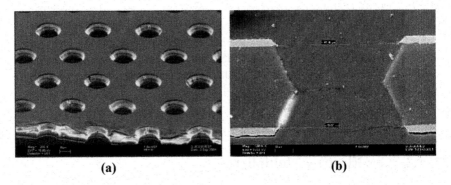

(a)   **(b)**

Fig. 175.1 a Microscopic picture of a GEM foil. **b** Cross section of a GEM hole showing its double-conical shape. There are two layers of copper on the top and on the bottom and a polymer layer in the middle (Reproduced from Sauli [3])

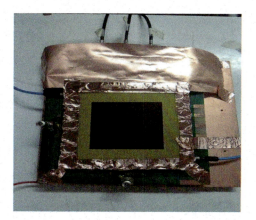

(a) Complete detector under gas flow

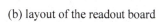

(b) layout of the readout board

Fig. 175.2 **a** Complete detector under gas flow **b** layout of the readout board

175.4 Results

In this study, the count rate is measured as a function of increasing bias voltage to the detector. It has been observed that the count rate increases with voltage and a plateau is observed from ΔV of 440 V onwards. The count rate as a function of bias voltage is shown in Fig. 175.3. The signal with Fe^{55} source obtained from the oscilloscope is shown in Fig. 175.4. The single anode readout pad used here is important for future application of GEM detector, which we will be performing soon.

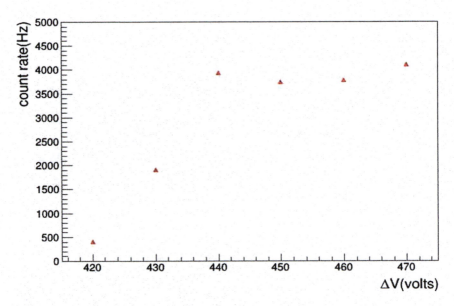

Fig. 175.3 Count rate as a function of the GEM voltage

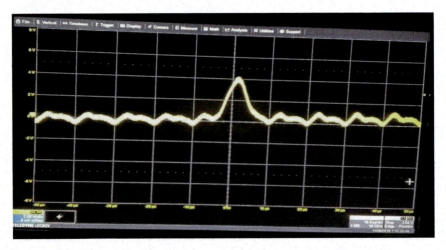

Fig. 175.4 Signal with Fe^{55} source after spectroscopic amplifier Gain $= 50$, Shaping time $= 2\,\mu s$)

Acknowledgements We would like to thank the members of IOP workshop for their invaluable cooperation.

References

1. F. Sauli, GEM: a new concept for electron amplification in gas Detectors. Nucl. Instrum. Methods A **386**, 531–534 (1997). https://doi.org/10.1016/S0168-9002(96)01172-2
2. F. Sauli, A. Sharma, Micropattern gaseous detectors. Annu. Rev. Nucl. Part. Sci. **49**, 341–388 (1999). https://doi.org/10.1146/annurev.nucl.49.1.341
3. F. Sauli, The gas electron multiplier (GEM): operating principles and applications. Nucl. Instrum. Methods A **805**, 2–24 (2016). https://doi.org/10.1016/j.nima.2015.07.060

Chapter 176
Testing of 10 cm × 10 cm Triple GEM Detector Using MUCH-XYTER v2.0 Electronics

C. Ghosh, G. Sikder, A. Kumar, Jogender Saini, A. K. Dubey, and Subhasis Chattopadhyay

Abstract Triple GEM (Gas Electron Multiplier) detector will be used as MUCH (Muon Chamber) in CBM experiment at GSI, to detect dimuons originating from the decay of low mass vector mesons (Friman et al. in Lect Notes Phys 841:1–980,[1]). CBM will operate at 10 MHz collision rate with maximum particle density of about 0.4 MHz/cm^2 at the innermost part of the MUCH detector layer located immediately after the carbon absorber (Technical Design Report for the CBM Muon Chambers in GSI, Darmstadt, [2]). Here we report the detailed test results of a 10 cm × 10 cm triple GEM detector (Dubey et al., Nucl Instr Meth A 718, 418-420, [3]) using a self triggered MUCH-XYTER v2.0 asic.

176.1 GEM Detector with Data Acquisition Chain

The experimental setup at VECC is shown in Fig. 176.1a consists of the entire DAQ chain including the FEB, AFCK, and FLIB (First level interface board). MUCH-XYTER v2.0 is a new type of asic produced at GSI for MUCH data acquisition and this has 128 analog front end channels with a dedicated 5 bit ADC for processing the charge pulses from the GEM detector. It can accept both positive and negative pulses with a data transmission rate of 320 Mbps.

The triple GEM detector [4, 5] has been tested with X-rays from Fe55 sources with Ar+CO$_2$ (70:30) gas mixture as an active medium. A typical pulse height spectrum of this triple GEM detector for Fe55 source is shown in Fig. 176.1b, this shows 32 comparators in the x-axis and hit counts in the y-axis; out of 128 channels, each

C. Ghosh (✉) · A. Kumar · J. Saini · A. K. Dubey · S. Chattopadhyay
Variable Energy Cyclotron Centre, 1/AF Bidhannagar, Kolkata 700064, India
e-mail: c.ghosh@vecc.gov.in

C. Ghosh · A. Kumar · A. K. Dubey · S. Chattopadhyay
Homi Bhabha National Institute, Mumbai 400094, India

G. Sikder
University of Calcutta, Bidhannagar, Kolkata 700106, India
e-mail: gitesh.sikder@gmail.com

© Springer Nature Singapore Pte Ltd. 2021
P. K. Behera et al. (eds.), *XXIII DAE High Energy Physics Symposium*,
Springer Proceedings in Physics 261,
https://doi.org/10.1007/978-981-33-4408-2_176

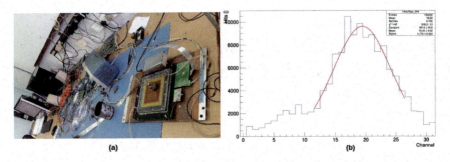

Fig. 176.1 **a** Experimental setup at VECC **b** Fe55 ADC spectra at detector high voltage = 4450 V

channel has 32 comparators, those were calibrated from 6 to 81 fC with a step of 2.5 fC. We observed the major peak at 5.9 keV and also the escape peak of Fe55 using the MUCH-XYTER electronics. With increasing high voltage, the peak position of the spectra is found to shift to higher ADC values.

176.2 Detector Gain Calculation

From Fig. 176.1b, one can get the mean value of the Fe55 ADC spectra peak, from that one can calculate the equivalent charge as, Q = 6 fC + mean * 2.5 fC. If this charge Q is divided by a single electron charge, then the total number of electrons can be estimated. Now if this number is divided by the number of primary electrons, i.e., 212, then we get the gain of the detector. Figure 176.2 shows the detector gain versus HV variation using Fe55 source.

$$Gain = \frac{[6 + (2.5 * mean)] * 10^{-15}}{212 * 1.6 * 10^{-19}}$$

176.3 Study of MUCH-XYTER Characteristics

To study the effect of calibration step on data acquisition, one FEB was calibrated at 3 different sets of calibration steps as 1.5, 2, and 2.5 fC, for each calibration step we measured the detector gain at various voltages. At any fixed HV, these three charge values are almost the same, this confirms that the FEB reads the correct charge from the detector irrespective of its calibration step, Fig. 176.2a confirms the same. We also studied the detector gain variation using different FEBs with identical parameters. we calibrated 2 FEBs at the same bias settings and the spectra from both FEBs were recorded, it was observed that both the FEBs showed almost the same ADC spectra

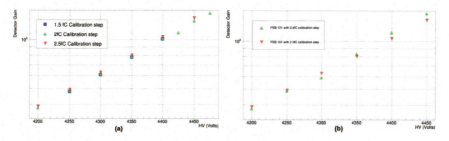

Fig. 176.2 a Detector Gain versus HV for different calibration step **b** Detector Gain versus HV for different FEBs with identical settings

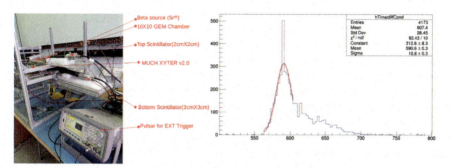

Fig. 176.3 a Coincidence setup **b** Time difference spectra at 5000 V

for the same bias settings and same detector parameters, Fig. 176.2b confirms the same.

We have studied the time resolution using MUCH-XYTER v2.0. A coincidence experiment has been setup, Fig. 176.3a, using 2 scintillator detectors and the 10 × 10 GEM detector. The detector is placed in between the 2 scintillators, the scintillator signals are sent through Leading Edge Discriminator (LED) and quad coincidence logic unit, then it is put into one channel of a FEB. The detector signal is recorded by the rest 127 channels of the same FEB.

176.4   Detector Efficiency and Time Resolution Scan

Detector efficiency has been measured at different High Voltage using a beta source. The efficiency has been calculated as Efficiency = (GEM counts *100)/2 F counts of Scintillators.

Time difference spectra of GEM with scintillator two fold (2 F) signal (Fig. 176.3b) peaks at 600 ns with a sigma of 10 ns which is in agreement with the previous nXYTER used at CERN test beam 2016. The spectra show an extended tail up to 750 ns, which results from time walk phenomena of the signal at lower ADC value.

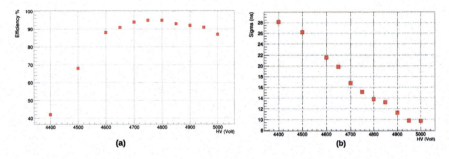

Fig. 176.4 a Detector efficiency versus HV **b** Detector time resolution versus HV

Figure 176.4a shows a plot of detector efficiency versus HV. Efficiency gets saturated at 4750 V with 95% value. After that, it goes down possibly due to the crosstalk hits from Detector channel to 2 F readout channel in XYTER. This crosstalk hit to 2 F readout channel can be eliminated if a separate FEB is used for 2 F, but due to time synchronization issue between 2 FEBs using an AFCK, we used only one FEB.

Figure 176.4b shows the variation of detector time resolution at different high voltage. We obtained resolution of the order of 17 ns at detector voltage of 4750 V and at 5000 V the value comes out in the order of 10 ns.

176.5 Conclusion

We have successfully tested a triple layer GEM detector with MUCH-XYTER v2.0 electronics and achieved an efficiency of 95% with a time resolution of 10 ns. All the characteristics of the new XYTER has been studied in detail, before using it in a mini CBM experiment [6]. Presently we are in process of testing the latest version v2.1 of the MUCH-XYTER asic for use in the main CBM.

References

1. B. Friman, C. Hohne, J. Knoll, S. Leupold, J. Randrup, R. Rapp, P. Senger, The CBM physics book: compressed baryonic matter in laboratory experiments. Lect. Notes Phys. **814**, 1–980 (2011)
2. Technical Design Report for the CBM Muon Chambers (2015) GSI, Darmstadt, https:// repository.gsi.de/record/161297
3. A.K. Dubey et al., Nucl. Instrum. Methods A **718**, 418–420 (2013)
4. A. Kumar et al., Proceedings of the DAE. Symp. Nucl. Phys. **62** (2017)
5. A.K. Dubey et al., Nucl. Instr. Meth. A **755** 62–68 (2014)
6. A. Kumar et al., Proceedings of the DAE. Symp. Nucl. Phys. **63** (2018)

Chapter 177
μSR with Mini-ICAL

N. Panchal

Abstract The India-based Neutrino Observatory (INO) is envisaged to house a 50 kton magnetized Iron-CALorimeter (ICAL) (Athar et al in INO Collaboration: Project Report, vol I, [1]) detector which is aimed to solve neutrino mass hierarchy (Ghosh et al in J High Energy Phys 2013(9), [2]). One of the vital aspects of the ICAL detector is to measure the magnetic field inside iron plates. The magnetic field is an essential parameter for reconstructing the momentum of neutrinos. An attempt to use the muon spin rotation technique to try to get information about the magnetic field in iron is made and discussed here.

177.1 Introduction

The ICAL detector will be comprising of 150 layers of 5.6 cm thick iron plates placed between two resistive plate chambers (RPCs) [3], as the active detector material. The presence of magnetic field in ICAL enables it to distinguish between positively and negatively charged particle trajectories. The measurement of B-field in ICAL is of paramount importance for the reconstruction of the momentum of neutrino. Muon Spin Rotation (μSR) is a non-destructive technique which can be used in ICAL to measure the B-field inside the iron plates. The feasibility studies for the same had been initiated at TIFR as discussed in [4]. A small magnet prototype (4 m $\times$ 4 m $\times$ 1.2 m) detector (mini-ICAL) comprising of 10 RPCs and 11 iron has been built at IICHEP, Madurai. The magnetic field in mini-ICAL is currently measured with two different techniques (a) by measuring the induced current in the sense coils around the plates and (b) by Hall probes measurements. The Hall probe measurements show

On behalf of mini-ICAL team.

N. Panchal ($\boxtimes$)
Homi Bhabha National Institute, Anushaktinagar, Mumbai 400094, India
e-mail: neha.dl0525@gmail.com

Tata Institute of Fundamental Research, Homi Bhabha Road, Colaba, Mumbai 400005, India

© Springer Nature Singapore Pte Ltd. 2021
P. K. Behera et al. (eds.), *XXIII DAE High Energy Physics Symposium*,
Springer Proceedings in Physics 261,
https://doi.org/10.1007/978-981-33-4408-2_177

almost 90% of the detector with B_{max} = 1.2 T at a current of 900 A. The signal from μSR at mini-ICAL will provide first direct measurement of B-field inside iron by using cosmic muons. This paper presents the preliminary results of the feasibility studies of μSR at mini-ICAL.

177.2   Mini-ICAL Detector

A small magnet prototype detector (mini-ICAL) has been commissioned at the Inter Institutional Center of High Energy Physics (IICHEP), Madurai. The dimensions of mini-ICAL are 4 m × 4 m × 1.2 m. The magnet assembly weighs 85 ton and comprises 11 layers of 5.6 cm thick iron plates. Presently, there are 10 layers of active detectors (RPCs) placed centrally inside the gaps available between the iron layers to track muons. The full 4 m × 4 m area of mini-ICAL detector is planned to be populated by 10 more RPCs in the second phase of detector installation. The main components of the mini-ICAL magnet assembly are the soft iron plates; the copper coils in 2 parts (U- and C-); sense coils wound around chosen plates on layers 0, 5, and 10; and aluminum strips fixed on the iron plates to guide the RPC trays. Four different sizes of iron plates (A, B, C, and D) were used and all the four iron plates were organized in order to make a square of dimension 4 m × 4 m. While placing the iron plates in a layer, gaps were left for the passage of copper coils through it. Two coils were placed symmetrically with 18 turns in each coil with a gap of 12 mm between each turn. The cross section of the copper conductor was 30 mm × 30 mm with a bore diameter of 17 mm. The maximum current needed for a field of 1.5 T was about 1000 A. Indeed this was also consistent with the rating of the DC power supply for the magnet. A 30 V 1500 A DC current supply (made by Danfysik) was used for powering the copper coils. The power supply is provided with both manual and computer control. A low conductivity water cooling system (LCWCS) was used for cooling the copper coils as well as the magnet power supply. The central $\sim$2 m × 2 m area was populated by RPCs in between 10 gaps of the magnet assembly. The gas mixture of 94.5% R134a ($C_2H_2F_4$ or Tetra fluoroethane), 0.3% Sulfur Hexafluoride (SF$_6$), and 4.3% of Isobutane (iC_4H_{10})) with pressure few mbar (above atmospheric pressure) was circulated in the RPCs. A closed-loop gas system was employed for the gas circulation in 10 RPCs. The mini ICAL electronics processes 1280 electronic channels and records the X-Y co-ordinates and time of flight of the charged particle trajectory in the detector on satisfying the trigger criteria. The FPGA-based HPTDC developed by CERN is used for storing the timestamps of the hits. The full electronics for the mini-ICAL is described in [5].

177.3   An Attempt to Observe μSR Signal at Mini-ICAL

An attempt was made to measure the magnetic field inside the iron plates using the μSR technique. The hardware trigger was set to be a logical AND of L9 & L8 & L7 & L6 (by convention L9 corresponds to topmost RPC layer) for this purpose. The trigger rate was observed to be ∼200 Hz. This trigger condition led to observe the decay of muon in 6 bottom (L5 to L0) Fe layers. The position (X, Y, Z) along with the TDC time was stored as hit information for every event, where (X, Y) was obtained from the strip number and Z was given by the layer number. The TDC trigger matching window (maximum range of the TDC) was set to 21.5 μs. The trigger signal was fanned out into two signals. While one of the signal was fed to one of the TDC channels as the reference signal, the other was delayed by 21.5 μs and used as the trigger for the TDC. When the TDC receives a trigger it latches all the timestamps which were available in the buffer. The reference signal was used as the timestamp of the muon trigger (t_m). A total of ∼98 h of data was acquired. The average magnetic field during the measurement was 1.3 T. The data was analyzed off-line with a ROOT-based C++ code. The analysis was done in three steps. In the first step, the noisy events are rejected by applying criteria on the event multiplicity and number of layers with hit. The filtered event should have

  a At least one hit in all layers above the iron layer at which muon was decayed,
  b No hit in all the layers below the iron layer at which muon was decayed,
  c Hit multiplicity in the layer <=3 with consecutive strip hits.

In the second step, the filtered muon tracks are fitted with both linear and circular fit. Since, magnetic field was applied in the Y-direction, there should be no curvature in the Y side hits. Hence, the Y side tracks are fitted only with linear fit. The muon tracks are further filtered on the basis of the goodness of the fit. A χ^2 error for the straight line fit and the circular fit was accepted to be 25 and 50, respectively. In the last step, an expected XY position ($X_0 Y_0$) of the muon track in the layer below the iron plate was extrapolated from the fitted tracks. The tracks in which the expected XY position, in the layer below the iron plate, lie between 0 and 64 and has atleast one TDC hit which is selected for further analysis. The hit within ±2 strips of the $X_0 Y_0$ was identified as a decay electron and the TDC time of which was stored as t_e.

The time difference between $\Delta t = t_e - t_m$ was plotted in a histogram with a bin size of 20 ns and the data was fitted with the equation $\Delta t = A \exp(\frac{t}{\tau}) + C$ and the lifetime for the muon was calculated. Since the applied magnetic field was in the Y-direction, the tracks with curvature in $-X$ and $+X$ direction are identified as μ^+ and μ^-, respectively. The distribution of Δt for μ^+ and μ^- along with the fit was shown if Fig. 177.1. The muon lifetime obtained for the μ^+ & μ^- are $(2.21 \pm 0.01)\mu$s and $(0.27 \pm 0.02)\mu$s, respectively.

In order to find the oscillation frequency (ω), a fast Fourier transform (FFT) of the Δt distribution with finer bin (500 ps) was calculated and is plotted in Fig. 177.2

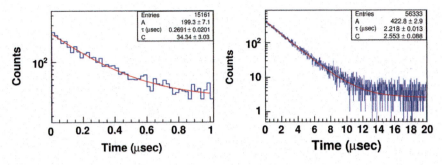

Fig. 177.1 Time distribution of μ^- (*Left*) and μ^+ (*Right*) scintillators

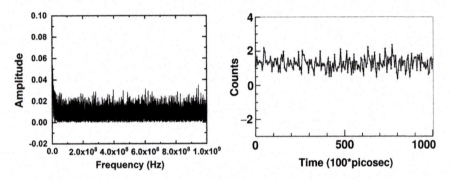

Fig. 177.2 *Left* The FFT of the timing spectrum for μ^+. *Right* The plot of the ratio between the number of counts in each bin and the value obtained from the fit function

(*Left*). No peak could be seen at the expected frequency $\nu = 135.5\frac{MHz}{T} \cdot 1.5\,T = 203\,\text{MHz}$. The oscillatory behavior of the early part (up to 100 ns) of the μ^+ lifetime spectra was closely observed by dividing the data in 500 ps bin with the corresponding fitted value and shown in Fig 177.2 (*Right*). However, no oscillation was observed.

177.4 Summary

The lifetime of muon in iron was measured as $(2.21 \pm 0.01)\,\mu$s and (0.27 ± 0.02) μs for μ^+ & μ^-, respectively. However, the modulation in the lifetime curve due to μSR signal could not be measured. This could be either due to less amount of data collected or due to the fact that the cosmic ray muon polarization is not retained in 5.6 cm thick iron plate or due to the ionization energy loss of the decay electron, it is not able to come out of the iron plate.

References

1. M.S. Athar et al., INO Collaboration: Project Report, vol. I (2006)
2. A. Ghosh et al., Determining the neutrino mass hierarchy with INO, T2K, NOvA and reactor experiments. J. High Energy Phys. **2013**, 9 (2013)
3. R. Santonico et al., Development of resistive plate counters. Nucl. Instrum. Meth. **187**, 377–380 (1981)
4. Neha et al., Can stopped cosmic muons be used to estimate the magnetic field in the prototype ICAL detector?, in *XXII DAE High Energy Physics Symposium-Proceedings*, vol. 203 (Springer Science and Business Media, LLC, 2018), p. 93
5. S. Achrekar et al., Electronics, trigger and data acquisition systems for the INO ICAL experiment, TIPP 2017. Springer Proc. Phys. **212** (2018)

Chapter 178
A Feasibility Study to Track Cosmic Ray Muons with Solid State Detectors Using GEANT4

S. Roy, S. K. Prasad, S. Biswas, S. Das, S. K. Ghosh, and S. Raha

Abstract We present the results of feasibility study for tracking cosmic ray muons with silicon detectors using GEANT4 simulation platform. The cosmic ray energy spectrum has revealed new fine structures and consensus about the origin. The composition of the primary particles has not yet been reached. Although enormous amount of data has been obtained both through direct measurements by means of satellite or balloon borne experiments as well as indirect method such as Extended Air Shower (EAS) experiments, there exist still a number of open questions. A fair knowledge of the atmospheric muon energy distributions at Earth is very important for understanding the physics of cosmic rays. A prototype detector setup is realized using GEANT4 package. Uniform magnetic field of varying magnitudes is applied across the detection volume of $(0.5\,m)^3$ to produce bending of the incident particle. At each magnetic field value, the lower and upper cut on energy of cosmic muons that can be measured using this detector setup is estimated. The detailed method and preliminary results are presented.

178.1 Introduction

At sea level, together with neutrinos, muons are the most abundant particles originated by the interactions of primary cosmic rays at the top of the atmosphere. In the primary interaction, many secondary particles are produced, in particular, pions. Such particles can decay into muons. The muons suffer a feeble interaction through the atmosphere. Therefore, they can transport direct information from the first hadronic interactions and consequently information about the primary particle. Cosmic ray flux measurements and air shower detection has been performed earlier and reported [1–4]. We are exploring the possibility of using silicon detectors for tracking cosmic ray muons. For this, we have designed a tracker using Silicon pad detectors in layers [5]. The tracker has been simulated using GEANT4 [6]. In the presence of magnetic

S. Roy (✉) · S. K. Prasad · S. Biswas · S. Das · S. K. Ghosh · S. Raha
Bose Institute, Department of Physics and CAPSS, EN-80, Sector V, Kolkata 700091, India
e-mail: shreyaroy2509@gmail.com

© Springer Nature Singapore Pte Ltd. 2021
P. K. Behera et al. (eds.), *XXIII DAE High Energy Physics Symposium*,
Springer Proceedings in Physics 261,
https://doi.org/10.1007/978-981-33-4408-2_178

field, charged particles like muon will bend and from the curvature of the track we can reconstruct the momentum of the particle. Our goal is to reproduce the nature of cosmic ray muon energy spectrum in the low momentum regime.

178.2 Tracker Simulation

The cosmic muon tracker has been simulated using GEANT4. The physical dimension of the tracker is 50 cm × 50 cm × 50 cm. The tracker has 10 layers and the gap between each layer is 5 cm. Each layer consist of 35 × 35 silicon pad detectors. The silicon pads are 300 μm thick and 1 cm × 1 cm in size. We have simulated a tracker of silicon pad detectors to track cosmic muons and to determine their energy in the low momentum regime. Keeping in mind the large size of the tracker, we have chosen a realistic value of the uniform magnetic field of ∼0.5 T, perpendicular to the direction of motion of the muons (cosmic muons are assumed to be vertically incident on the detector setup, although practically there will be muons from all directions). In a particular case, if the cosmic muons are incident vertically (along-Z axis) on our setup, then the magnetic field will be applied along X-axis. The incident muon hits the silicon pad of the tracker layers. The hit points (x, y, z) and pad numbers on different detector layers are recorded and used for track reconstruction.

178.3 Muon Reconstruction

The tracks of the incident muons are reconstructed using the recorded hit points information of the muons on different detector layers. The radius of the track curvature is estimated by fitting the reconstructed tracks. To reconstruct the momentum of the muon passing through the tracker, the following relation has been used. The equation is developed as shown below. We start with the Lorentz force.

$$\mathbf{F} = q.\mathbf{v} \times \mathbf{B} \qquad (178.1)$$

where B is the magnetic field, q is the electric charge of the particle, and v is the velocity of the incident particle If $v \perp B$, then the particle moves in a circle. For circular motion,

$$F = \frac{mv^2}{r} \qquad (178.2)$$

where r is the radius of the circular path Substituting (178.2) in (178.1), we get

$$\frac{mv^2}{r} = qvB \tag{178.3}$$

$$mv = p = qBr \tag{178.4}$$

$$p_{reconstructed} = 0.3qBr \tag{178.5}$$

in units of MeV/c, kilogauss, cm.

Using this relation and putting q = 1e, B = 0.5 T, and the value of r from fitting the muon tracks, we have calculated the momentum of the incident muons.

178.4 Results

We have studied the lower and upper bound of the incident muon energy which can be detected using this setup. The table below shows the lower and upper energy cut off for cosmic muons that can be measured using this muon tracker operating at different magnetic field values. It has been observed that below 15 MeV, the track is distorted due to multiple scattering. At high energies, the track is nearly a straight line and the upper cut-off is determined by finding out the energy for which the bending of the track is smaller than the granularity of the detector. One of the important parameters for a tracking detector is the resolution of the momentum reconstruction because this would limit its use to a particular momentum range of the incoming particle. The momentum resolution has been calculated using the following formula

$$\frac{\Delta p}{p} \% = \frac{p_{reconstructed} \sim p_{incident}}{p_{incident}} \times 100\% \tag{178.6}$$

Figure 178.1 shows the momentum resolution of the muon tracker. It is found that the tracker has a very good momentum resolution (<3%) in the energy range 200–1000 MeV/c. It is feasible to use this tracker in this particular energy range for cosmic muons (Table 178.1).

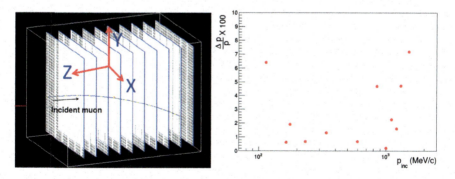

Fig. 178.1 Muon tracker simulated in GEANT4 (left), Momentum resolution of the tracker (right)

Table 178.1 Lower and upper cut on muon energy that can be measured with the tracker for different magnetic field intensities

Magnetic field (Tesla)	Lower cut (MeV)	Upper cut (MeV)
0.1	15	250
0.25	15	750
0.5	15	1850
0.75	15	2600
1	15	4000

178.5 Summary and Outlook

A prototype silicon detector setup is realized using GEANT4 simulation package. Uniform magnetic field of varying magnitudes is applied across the detection volume of $(0.5\,\text{m})^3$ to produce bending of the muon. The lower and upper cut on energy of cosmic muons that can be measured using this tracker is 15 MeV and 1850 MeV, respectively, for a magnetic field of 0.5 T and the tracker has a very good momentum resolution ($<3\%$) in the momentum range 200–1000 MeV/c. Feasibility of the tracker to measure cosmic muon energy by momentum reconstruction technique is studied, and in future this may be implemented as well. Finding momentum resolution at various magnetic fields for muons of different incident energy and for different number of detector layers is in the future plan.

References

1. S. Roy et al., Plastic scintillator detector array for detection of cosmic ray air shower, https://doi.org/10.1016/j.nima.2018.09.109
2. S. Biswas et al., JINST 12 C06026.120 (2017)
3. S. Roy et al., Proceedings of ADNHEAP 2017. Springer Proc. Phys. **121**, 201. ISBN 978-981-10-7664-0
4. S. Shaw et al., Proceedings of the DAE. Symp. Nucl. Phys. **62** (2017)
5. S. Muhuri et al., arXiv:1407.5724v1
6. S. Agostinelli et al., NIM A **506**, 250–303 (2003)

Chapter 179
Update on Muon Reconstruction for INO-ICAL

A. D. Bhatt and Gobinda Majumder

Abstract The INO-ICAL is a proposed neutrino physics experiment, which will be made of 50 kTon of magnetized iron layers, formed of with 150 layers of $4\,m \times 2\,m \times 56\,mm$ thick iron plates, interleave with of RPCs as a sensitive detector. To study the detector's response to various particles, the GEANT4 toolkit is used to simulate the matter particle interactions. The GEANT4 output is digitized for detector properties based on the data collected in the prototype detector at Madurai. The momentum and direction of charged particle tracks are estimated using the Kalman filter technique. In this work, the concept of fiducial volume for ICAL detector is discussed. Also, the measurement of the momentum of fully contained charge particles using the total pathlength of the trajectory is explained and the results are compared with the Kalman filter technique.

179.1 Introduction

The India-based Neutrino Observatory (INO) project is a proposed multi-institutional effort aimed at building an underground laboratory with a rock cover of approximately 1.3 km at Pottipuram in Bodi West hills of Theni District of Tamilnadu in India, with the main focus on research in non-accelerator-based high energy physics and nuclear physics [1]. The primary focus of the programme is to study atmospheric neutrinos with a magnetized Iron CALorimeter (ICAL) detector with the main goal to measure the sign of the 2–3 mass-squared difference, $\Delta m_{32}^2 \left(= m_3^2 - m_2^2\right)$ precisely through matter effects. The physics capabilities of ICAL are mainly derived

A. D. Bhatt and G. Majumder on behalf of INO mini-ICAL group.

A. D. Bhatt (✉)
Institute of Nuclear Physics Polish Academy of Sciences, PL-31342 Krakow, Poland
e-mail: apoorva.bhatt@tifr.res.in

A. D. Bhatt · G. Majumder
Tata Institute of Fundamental Research, Mumbai, India
e-mail: gobinda@tifr.res.in

© Springer Nature Singapore Pte Ltd. 2021
P. K. Behera et al. (eds.), *XXIII DAE High Energy Physics Symposium*,
Springer Proceedings in Physics 261,
https://doi.org/10.1007/978-981-33-4408-2_179

from detection of muons through Charged-Current (CC) events produced from ν_μ and $\bar{\nu}_\mu$ interactions.

The muon response of ICAL detector was previously discussed in [2, 3]. In those studies, both fully contained and partially contained events were considered for analysis. Also, the muon response in the central region [2] of the central module of the detector where the magnetic field is almost uniform and in the peripheral regions, [3] where it is non-uniform were studied separately.

In the present work, the muons are studied with fully and partially contained event separately irrespective of the region of the detector. If the muon event is fully contained, then the range of the muon is precisely known. The momentum measurement from the range of the muon inside the detector gives a better estimation than curvature measurement. This is applicable only for the fully contained event. In the next subsection, an algorithm (for ICAL detector) used to tag Fully Contained (FC) muon event and the calibration scheme for momentum estimate of these fully contained events are discussed.

179.2 ICAL Detector Simulation

The ICAL detector is simulated using the GEANT4 toolkit [4]. Iron layers, 5.6 cm thick, are interleaved with the active RPCs with a 2 mm gas gap. The spacing between two consecutive RPCs is 9.6 cm. The layout of the ICAL detector geometry is shown in 179.1a. When a charged particle propagates through the ICAL detector, hits in the X and Y strips of the RPC layers are recorded. The layer number provides the z-coordinate. The details of the geometry in simulations have been discussed in [2].

In the previous work, the total RPC dimension in the simulation detector geometry was 184 cm × 184 cm × 1.88 cm. As per the detector design at that time, elec-

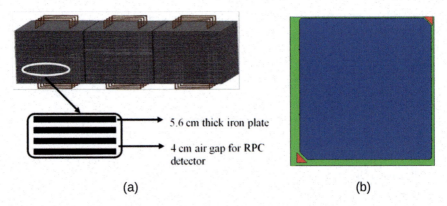

5.6 cm thick iron plate

4 cm air gap for RPC detector

(a)   (b)

Fig. 179.1 a Schematic view of the 50 kt ICAL detector and **b** A basic RPC unit along with the tray

tronics board was to be kept on top of the RPCs. To accommodate that in the detector geometry, the electronic board material (G10) was uniformly spread over the entire RPC. Hence, the thickness of RPC (in the geometry) was larger by an extra cm. In the new RPC design implemented in the current geometry, the RPC size is 174 cm × 183.5 cm × 1.8 cm. There are two triangular cuts on two diagonal corners to accommodate two electronic boards (one for DAQ and another for H.V. power supply) as shown in Fig. 179.1b. The RPC is kept in a tray made of Fibreglass Reinforced Plastic (FRP) material along with the two triangular electronic boards. The RPC is not kept at the centre of FRP Tray but slightly off centre. The DAQ Board is right angle triangle with side length 12.5 cm and thickness 0.5 cm, and H.V. Power Supply Board is also right angle triangle with side length 10 cm and thickness 0.5 cm. The RPC tray size is 191 cm × 194 cm × 3.4 cm. This is the main difference in geometry in the simulations framework as compared to previous works. The width of the readout strip of RPC is 3 cm. The other aspects of the geometry like the magnetic field, etc. are as described in the previous work.

179.3 Fully and Partially Contained Events

For a "fully" contained event, the momentum measurement from the range of muon inside the detector gives a better estimate than the one measured using the curvature. The simple assumption for an FC event is that the vertex and end points of the trajectory are inside the fiducial volume. Also, the Fiducial Volume is needed to distinguish events produced in the rock from the events produced inside the ICAL detector. The fiducial volume for ICAL detector is taken by excluding 2 RPC layers from top and bottom and 3 strips on the either side of X- and Y-direction. To check the validity of these criteria, single μ^- events are generated in simulation. In the generated dataset, there are two types of events, (i) Generated within Fiducial volume (FGEN) and (ii) Generated outside of Fiducial volume (OGEN). After reconstruction, extrapolation is needed to analyse the reconstructed position of the vertex of the tracks to check whether it is within the fiducial volume or not. The events are observed in the following criteria:

- FF—Generated fiducial and reconstructed fiducial—0.97 ± 0.01.
- FO—Generated fiducial and reconstructed outside—0.025 ± 0.001.
- OF—Generated outside and reconstructed fiducial—0.035 ± 0.001.
- OO—Generated outside and reconstructed outside—0.964 ± 0.007.

Now, the number of FO and OF combinations is quite small which implies the reconstruction is quite accurate. But, it is not as simple as it is assumed. In Fig. 179.2, a few types of possible events, which can be observed in ICAL detector, are depicted:

Event 1 Clearly coming from the outside and the criteria mentioned above will clearly tag this as coming from outside the Fiducial volume.

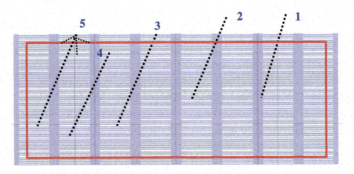

Fig. 179.2 Fiducial volume for ICAL detector

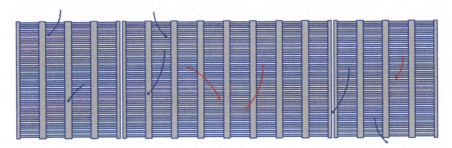

Fig. 179.3 Various possible FC and PC events in ICAL

Event 2   Coming from outside, but as the first hit is well within the detector volume,
            it will be tagged as the event inside the Fiducial volume.

Event 3   Similar to *Event 1*

Event 4   Will be tagged as well as within the Fiducial volume.

Event 5   Starts from the 1st layer of ICAL, but along with hadron shower. In prin-
            ciple, this event should be considered within my Fiducial volume but just
            applying the criteria will not permit it.

Looking at these events, it is clear that the simple criteria of excluding some fixed
number of strips and layers from the boundaries of the ICAL detector is insufficient
to tag an event as *Fully* or *Partially* contained event.

The algorithm presented here can only be applied to muon tracks. Figure 179.3
presents some examples of fully and partially contained muon tracks possible in the
ICAL detector. In the algorithm, the track is extrapolated back to two layers in the
backward direction from the vertex and in the forward direction from the endpoint.
If the extrapolated points from one of the ends of the track are in the dead region of
the detector, then it is tagged as *Partially Contained* (PC) track. If the extrapolated
points are in the sensitive region RPC, i.e. the gas volume, then it is tagged as a *FC*
track. In Fig. 179.3, the red track is only fully contained, remaining others are all
partially contained. For FC tracks, the range of the muons is calculated easily as the
vertex and end point are within the sensitive region of the RPC. Using the range of

muons to estimate the momentum will provide a better estimate than the curvature in the magnetic field. For the current analysis, the description of simulation and the conditions on event selection and cuts are discussed in the following section.

179.4 Calibration and Analysis of Fully Contained Tracks

In this study, one million muon events are simulated in ICAL detector. The vertices of these events are uniformly spread over the entire volume of detector. The events were generated in momentum range from 0.5 to 20.5 GeV and its momentum direction was smeared uniformly in polar ($cos\theta_{gen}$) and azimuthal (ϕ) angle. In the calibration and analysis, only the events with one reconstructed track having minimum of five layers in the track fit are selected. The events are segregated in different $|cos\theta_{gen}|$ bins for calibration and analysis.

To obtain the calibration relation of momentum as a function of length of the track, the first step is getting the scatter plot of the true momentum of muons vs the density weighted pathlength of fitted tracks in the detector for selected events. In Fig. 179.4a, for the events with all $|cos\theta_{gen}|$ bins ($|cos\theta_{gen}|\epsilon$ [0.4, 1.0]) the scatter plot is presented. As for events with $|cos\theta_{gen}| < 0.4$, the reconstruction efficiency is low, these events are not considered in these analysis [2].

To observe the calibration curve, the profile histogramme of this scattered plot is constructed as shown in Fig. 179.4b. This plot shows the correlation between muon momentum and density weighted range. It is fitted by a second-order polynomial function. In this study, for each $|cos\theta_{gen}|$ bin, a different calibration is obtained as well as for the events will all $|cos\theta_{gen}|$ bins to observe if there is any bias in the calibration due to the zenith angle.

From the range information obtained from fitted tracks and the polynomial functions obtained in the previous step, the muon momentum is reconstructed for each event for two different calibrations ($cos\theta_{gen}$ dependent and all events). The distribution of $(p_{reco}/p_{gen} - 1)$ is plotted for different ranges of p_{gen} and fitted by Gaussian function as shown in Fig. 179.4c, d. The sigma (σ) of the fitted Gaussian function is treated as the resolution for the mean muon momentum of the particular momentum

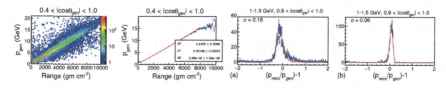

Fig. 179.4 a Scatter plots of muon momentum with range of muons in the detector. **b** Fitted calibration function for relation between range of muons in the detector and momentum where p_0, p_1 and p_2 are parameters of the second-order polynomial function. Resolution plots for a p_{gen} of 1–1.5 GeV and $|cos\theta_{gen}|\epsilon$ [0.9, 1.0] **c** for momentum estimated using curvature fit and **d** for momentum estimated from range calibration

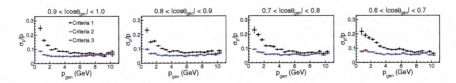

Fig. 179.5 Momentum resolution as a function of momentum for fully contained muon tracks. Criteria 1 is for momentum estimated using curvature fit. Criteria 2 is for momentum estimated using range calibration in different $|cos\theta_{gen}|$ bins. Criteria 3 refers to momentum computed using range calibration of muon tracks with all $|cos\theta_{gen}|$ bins

range. In Fig. 179.4c, d, the $(p_{reco}/p_{gen} - 1)$ distributions for p_{gen} range 1–1.5 GeV and $|cos\theta_{gen}|\epsilon$ [0.9, 1.0], for two different methods of estimating the momentum is presented. In Fig. 179.4c, the momentum is estimated from the curvature fit, and in Fig. 179.4d, momentum is computed using range calibration of muon tracks with all $|cos\theta_{gen}|$ bins.

In Fig. 179.4e–h, the momentum resolution is plotted as function of momentum for different $|cos\theta_{gen}|$ bins. The black points in the plot represent resolution for momentum reconstructed using curvature fit. The red and blue points in the plot are for momentum computed using the calibration described here for fully contained events. The small difference between Criteria 2 and 3 means that the impact of the more precise calibration using individual $cos\theta_{gen}$ bins is quite small. Also, at lower momentum, resolution is poorer because of multiple scattering, whereas in the higher momenta bins, the number of events is less hence fluctuations are high. It is clearly observed that for fully contained events momentum reconstructed using the range of muons through the detector gives a better estimate than the curvature fit for a large range of momentum (Fig. 179.5).

179.5 Summary

Here, an algorithm to tag "fully contained" muon tracks is described. Also, a method of calibration to estimate momentum from the pathlength of the track for fully contained muon trajectories is discussed. For the fully contained charged particle, there is a substantial improvement of momentum resolution from the measurement of pathlength in comparison with the result from the Kalman filter technique and the ICAL experiment will use the pathlength technique for the fully contained events.

References

1. A. Kumar et al., Physics potential of the ICAL detector at the India-based Neutrino Observatory (INO). Pramana—J. Phys. **89**, 79 (2017)
2. A. Chatterjee et al., A simulations study of the muon response of the iron calorimeter detector at the India-based neutrino observatory. JINST **9**, P07001 (2014)
3. R. Kanishka et al., Simulations study of muon response in the peripheral regions of the Iron calorimeter detector at the India-based neutrino observatory. JINST **10**, P03011 (2015)
4. GEANT4 Collaboration., GEANT4: a simulation toolkit. Nucl. Instrum. Meth. A **506** 250–303 (2003)

Author Index

Printed by Printforce, United Kingdom